Efficiency and Sustainability of the Distributed Renewable Hybrid Power Systems Based on the Energy Internet, Blockchain Technology and Smart Contracts-Volume II

Efficiency and Sustainability of the Distributed Renewable Hybrid Power Systems Based on the Energy Internet, Blockchain Technology and Smart Contracts-Volume II

Editors

Nicu Bizon
Mamadou Baïlo Camara
Bhargav Appasani

MDPI • Basel • Beijing • Wuhan • Barcelona • Belgrade • Manchester • Tokyo • Cluj • Tianjin

Editors

Nicu Bizon
Faculty of Electronics,
Communication and
Computers
University of Pitesti
Pitesti
Romania

Mamadou Baïlo Camara
Laboratoire GREAH
Université Le Havre
Normandie
Le Havre
France

Bhargav Appasani
School of Electronics
Engineering
Kalinga Institute of Industrial
Technology
Bhubaneswar
India

Editorial Office
MDPI
St. Alban-Anlage 66
4052 Basel, Switzerland

This is a reprint of articles from the Special Issue published online in the open access journal *Sustainability* (ISSN 2071-1050) (available at: www.mdpi.com/journal/sustainability/special_issues/Renewable_Hybrid_Power).

For citation purposes, cite each article independently as indicated on the article page online and as indicated below:

LastName, A.A.; LastName, B.B.; LastName, C.C. Article Title. *Journal Name* **Year**, *Volume Number*, Page Range.

ISBN 978-3-0365-6371-8 (Hbk)
ISBN 978-3-0365-6370-1 (PDF)

Contents

Preface to "Efficiency and Sustainability of the Distributed Renewable Hybrid Power Systems Based on the Energy Internet, Blockchain Technology and Smart Contracts-Volume II"

The very fast increase in the world's energy demand over the last decade, and the request for sustainable development, can be approached through micro- and nano-grids using hybrid power systems based on the energy internet, blockchain technology, and smart contracts.

This book is the second volume in these topics and includes innovative solutions and experimental research, as well as state-of-the-art studies, in the following challenging fields:

- Microgrids, nanogrids, smart grids, smart cities, and smart associations of farms, buildings, parking, and homes;
- Smart grid cyber security using the energy internet, blockchain, and smart contract-based applications;
- Operations of smart associations integrated with distributed generation;
- Smart grid architecture and energy management models;
- Fuel cell (FC) systems: modeling, control, optimization, and innovative technologies to improve the fuel economy, lifetime, reliability, and safety in operation;
- Hybrid power systems (HPSs) based on renewable energy sources (RESs) (RES HPS): optimized RES HPSs architectures; global maximum power point tracking (GMPPT) control algorithms to improve the energy harvesting from RESs; advanced energy management strategies (EMSs) to optimally ensure the power flow balance on DC (and/or AC bus) for standalone RES HPSs or grid-connected RES HPSs (micro-, nanogrids, etc.);
- RES HPS with FC system as a backup energy source (FC RES HPS): innovative solutions to mitigate the RES power variability and load dynamics of energy storage systems (ESSs) by control of the generated FC power; DC voltage regulation and/or load pulse mitigation by active control of the power converters from hybrid ESS;
- FC vehicles (FCVs): FCV powertrain, ESSs topologies, and hybridization technologies and EMSs to improve the fuel economy;
- Optimal sizing of FC RES HPSs and FCVs;
- Business opportunities, open issues, and future trends.

The climate changes that are becoming visible today are a challenge for the global research community. The stationary applications sector is one of the most important energy consumers. Harnessing the potential of renewable energy worldwide is currently being considered to find alternatives for obtaining energy by using technologies that offer maximum efficiency and minimum pollution. In this context, renewable energy sources, fuel cell systems and other energy generating sources must be optimally combined and connected to the grid system using advanced energy transaction methods.

As this book presents the latest solutions in the implementation of fuel cell and renewable energy in mobile and stationary applications, such as hybrid and microgrid power systems based on the Energy Internet, Blockchain technology, and smart contracts, we hope that they will be of interest to readers working in the related fields mentioned above.

Nicu Bizon, Mamadou Baïlo Camara, and Bhargav Appasani
Editors

sustainability

MDPI

Review

Blockchain-Enabled Smart Grid Applications: Architecture, Challenges, and Solutions

Bhargav Appasani [1], Sunil Kumar Mishra [1], Amitkumar V. Jha [1], Santosh Kumar Mishra [1], Florentina Magda Enescu [2], Ioan Sorin Sorlei [3], Fernando Georgel Bîrleanu [4], Noureddine Takorabet [5], Phatiphat Thounthong [5,6] and Nicu Bizon [2,3,4,*]

1. School of Electronics Engineering, Kalinga Institute of Industrial Technology, Bhubaneswar 751024, India; bhargav.appasanifet@kiit.ac.in (B.A.); sunil.mishrafet@kiit.ac.in (S.K.M.); amit.jhafet@kiit.ac.in (A.V.J.); 2081108@kiit.ac.in (S.K.M.)
2. Faculty of Electronics, Communication and Computers, University of Pitesti, 110040 Pitesti, Romania; florentina.enescu@upit.ro
3. ICSI Energy, National Research and Development Institute for Cryogenic and Isotopic Technologies, 240050 Ramnicu Valcea, Romania; sorin.sorlei@icsi.ro
4. Doctoral School, University Politehnica of Bucharest, Splaiul Independentei Street no. 313, 060042 Bucharest, Romania; birleanu.fernando@gmail.com
5. Group of Research in Electrical Engineering of Nancy (GREEN), University of Lorraine, 2 Avenue de la Forêt de Haye, 54518 Vandeuvre lès Nancy, CEDEX, F-54000 Nancy, France; noureddine.takorabet@univ-lorraine.fr (N.T.); phatiphat.t@fte.kmutnb.ac.th (P.T.)
6. Renewable Energy Research Centre (RERC), Department of Teacher Training in Electrical Engineering, Faculty of Technical Education, King Mongkut's University of Technology North Bangkok, 1518 Pracharat 1 Road, Wongsawang, Bangsue, Bangkok 10800, Thailand

* Correspondence: nicu.bizon@upit.ro

Citation: Appasani, B.; Mishra, S.K.; Jha, A.V.; Mishra, S.K.; Enescu, F.M.; Sorlei, I.S.; Bîrleanu, F.G.; Takorabet, N.; Thounthong, P.; Bizon, N. Blockchain-Enabled Smart Grid Applications: Architecture, Challenges, and Solutions. *Sustainability* **2022**, *14*, 8801. https://doi.org/10.3390/su14148801

Academic Editor: Thanikanti Sudhakar Babu

Received: 11 April 2022
Accepted: 14 July 2022
Published: 18 July 2022

Abstract: The conventional electrical grid is undergoing substantial growth for reliable grid operation and for more efficient and sustainable energy use. The traditional grid is now metamorphosing into a smart grid (SG) that incorporates a diverse, heterogeneous blend of operating measures such as smart appliances, meters, and renewable energy resources. With better efficient results and dependability, the SG can be described as a modern electric power grid architecture. The SG is one of the greatest potential advances as a promising solution for the energy crisis. However, it is complex and its decentralization could be of tremendous benefit. Moreover, digitalization and integration of a large number of growing connections make it a target of cyber-attacks. In this sense, blockchain is a promising SG paradigm solution that offers several excellent features. There has been considerable effort put into using blockchains in the smart grid for its decentralization and enhanced cybersecurity; however, it has not been thoroughly studied in both application and architectural perspectives. An in-depth study was conducted on blockchain-enabled SG applications. Blockchain architectures for various applications, such as the synchrophasor applications, electric vehicles, energy management systems, etc., were proposed. The purpose of this article is to provide directions for future research efforts aimed at secure and decentralized SG applications using blockchain.

Keywords: smart grid; blockchain; smart contracts; cybersecurity; microgrids; electric vehicles; energy transactions; energy management; smart cities; advanced metering infrastructure; home automation; smart homes

1. Introduction

The power grid is a complex engineering marvel, which is undergoing rapid changes due to the proliferation of renewable energy resources, high-speed signal processors, and intelligent sensors, etc. The present requirement involves bi-directional flow energy and information between the power generators and the power consumers. So, the traditional power grid is evolving into a smart grid (SG), a grid that is capable of dynamically monitoring and controlling the flow of power, providing reliable power to the consumers [1].

The SG connects heterogeneous components that vary in their functionality and requirements. These components include renewable and non-renewable energy sources, intelligent sensors, controllers, etc. The statistics on research publications related to the SG are shown in Figure 1. These statistics were obtained from the Scopus database. The various applications that the SG caters to are shown in Figure 2. In Figure 2, the share of research publications from the application's perspective is shown. From this figure, it can be observed that the main applications in an SG are the energy management systems (EMS), electric vehicles (EVs), microgrids (MGs), smart cities (SCs), home automation (HA), advanced metering infrastructure (AMI), and synchrophasor applications (SPAs) [2].

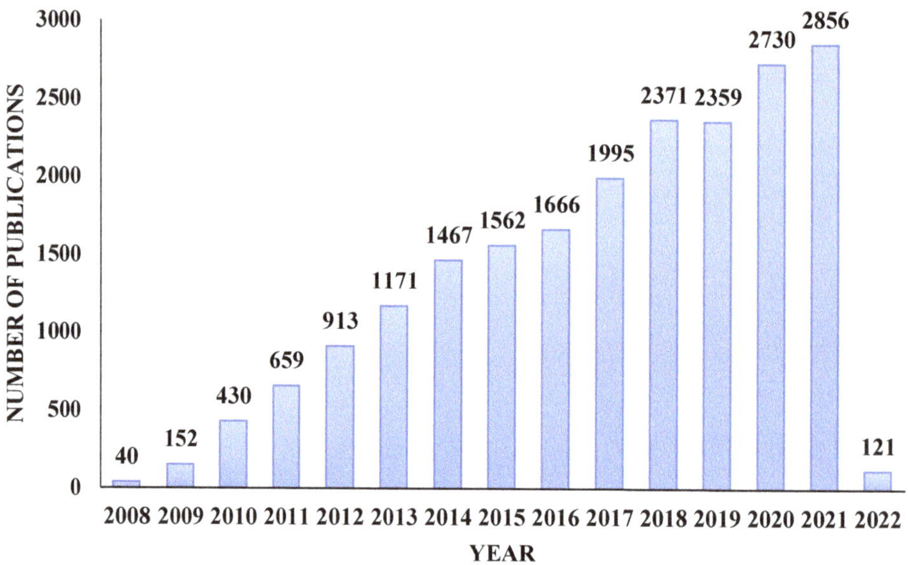

Figure 1. Publication statistics on SG.

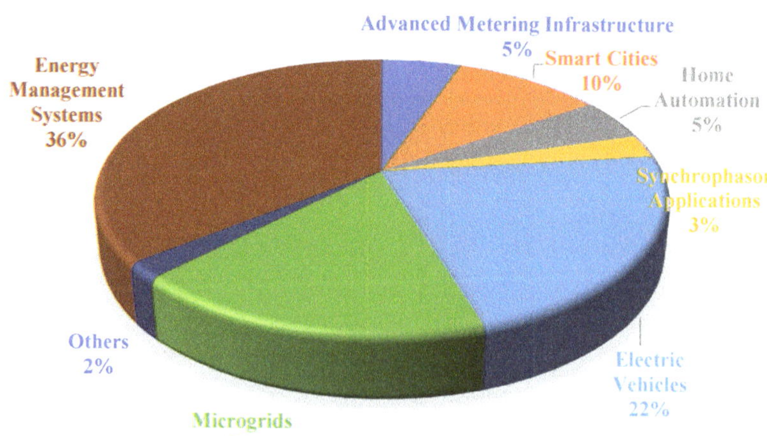

Figure 2. Distribution of research related to SG.

SG enhances the reliability of power supply and materializes several applications at the cost of increased complexity [3]. In this complex network, at a given instance, there are several entities in the grid that carry out transactions. An important concern is validating a transaction between the various entities involved in a particular SG application. A promising and secure solution for this problem is the use of Blockchain technology.

Blockchain technology, first introduced by Satoshi Nakamoto, helps achieve consensus about the authenticity of a particular transaction and helps maintain trust between various entities involved [4]. The number of papers published on blockchain technology every year is shown in Figure 3. Additionally, the corresponding number of papers published on Blockchain for SG is shown in this figure. The publication statistics were obtained from the Scopus database.

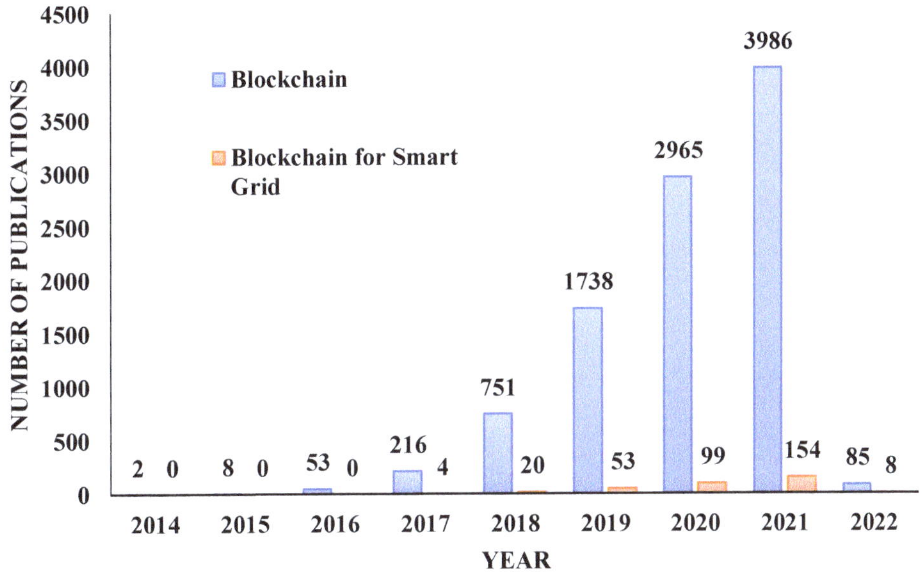

Figure 3. Publication statistics on blockchain and blockchain for SG.

The statistics indicate that blockchain technology is not being exploited for SG applications. Only 3.5% of publications on the blockchain are related to the SG applications. The motivation for this review was to explore the research available on the blockchain for SG, categorize it based on the application, propose the blockchain architectures for the various SG applications, identify the challenges in this regard, and suggest suitable solutions. The review papers and surveys on blockchain for SG are summarized in Table 1. Contrary to these works, the present work presents a boarder perspective on different SG applications with the blockchain. Moreover, the present work also describes the architecture of the blockchain-enabled SG applications. A wide range of potential applications of SG is considered, such as EV, AMI, SPA, MGs, SCs, HA, and EMS.

The paper is organized in the following sections, as represented in Figure 4. Section 2 discusses the basic concepts of a blockchain. It presents the terminology related to the blockchain and its general architecture. Section 3 presents a review of the blockchain-enabled SG applications. Different applications are discussed, and their architectures are presented. The security concerns pertaining to these applications are discussed in Section 4, and Section 5 is the conclusion of this review.

Table 1. Existing reviews on blockchain for SG. "✓" and "✕" indicates "included" and "excluded" respectively in literature.

Reference	Blockchain from an SG Application Perspective			SG Applications Considered						
	Architecture	Security	General	EVs	AMI	SPA	MGs	SCs	HA	EMS
[4]	✕	✕	✓	✕	✕	✕	✓	✕	✕	✓
[5]	✕	✕	✓	✓	✕	✕	✕	✕	✕	✕
[6]	✕	✕	✕	✕	✕	✕	✕	✕	✕	✓
[7]	✕	✕	✓	✕	✕	✕	✕	✕	✓	✕
[8]	✕	✓	✕	✓	✕	✕	✕	✕	✕	✓
[9]	✕	✓	✓	✕	✕	✕	✕	✕	✕	✕
[10]	✕	✓	✓	✕	✕	✕	✕	✕	✕	✕
[11]	✕	✕	✓	✕	✕	✕	✕	✕	✕	✕
[12]	✕	✕	✓	✕	✕	✕	✕	✕	✕	✓
[13]	✕	✓	✓	✕	✕	✕	✓	✕	✕	✓
[14]	✕	✓	✓	✓	✕	✕	✓	✕	✕	✓
[15]	✕	✕	✓	✕	✕	✕	✕	✕	✕	✓
[16]	✓	✕	✓	✕	✕	✕	✓	✕	✕	✓
[17]	✕	✕	✓	✕	✕	✕	✕	✕	✕	✕
[18]	✕	✕	✓	✕	✕	✕	✕	✕	✕	✓
[19]	✕	✕	✓	✕	✕	✕	✕	✕	✕	✕
[20]	✕	✕	✓	✕	✕	✕	✓	✕	✕	✕
[21]	✕	✓	✓	✕	✕	✕	✕	✕	✕	✕
[22]	✕	✕	✓	✕	✕	✕	✓	✕	✕	✓
This survey	✓	✓	✓	✓	✓	✓	✓	✓	✓	✓

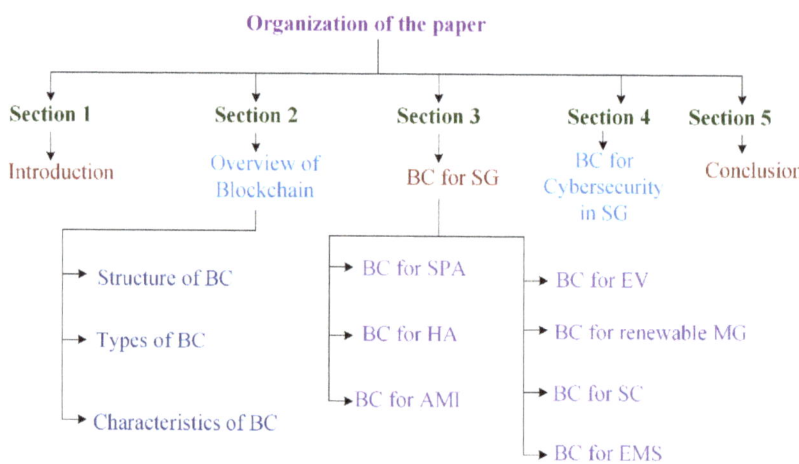

Figure 4. Organization of the paper.

2. Overview of Blockchain

In the past few years, blockchain technology has received tremendous attention worldwide. At the beginning of the technology's inception for application in digital currency, or cryptocurrency, blockchain was considered a cryptocurrency [23]. Bitcoin, the most popular cryptocurrency, was considered to be the blockchain. However, blockchain is the backbone of these cryptocurrencies. It is a distributed ledger for transactions in a decentralized network. Initially, the researchers were skeptical about this technology, but the popularity of Bitcoin changed their perception. This can be corroborated in the sudden growth in the number of published articles on the blockchain after 2016 as shown in Figure 3. Blockchain is being considered in various other domains such as banking, healthcare, healthcare, industries, etc. These various applications are depicted in Figure 5.

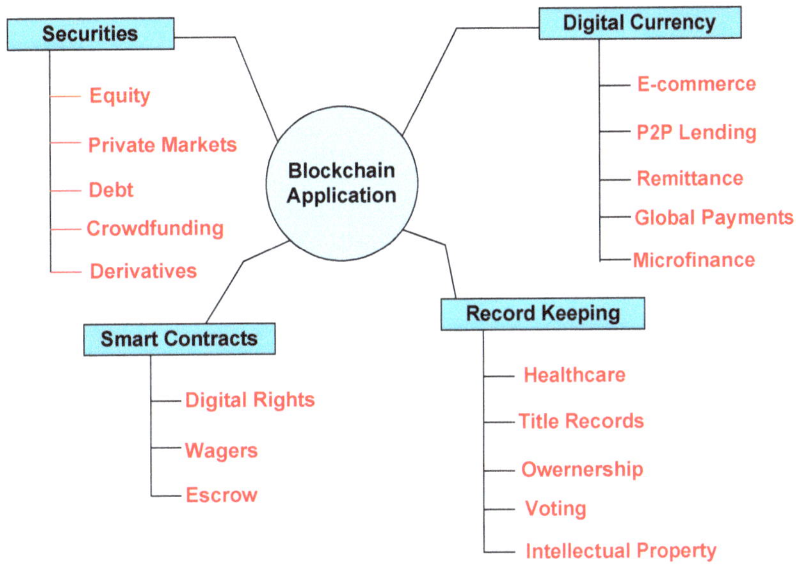

Figure 5. Applications of blockchain.

2.1. *Structure of Blockchain*

The blockchain comprises a series of blocks of transactions linked together in a chain, as shown in Figure 6. Client/server architecture is used in traditional client/server systems, and various administrators are in charge of them. On the other hand, blockchain is a distributed, decentralized peer-to-peer (P2P) network [24]. Each and every network participant can control the network. The network is made up of many connected computers or nodes, and the blocks in the chain cannot be changed without the network's approval. Each node in the network has its copy of the digital ledger.

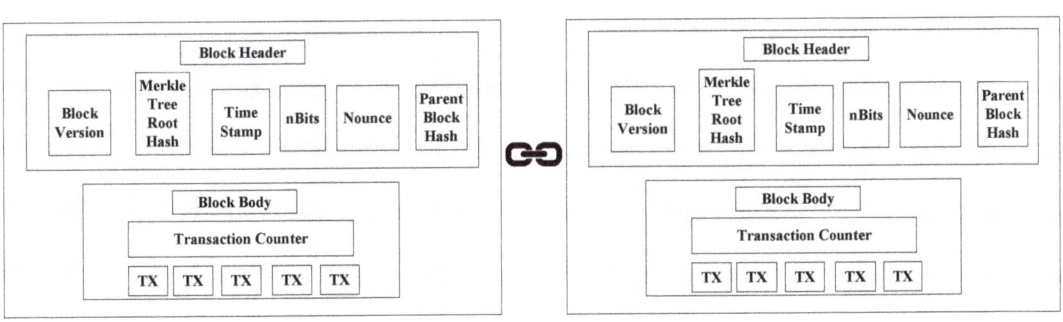

Figure 6. Structure of a blockchain.

The main constituents of a blockchain and the associated terminology are described as follows:

1. Block: In a blockchain, pointers and linked list data structures are utilized to represent blocks. Using a linked list, the blocks are sorted in a logical order and aligned up with one another. A block is a data set containing transaction information like timestamps and links to previous blocks and is produced using a secure hash technique.

The location of the next block is indicated via pointers. Every block is divided into two sections: the block header and the block body.

The block header has the following fields:

(i.) Block version: specifies which set of block validation criteria should be used.
(ii.) Merkle tree root hash: the sum of all transactions in the frame's hash value.
(iii.) Timestamp: from 1 January 1970, the current time is expressed in seconds in universal time.
(iv.) nBits: a valid block hash's goal threshold.
(v.) Nonce: a 4-byte field that starts with 0 and rises for each hash computation.
(vi.) Parent block hash: a 256-bit hash value that refers to the block before it.

A transaction counter and transactions make up the block body. The maximum number of transactions stored in a block is determined by the block size and the transaction size.

2. Public and Private keys: Blockchain is a constantly increasing network of interconnected and secured blocks using cryptographic processes [25]. To validate transactional authentication, blockchain employs an asymmetric key technique. The transactions in the block are encrypted using a private key. Every other node in the network can access these transactions. These nodes can decrypt the data using a public key available to all the nodes in the network.
3. Hash function: Every block has a cryptographic hash related to the previous block. Hashing creates a unique fixed-length string to identify a piece of data. The length of the string is independent of the size of the data.
4. Consensus process: A set of protocols and consensus from all network participants are used to validate new blocks. Consensus is needed to decide on the validity of the block. Several approaches are available for the consensus process, such as proof of work, proof of stake, practical byzantine fault tolerance, etc.
5. Smart Contracts: Smart contracts are programs that execute automatically and control the transactions between the distributed nodes in the blockchain network.

2.2. Types of Blockchain

The type of a blockchain depends on the nature of the application. There are three types of blockchains: public, private, and consortium [26]. These three types of blockchains are represented along with their properties in Figure 7.

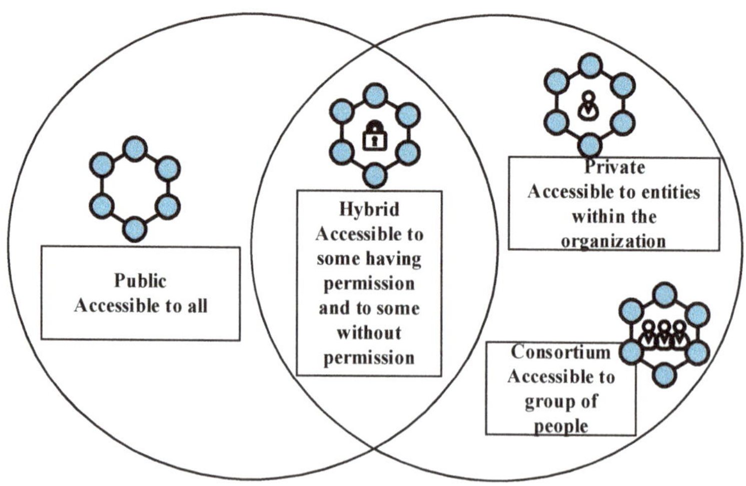

Figure 7. Types of blockchains and their properties.

There is no control over a permissionless or public blockchain. Anyone may access the network and read or write data. Permissioned ledgers, on the other hand, are only accessible to network users who have been authenticated. Since they are encrypted with a private key, everyone cannot read the blocks. The properties of public and private blockchains are combined in consortium blockchains.

2.3. Characteristics of Blockchain

A blockchain is a decentralized network, and unlike a centralized system, the transactions are validated by the nodes in the network [27]. The identity of the nodes in the network remains unanimous, and once a transaction is validated by the nodes and added to the blockchain, it is impossible to reverse the transaction. Thus, the blockchain is immutable. The various other characteristics of a blockchain are depicted in Figure 8.

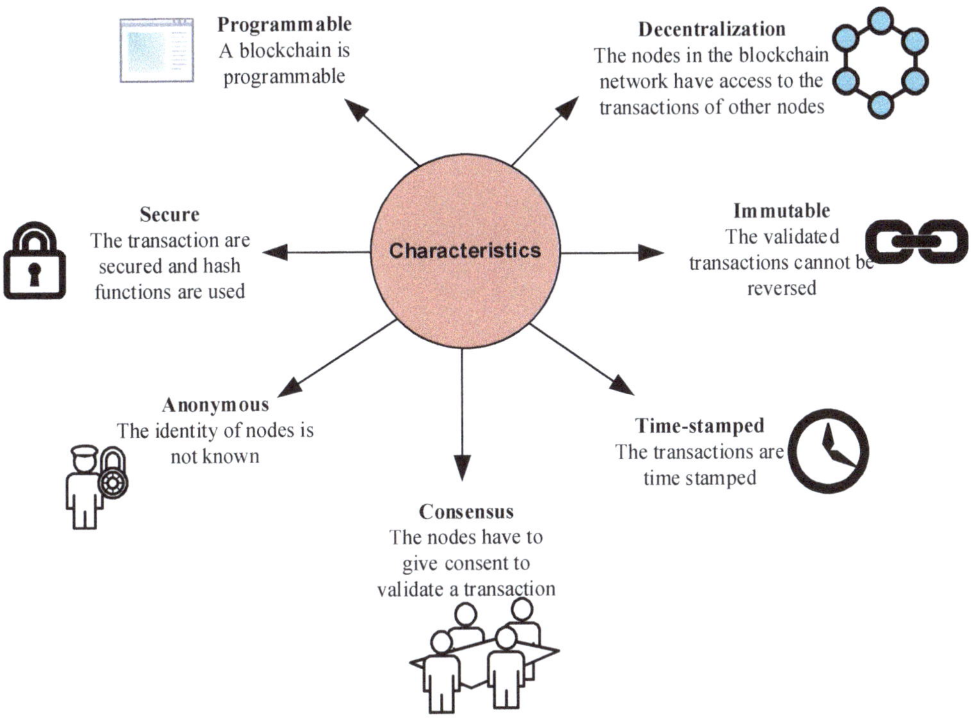

Figure 8. Characteristics of a blockchain.

Although blockchain technology has gained traction in future Internet systems, several difficulties must be properly addressed. Expertise in blockchain technology is critical, as the technology is still in the nascent stages. Adoption of BCT provides promised benefits in various fields, but the high initial infrastructure costs are a big worry for businesses. The deployment of blockchain technology is also influenced by privacy and security concerns. Scalability and legal requirements are also significant obstacles to its implementation.

3. Blockchain for Smart Grid

Blockchain technology has much potential to transform applications by creating more trust and increasing decentralization. Despite its rapid growth, its advantages are not being aggressively exploited by the SG applications. The number of articles published on blockchain from the perspective of the various SG applications is shown in Figure 9. These statistics were taken from the Scopus database and considered only articles published in journals.

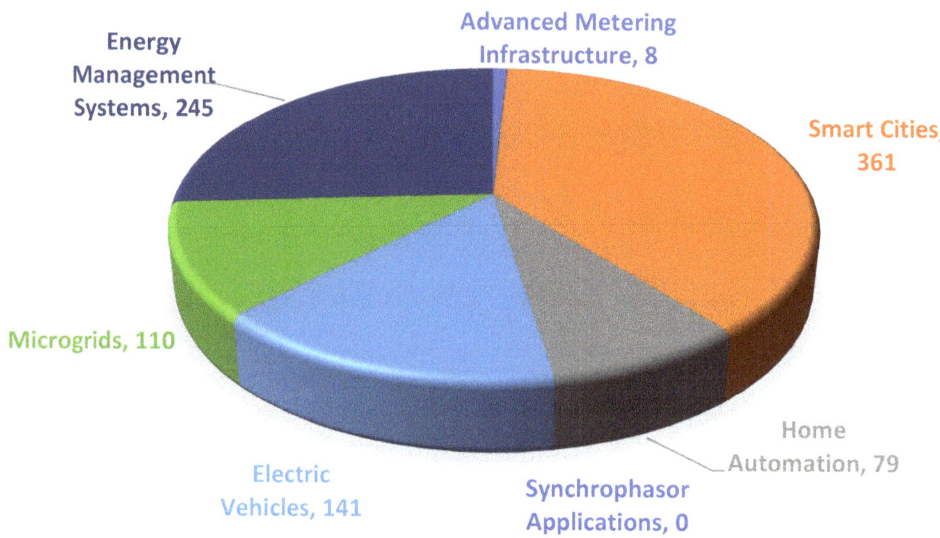

Figure 9. Publication statistics on blockchain for SG application.

Blockchain is widely adopted for energy management applications in an SG. Blockchain is also widely used for SCs, EVs, and MGs. SPAs, responsible for the wide area monitoring and control of the grid, are not employing blockchain technology for decentralizing the process. Only four conference articles reported the use of blockchain technology for SPA. In this section, blockchain technology will be explored from the perspective of these applications.

3.1. *Blockchain for Synchrophasor Application*

The major outages across the globe, such as those in Brazil in February 2011, the Pacific Southwest in September 2011, India in July 2012, Vietnam in May 2013, the Philippines in June 2013, Bangladesh in November 2014, etc., have necessitated the wide-area measurement system (WAMS) in the SG [28,29]. The WAMS is a comprehensive solution to monitor, control, and maintain the SG by incorporating the state-of-the-art infrastructure, emerging technology, and tools.

Recently, synchrophasor technology emerged as a viable solution for the WAMS. The synchrophasor technology enables WAMS to monitor, control, and coordinate the SG in real-time and precisely [30]. The fundamental architecture of the synchrophasor measurement system involves a phasor measurement unit (PMU), phasor data concentrator (PDC), and the communication network [31]. The PMUs are high-speed sensors that monitor the grid in real-time by measuring the grid voltages and currents. These measurements are time-synchronized using the global positioning system (GPS) and communicated to the PDC, which acts as an aggregator. The time-synchronized measurements of PMUs are referred to as synchrophasor data.

The communication network acts as a backbone since it provides the infrastructure for communicating synchrophasor data between PMUs and PDCs [32]. The more generic architecture of WAMS comprises decentralized architecture where the devices are hierarchically arranged. The decentralized hierarchical architecture of the WAMS with three levels of hierarchy is shown in Figure 10. A local PDC may be located close to the microgrids, aggregating synchrophasor data from several PMUs in a power grid. Further, there may be a master PDC that aggregates data from several local PDCs. Finally, the data from several master PDCs may be aggregated by a PDC known as a super PDC located at the regional level, which is the highest level in the proposed hierarchy.

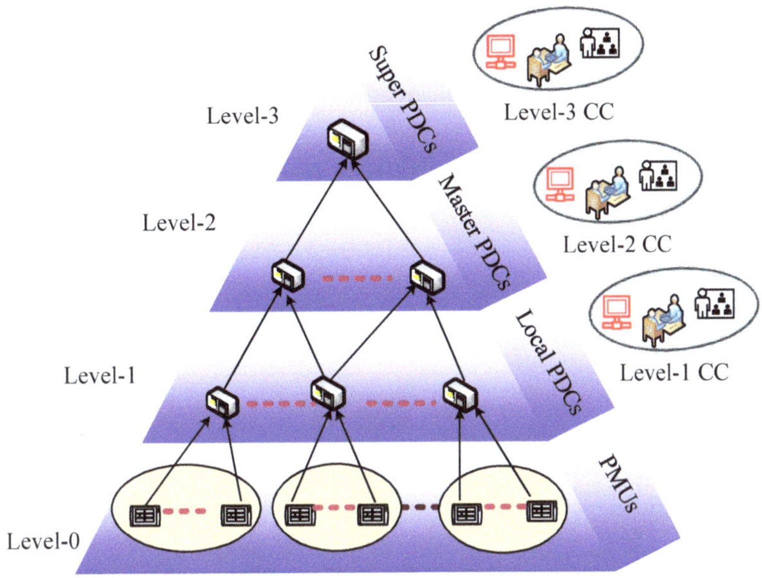

Figure 10. Hierarchy in a WAMS.

The data pertaining to the health of the grid can be used in WAMS for state estimation, stability analysis, situational awareness, etc., of the SG and its other operational-related functionalities. However, such data, typically referred to as synchrophasor data, can be exploited by cyber-attacks such as denial of service (DoS), distributed denial of service (DDoS), false data injection, spoofing, data tampering, etc. [33]. These attacks put the WAMS at risk, and its efficacy becomes questionable. The risk identification and assessment of smart grids is thoroughly discussed by Jha et al. in [34], where the authors considered risk assessment analysis of smart grid communication networks. The blockchain can be used with synchrophasor technology to mitigate the risk of cyber-attacks in a WAMS. Additionally, blockchain technology can simultaneously enhance the robustness, reliability, and integrity of the synchrophasor data by incorporating a decentralized peer-to-peer approach to communicate synchrophasor data in a WAMS.

3.1.1. Blockchain Architecture for SPA

The blockchain architecture for the SPA in an SG will consist of three fundamental components:

1. The member nodes, which are the PMUs or the PDC. Each node generates its synchrophasor data and shares it using the IEEE C37.118-2 [35].
2. A shared ledger containing the synchrophasor data collected by all the member nodes.
3. A peer-to-peer distributed network between the member nodes.

The architecture of a blockchain for SPAs is shown in Figure 11. As shown in the figure, the PMUs are connected in a fashion to create a distributed peer-to-peer network where all PMUs are enabled as member nodes. Each PMU is responsible for collectively updating the shared ledger. The synchrophasor data from a PMU is referred to as a synchrophasor transaction. The synchrophasor transactions are generated by PMUs which can be verified using authentication methods such as the elliptic curve digital signature algorithm. Despite this authentication, it is quite possible that the false identity of a PMU can be created to obtain access to the network causing danger to the resources. Such an attack can be mitigated using device identity validation methods such as the Bloom filter-based PMU identity validation approach.

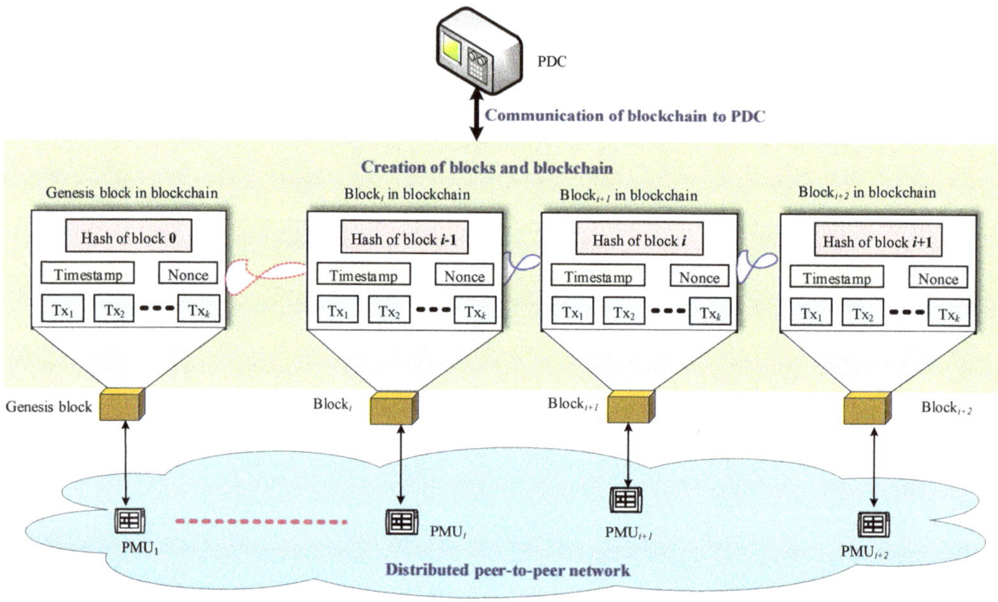

Figure 11. Blockchain architecture for SPA at the level of PMUs.

The PMUs are connected in a fashion to create distributed peer-to-peer network. Each PMU in the distributed peer-to-peer networks acts as a member which mines the block, where the synchrophasor transactions are included in a block. The contents of the block are shown in Figure 12.

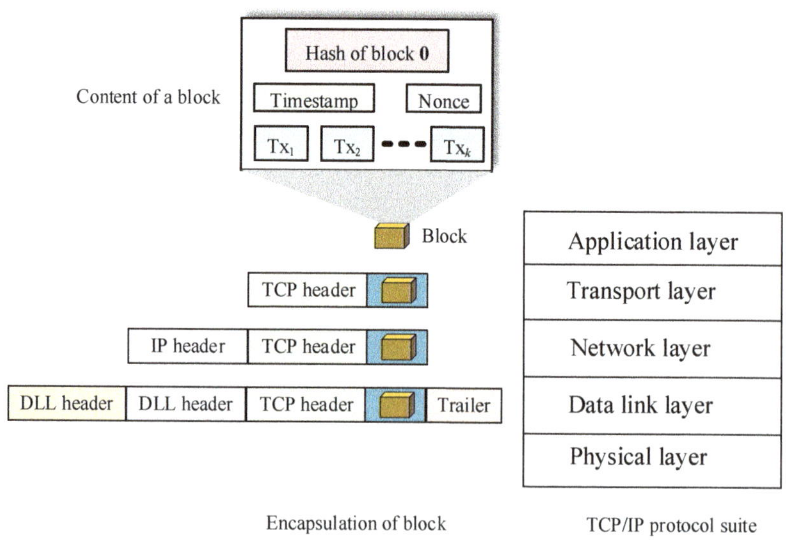

Figure 12. Contents of the block in a blockchain-based SPA.

Each block is generated using the IEEE C37.118-2 standard. The block is encapsulated with other protocols for communication over the TCP/IP network. The PMUs are responsible for consensus execution and block validation. There are several approaches for consensus execution and validation in blockchain technology. For SPA, the Markel tree-based approach can be used for consensus execution as it converges quickly without compromising the integrity of the synchrophasor transactions. Further, PMUs can follow consensus based on proof-of-work (PoW), where nonce is searched, which is a random number. When all the synchrophasor transactions grouped in a block are validated and PoW is completed, only a block is considered successfully mined by the PMU. On validating with PoW, the newly created block is appended to the existing chain to update the blockchain. The first block in the blockchain is a genesis block, which a PMU in the network can generate. It is imperative that any PMU can validate any number of blocks and receives the whole existing blockchain from executing the consensus and PoW. The decentralization can also help remove the PDC, and the PMUs themselves can take commensurate actions based on the measurements available from other PMUs.

3.1.2. Challenges and Solutions for the Implementation of Blockchain-Based SPA

PMUs operate at a very high rate, typically 30–60 samples per second in a time-synchronized manner. Hence, the additional functionalities of creating the blocks and validating burden the device and hampers the granularity of its measurements. An alternative solution to this problem is implementing the blockchain at a higher level in the hierarchy, i.e., at the local PDCs. The architecture for implementing the blockchain at the level of local PDCs is shown in Figure 13.

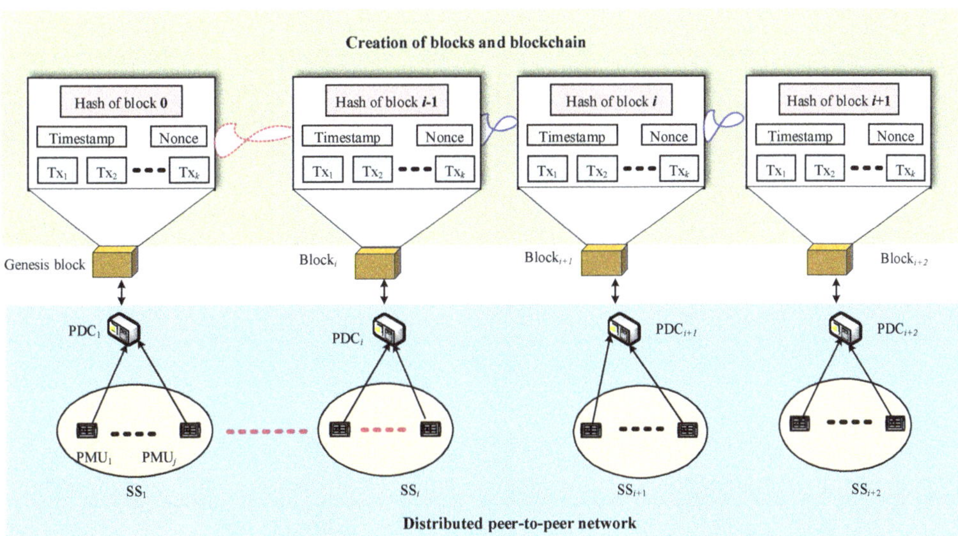

Figure 13. Blockchain architecture for SPA at the level of local PDCs.

SPAs are mission-critical, so it becomes computationally intensive to validate all the transactions. A solution to this problem is to terminate the chain at periodic intervals and start a new chain. This reduces the secureness of the chain, so additional measures will be needed to maintain the security. Because of the problems of the mission-critical nature of the application and the high data rate of the PMU, not many works are available on this topic.

3.2. *Blockchain for Home Automation*

A smart house is an integrated Internet of Things (IoT) domicile that provides users security, health, comfort, and a higher standard of life, among other benefits. People's life and independent living are made easier with smart home solutions. They provide valuable capabilities such as behavior tracking and safety evaluations, which have drawn the attention of consumers and device makers. Although intelligent homes provide significant benefits to homeowners and other interested parties, they are vulnerable to harmful cyber-attacks that risk users' safety and privacy [36]. Traditional solutions to such dangers exist, but they are extremely centralized and prone to large-scale attacks. As a result, the adaptability and scalability needed for effective utilization in the cutting-edge field of autonomous smart home applications and facilities are absent. Several clever technologies make life easier for individuals. Such programs generate enormous volumes of data. The archiving of this ever-changing material into repositories raises security problems. In cybersecurity technologies with remote connectivity and data transmission, blockchain has performed well. Thus, it is being employed for home automation applications.

3.2.1. Blockchain Architecture for HA

Home automation involves several smart devices, such as smart TVs, lights, etc. These devices monitor and control the various parameters of the house, which operate independently or are coordinated by a user. The interconnectivity of these smart devices is required to achieve the objective of HA. The interoperability challenges between the smart devices are handled using an IoT gateway. Users from one home cannot control the devices of another home to avoid a security breach. The service provider is responsible for providing necessary recommendations to the user for controlling smart devices based on prediction algorithms. The service provider can use machine learning algorithms for better recommendations or predictions. The blockchain network is used to connect different users and service providers to enhance security in the HA [37]. The blockchain network may be built using Ethereum or Hyperledger. The general architecture of blockchain for HA is shown in Figure 14.

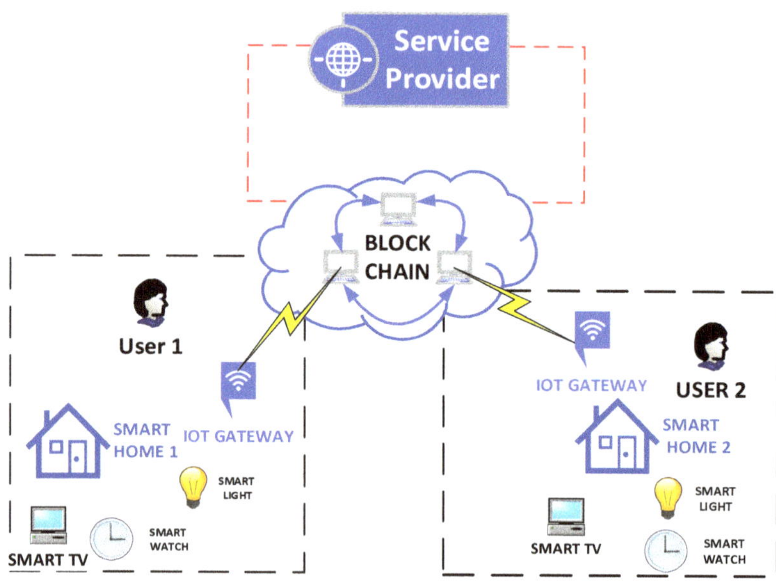

Figure 14. The architecture of blockchain for HA.

The user within the house can control the entities within his home; he cannot have access to the entities present in another smart home. The various devices in the home can be directly connected to the blockchain network through the gateway. The data from the devices can be placed into the blocks, which are then chained together using the hashing mechanism of the blockchain. The service provider can analyze the data and send suggestions to the users, but he cannot directly control the devices in the smart home. This architecture can be customized based on the user's specific requirements by the service provider. The various devices in the home can be directly connected to the blockchain network through the gateway. The data from the devices can be placed into the blocks, which are then chained together using the hashing mechanism of the blockchain.

3.2.2. State-of-the-Art on Blockchain for HA

The literature on blockchain for HA applications discusses access control mechanisms, homecare systems, utility payment services, etc. These works are summarized in Table 2.

Table 2. Review of the works on blockchain for HA applications.

Reference	Domain	Blockchain Mechanism Used	Summary
[38]	Access control	Private blockchain	For access control in smart homes, which is computationally fast and economical but is susceptible to malicious attacks.
[39]	Home care	Ethereum blockchain	Provides a secure means of sending healthcare data to the healthcare center, but has increased overhead.
[40]	Home care	Private blockchain	Reduces communication overhead for sending patient data but has more overhead.
[41]	EV charging bill payment	Lightweight basic blockchain	Reduces the size of block for payment of charging bill. This is also vulnerable to security attacks.
[42]	Home care	Consortium blockchain	Data of the aged people is stored efficiently with enhanced quality but is susceptible to DoS attack.
[43]	Authentication mechanism	Ethereum blockchain	A scalable but expensive mechanism for authentication of IoT devices.
[44]	Automated payment	Bitcoin blockchain	A highly scalable automated payment system that also allows off-chain transactions.
[45]	Lightweight payment system	NA	A low-power and fast payment system. This may be susceptible to malicious attacks.

3.2.3. Challenges and Solutions for the Implementation of Blockchain-Based HA

Various blockchain systems are being used for HA applications [46]. These systems have their specific data format, and their interoperability is challenging. Additionally, the consensus algorithms used by these systems are different. For seamless interaction, standardization of blockchain systems is required. Another challenge to implementing blockchain for HA applications is the real-time analytics of streaming data. The data have to be processed and analyzed in real-time. For example, an intruder detection system requires real-time face detection. Processing blockchains for real-time applications is challenging. A possible solution is to use a lightweight framework for this application.

3.3. *Blockchain for Advanced Metering Infrastructure*

The heart of the AMI is a smart meter used to collect, monitor, and communicate the data related to energy consumption corresponding to every user. The meter data are used differently by different entities. For example, the grid operator can use this data for load forecasting and planning, and the market operator can use smart meter data for dynamic pricing and billing. On the other hand, the users can use such data to manage their electricity usage. Whereas AMI provides ample advantages, secure AMI data transaction is challenging. The blockchain-based AMI plays an important role in achieving this objective. A generic framework for implementing AMI using blockchain is shown in Figure 15.

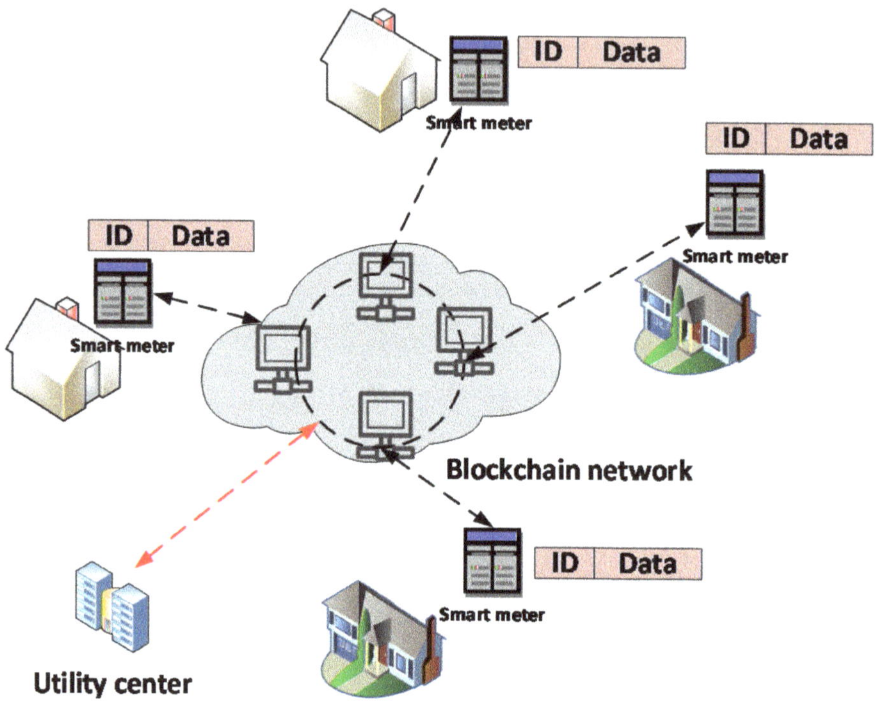

Figure 15. The architecture of blockchain for AMI.

The smart meters can be directly connected to the blockchain network through the gateway [47]. The data from the meters contains meter IDs and other utility-related information as per the IEC 62056 protocol. These meters are connected to the servers or nodes inside the blockchain network that create the blocks using the data received from the AMIs. These blocks are then shared with all other nodes inside the blockchain-enabled network. This network can only be accessed by the nodes related to the utility center and so should be a private blockchain network. The private blockchain can be used for smart contracts and validations to provide energy utilization transparency without compromising security and privacy.

Challenges for the Implementation of Blockchain for AMI

Blockchain has not been used widely for this SG application despite its utility. Researchers have used it to enhance the security of AMI applications. In ref. [48], a lightweight blockchain-based framework was proposed to enhance AMI's security. The framework was secure against attacks, and its energy consumption was low. In ref. [49], blockchain was used to preserve the integrity of the customers using AMI. As with the blockchain for HA applications, the blockchain for AMI is also plagued with interoperability and real-time constraints.

3.4. Blockchain for Electric Vehicles

The technological evolution of electric vehicles (EVs) and the rapid growth of the smart grid have led to the emergence of new connectivity structures—vehicle-to-grid (V2G) [50]. In the future, the importance of EVs using technologies such as the Internet of

Vehicles (IoV) [51] or the Internet of Things (IoT) [52] will increase, as it offers innumerable advantages, for example logistics companies provide fixed charging stations (CSs) for their fleet of vehicles.

Interconnectivity requirements with all technology systems in the real world have led to the emergence of vehicle-to-everything (V2X) technology [53], using integrated vehicle sensor platforms that use the centralization of various functions through an integrated EV server, connected by a series of connectivity devices such as CAN, LIN, Wi-Fi, and Bluetooth technology [54]. The results of V2X performances are based on a series of information on the collection and dissemination of multi-networks and technological capabilities between electric vehicles.

The security factor, the speed of data transfer between interconnected vehicles, and the wide coverage of telecommunications systems led to the emergence of 5G networks and their distribution very quickly in the world [55]. The infrastructure of multi-networks communication systems through 5G technology has the power to process applications at a superior level. The 5G network drives the V2X protocol, generating many scenarios for data management by promoting the development and integration of blockchain applications [56]. The implementation of blockchain systems in the vehicle-to-everything protocol tends to reinvent intelligent transport systems, leading to high efficiency of transport and road safety services [57].

3.4.1. Architecture of Blockchain for EVs

The general blockchain architecture for the EV application is shown in Figure 16. The blockchain-based EVs infrastructure requires regular nodes to capture mobile cars' dynamics. These nodes are responsible for smart contracts and block validations, forming the basis of the blockchain. The mobile cars send their data to such regularly placed blockchain nodes or access points. The interconnectivity between mobile electric vehicles and nodes is through WiFi. An ID number uniquely identifies each EV. The data that an EV sends to the access point involve the battery status, vehicle status, bill payment for charging, etc. The data are placed into the blockchain network by the access points as blocks. The various nodes in the blockchain validate the transactions. The blockchain network is also accessible to the transport authority, who can continuously monitor the status of the EVs and send personalized recommendations or warnings to the EV user. However, the transportation authority cannot change the parameters of the EV.

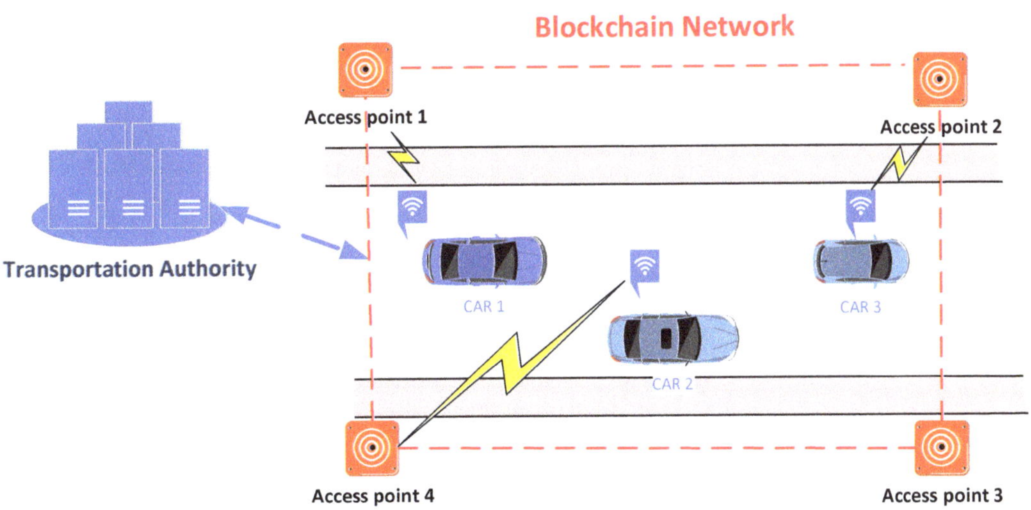

Figure 16. Architecture of blockchain for EV applications.

3.4.2. A panoramic Overview of Blockchain for EV

Despite abundant opportunities to incorporate blockchain in EV applications, some of the challenges are inherently present in the blockchain-enabled EV system, such as mining of the data, cybersecurity, handling of voluminous data, etc. Furqan Jameel et al. [58] provided a solution for unloading mining tasks in vehicle-to-everything cellular networks. A short block length transmission architecture has been proposed to meet the low-latency requirements for cybersecurity applications of EVs. In practice, finite block length architecture is a fairer approach to modeling blockchain networks. The inspiration for the theoretical application of adopting games defines a type of challenge for solving mining tasks and efficiently unloading them to clusters. The advantage of using blockchain databases ensures good data transfer rates and maintains the vehicles' fairness in the unloading process.

However, a significant disadvantage is the scalability of the data chains within the blockchain, which can be a design problem. Because data security is the main issue in conventional blockchain networks, the impact it has on the process of downloading data into electric vehicles is a real challenge. However, in ref. [59], the authors proposed a new coding sequence—Secure V2X—that capitalizes on the characteristics of the blockchain and the data networks protecting the confidentiality and security of the V2X protocol.

In addition to the benefits that blockchain initiates in low-security areas, confidentiality is the main issue in trading energy in a collective network type peer-to-peer (P2P) (E-trading). In recent years, electric vehicles have received worldwide recognition due to their potential in the green transportation system. The rapid development of technologies in smart communication networks has allowed EVs to relate to the environment. The electricity production costs are constantly decreasing through the implementation of renewable energy sources and smart grids [60]. Thus, the major challenge of peer-to-peer technology, E-trading and D-trading [61], and integration for electric vehicles is the development of a secure communication architecture that maintains data confidentiality and information anonymity. In addition, the objective of the blockchain is to mask trading relationships without compromising data integrity [62].

Various review papers in the literature focus on blockchain technologies applied in the Future Smart Grid [63,64]. Although the technology is considered one with a wide range of advantages, security needs to be assessed systematically to enhance reliability of the SG [65].

Motivated by previous development, Marina Dorokhova et al. [66] proposed integrating electric vehicle charging systems based on blockchain technology. The study is based on a popular blockchain platform, Ethereum, for interconnecting EV infrastructure and real-world infrastructure [67]. The advantage it offers is the crediting in the safety zone of the energy flows between the owners of electric vehicles and the companies that own charging stations. The only barriers that could be removed in the future are the limitations of the blockchain-high transaction costs due to network loads, high power consumption, or transactions that do not change in case of errors.

A case study by Shivam Saxena et al. [68] further demonstrated the need for techno-economic evaluation of residential energy trading systems. The EV is a part of such system, which can be enhanced through the blockchain. Using blockchain in EVs not only improves the household's participation in the electricity markets but also drastically reduces the negative impact on the energy distribution network [69]. These seminal works are comprehensively summarized in Table 3.

Table 3. Summary of works related to blockchain for EVs.

Reference	Subdomain	Objectives	Solutions/Results	Technologies	Advantages/Opportunities	Challenges
[58]	V2X Communications	Efficient solutions for unloading mining tasks in cellular vehicle-to-everything networks	Adopting a game-theoretic approach to efficiently unload the mining tasks to the mining clusters	Blockchain-based cellular V2X networks	Good data transfer rates and maintain the fairness of the vehicles in the unloading process	Scalability of the data chains within the blockchain and the impact of data security in the process of downloading data into EVs
[59]	Secure V2X Communications	Network performance	Deploying a novel framework (Secure V2X)	Blockchain and NDN (named data networking)	Protecting the confidentiality and security of the V2X protocol	Without the right cluster, the Secure V2X sequence do not helps to improve network performance
[61]	Energy trading and charging payment system for EVs	Employing blockchain technology to provide trust between users	Maintaining the data confidentiality and information anonymity	Private blockchain	Improves the distribution network and renewable energy network	Development of a secure communication architecture
[65]	Charging Management	Integration of EVs charging systems interconnected with real world infrastructure	Charging management framework	Ethereum blockchain platform	Crediting in the safety zone of the energy flows between the owners of electric vehicles and the companies that own charging stations	Limitations of the blockchain-high transaction costs and high power consumption
[67]	Residential communities	Technical -economic evaluation for residential energy trading systems	Residential energy trading systems	Blockchain platform	Reduces the impact on the energy distribution network	Peak energy demand is very high

3.4.3. Challenges and Solutions for the Implementation of Blockchain for EVs

The scalability of blockchain data chains, data security in the download process, and confidentiality are challenges that are yet to be addressed. The major challenge of peer-to-peer technologies is the processing of energy transactions and the anonymity of information. The high resource requirement and transaction cost in terms of energy consumption plagued the use of blockchain technology for EV applications with WSN infrastructure. Overcoming these limitations would make blockchain technology the main key factor for EVs. The development of lightweight blockchain algorithms for reaching consensus in real-time can be a probable solution.

3.5. Blockchain for Renewable Microgrids

With every day passing, there is a continuous transition and evolution to a renewable grid that is based on various distributed energy resources such as photovoltaics, fuel cells, microturbines, batteries, etc. These transitions rely on the successful deployment of blockchain technology.

3.5.1. Architecture of Blockchain for MGs

The generalized blockchain architecture for the MG application is shown in Figure 17. In general, the power grid of a zone is sprawled over a large geographical area where different MGs are considered. The different MGs are interconnected using the blockchain network. The blockchain network aims to enhance security and privacy in the MG operation without hampering transparency and data integrity. The data block carries information regarding the energy generated, energy to be shared with other microgrids, etc. The data pertaining to the MG are grouped into the blocks where each newly generated block is validated using a consensus algorithm. The block is then placed onto the blockchain network and is added to the blockchain after being validated. The nodes in the blockchain need proper algorithms to reach a consensus on the energy being traded, the price at which the electricity is being traded, etc.

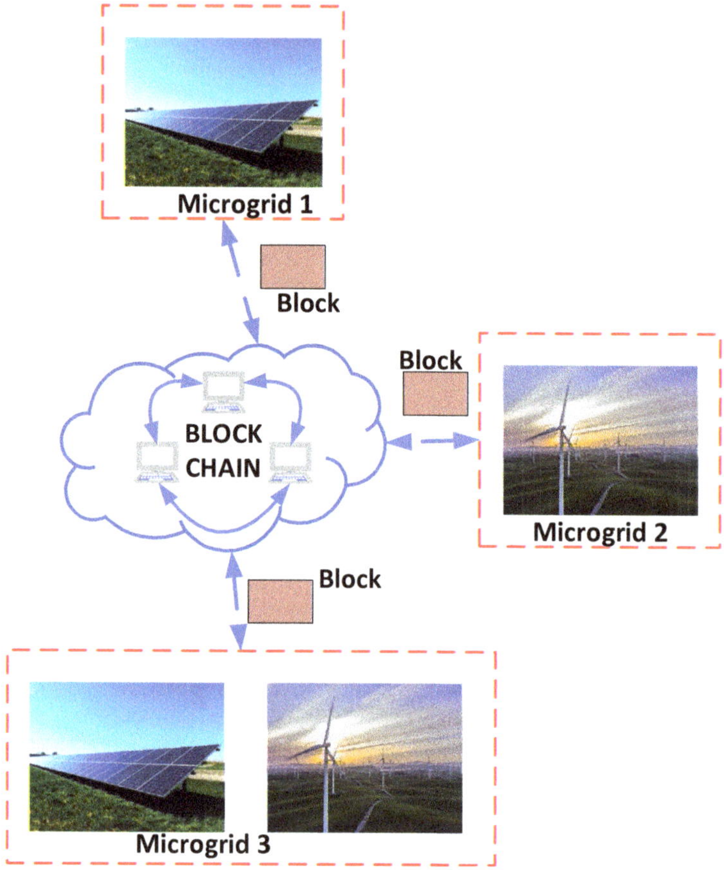

Figure 17. Blockchain for microgrids.

3.5.2. A Panoramic Overview of Blockchain for MG

Early inquiries about the energy sector with the accent on the smart grid and microgrids are mainly found in refs. [70–73], where different requirements, technologies, architectures, trends, and cyber security issues are largely debated.

With rising social, economic, political, and environmental concerns and strategies such as increasing power consumption, dealing with the middleman, market liberalization, pollution, etc., blockchain is seen as a promising solution in renewable microgrids for efficient operation such as complex point-to-point transactions between producers, traders, and users using elaborate algorithms in order to validate, secure, and record these transactions.

The different authors reviewed blockchain in the context of microgrids from several perspectives. In ref. [74], the need for blockchain, benefits, and challenges was reviewed. In ref. [11], real solutions such as the Brooklyn Micro Grid based on the blockchain environment with the Proof of Work (PoW) mechanism were presented. Other comprehensive reviews can be found in ref. [75] that can serve as quality background research for those who want to propose and implement feasible solutions and methodologies for renewable microgrids based on blockchain technology.

On the other hand, many works propose different solutions and approaches that use blockchain technology to enhance and improve microgrids and their applications. To start, ref. [76] proposed an approach for using blockchain on the Dominican Republic's electricity market, referred to as the main step in empowering automatic management of economic transfers with funds authenticating and supplier's guarantee. The approach presents an economic and energy blockchain-based flow to decentralize the current flows that involve total control through banking operations. Of course, such an approach must face serious economic interests, political regulations, and technological limitations to achieve its goals, but it is seen as a first step in applying blockchain in the electricity sector.

In ref. [77], a local energy market model using private blockchain via home energy management and demurrage mechanisms was presented. In the proposed model, there are three major actors: a small community (several microgrids) that uses photovoltaic systems as renewable energy as the prosumers; the consumers; and the main grid. Using Critical Peak Price (CPP) and Real-Time Price (RTP) schemes, simulations showed that costs were significantly reduced. Moreover, ref. [78] proposed a blockchain-based decentralized market mechanism to establish the price using auction methods. However, this is plagued with two limitations: difficulty in selling the oversupply of energy through auction and big challenges in ensuring privacy and security. Another solution for P2P energy trading was presented in ref. [79] by implementing the blockchain-based decentralized energy flexibility market for P2P transactions among prosumers. Two additional frameworks, first as decentralized blockchain-based and second as semi-decentralized, can be found in ref. [80], where the P2P energy trading subject was analyzed.

A more applied approach is in ref. [81], which proposed a method for an effective P2P blockchain-based energy market between a microgrid and the smart grid (IEEE 24-bus test system) where the function of distributed consensus algorithm was evaluated in the presence of Fault Data Injection Attack (FDIA). The main findings of this paper showed that the consensus process keeps running even if the case of a cyber-attack and the output response of the P2P market is very close to the centralized energy market. Maintaining the idea of the applied solution, the authors in ref. [82] suggested a model for an integrated energy management platform based on blockchain technology and, at the same time, implement a bilateral trading mechanism with simulation results showing significant optimization of the energy flow in a microgrid. Another model for blockchain-based energy management was suggested in ref. [83], where a Pythagorean fuzzy method was used in choosing the best solution for energy production, distribution, and waste control.

Further, in ref. [84], another P2P energy trading mechanism between microgrids based on the same technology using a fuzzy meta-heuristic approach as a pricing solution was presented with results showing increased profitability and reduced CO_2 emissions. Additionally, the fusion of the electricity market and blockchain was studied in ref. [85], where transactions were highlighted using multi-agent cooperation and sharing platform based on the Ethereum private blockchain, with results revealing several benefits such as transparent transactions and intelligent mutual trust.

Going deeper and deeper into the heart of the topic of this section, we arrive at the point where blockchain applications variates in terms of the constructive technology that microgrids are built on, this referring to AC microgrids, DC microgrids, or hybrid AC-DC MGs [86–90]. First, blockchain was used in ref. [86] to increase the security for interconnected hybrid AC-DC microgrids using a modified sine cosine algorithm to achieve the optimal decision in the shortest time and with high accuracy. The approaches in refs. [87] and [88] are based on the blockchain technology for energy management concerning DC and hybrid AC-DC microgrids using different strategies such as fuzzy logic control or the whale optimization algorithm. These seminal works are comprehensively summarized in Table 4.

Table 4. Material summary—blockchain for renewable microgrids.

Reference	Subdomain	Objectives	Solutions/Results	Technologies	Advantages/Opportunities	Challenges
[76]	Local energy market	Replacing transactions based on banking entities with cryptocurrency-based transactions	Economic and energy blockchain-based flow	Public blockchain	Funds authenticating and automatic control of transactions	Political regulations, economic interests and technological limitations
[77]	Local energy market/microgrid/smart grid	Optimizing energy consumption and minimizing electricity costs.	Reduced electricity cost and optimized energy consumption, especially at peak hours	Private blockchain with PoW mechanism	Optimal electricity cost for each time slot and local energy demand and generation balance	Implementing penalty policy
[78]	Microgrid/smart grid	Minimizing electricity costs	Decentralized market mechanism	Private blockchain	-	Selling oversupply
[79]	Local energy market/microgrid	P2P energy transactions	Electricity costs reduced	Public blockchain	Control of energy generation and flows and full ratio of self-consumption from renewable energy	Political regulations, economic interests, and technological limitations
[80]	Local energy market/microgrid	P2P energy transactions	Decentralized proposed framework and semi-centralized proposed framework	Solc, Mocha, React.js, Next.js, Ganachecli, Metamask, Ganache-cli, and Web3	Framework 1 uses more transactions, is less flexible and more secure/Framework 2 uses less transactions is more flexible and less secure	Smart contract limitations
[81]	Microgrid/smart grid	Ensure security and achieve consensus when cyber-attacks occur	Proposed architecture	Either public or private blockchain	Efficiency against attacks	Transaction security
[82]	Microgrid	Optimize the energy flow in a microgrid	Proposed model/optimized energy flow	Private blockchain	Reduced import costs	Security and communication efficiency
[83]	Renewable energy	Energy management	Proposed methodology and framework	Either public or private blockchain	-	Technological infrastructure and investment prices
[84]	Local energy market/microgrid	P2P energy transactions	Proposed fuzzy meta-heuristic approach	Either public or private blockchain	Encourage P2P energy transactions	Security and risks concerns
[85]	Local energy market/microgrid	P2P energy transactions	Proposed trading platform	Private blockchain	Transparent transactions	Political regulations, economic interests, and technological limitations
[86]	Hybrid AC-DC microgrid	Increase security	Proposed framework	Public blockchain	Increased security	Power injection limitations
[87]	DC microgrid	Energy management	Proposed framework	Either public or private blockchain	Maximum utilization of renewables	Political regulations, economic interests, and technological limitations
[88]	Hybrid AC-DC microgrid	Energy management	Proposed framework	Private blockchain	Optimal energy management and secured transactions	Political regulations, economic interests, and technological limitations

3.5.3. Challenges for Implementation of Blockchain for Microgrids

Like the numerous advantages, many challenges must be overcome in the blockchain-based renewable microgrids [91,92]. These challenges refer to technological constraints, economic aspects, social uncertainties, environmental concerns, political and institutional limitations, and law, regulations, norms, or end-to-end privacy and security.

A feasible and efficient balance between key features such as security, energy management, constraints, and costs is still challenging. Different consortiums operate different microgrids, so it is important to analyze and decide on the correct algorithm or methods to use, the best technology, the most suitable investor, and a very well-trained team.

3.6. *Blockchain for Smart City*

With the development and use of blockchain technology, the Internet of Things (IoT), and Cloud Computing, rapid evolution can be observed in the smart city paradigm.

3.6.1. A panoramic Overview on Blockchain for SC

In refs. [93,94], some of the problems related to smart city transportation were debated. These works demonstrated that there are concerns in rethinking the transformations of localities in terms of improvement of public transport and logistics [95,96], water supply [97], green energy [98], environment [99], health [100,101], education [102–105], and economics [106–109] by using the blockchain, which offers the possibility to use distributed stored data, and performs transactions without intermediaries between producers and beneficiaries [106,109] without data security problems [107]. The blockchain architecture [93,94] is the one that will strengthen the importance of using smart contracts in the development of transactions between the parties. These contracts are triggered by operations (agreements) between the parties or are determined by sensors, actuators, or IoT tags [97]. So, the blockchain and smart contracts are the ones that contribute to the transformation of localities into smart cities, finding the optimal adequacy in the development of logistics, energy, environment, water quality, health, etc. Some seminal results of the prospective of blockchain on health care are summarized in Table 5, whereas its applications in other smart city domains are summarized in Table 6.

Table 5. Summary of literature on blockchain for smart city health care system.

Reference	Objectives	Solutions/Results	Advantages/Opportunities
[93]	Smart village architecture	Blockchain in healthcare	Raising the standard of living of citizens
[94]	Application of BC in the health system	Implementation of BC in healthcare	Data storage security, privacy, and integrity in online consultation
[95]	Application of BC technology in the healthcare	How to apply BC technology in health to monitor the patient's health	Real-time patient monitoring, efficient data handling
[96]	Public health in the smart society	Prediction regarding the health status of the population using BC	Modernization of the healthcare system with enhanced data integrity, security, and privacy
[97]	Development of a BC based platform for healthcare	Model-based platform as a solution for healthcare information exchange	Enhanced privacy and security using a combined approach based on off-chain storage and on-chain verification

Table 6. Summary of literature on Blockchain for Smart City.

Reference	Objectives	Solutions/Results	Advantages/Opportunities
[98]	Green energy marketing	Utilization of photovoltaic parks	use of green energy, reduction of pollution, sale of surplus energy, decrease the production price
[99]	Energy management	Low-cost solution to the energy system	Efficient trading, production quality, capitalization of energy surplus
[100]	Incorporation of green energy in irrigation	Smart irrigation system based on photovoltaic parks	Efficient trading, management, and utilization of energy for irrigation systems
[101]	Scalable network of smart cities with hybrid architecture	Development of a model for real-time processing of the edge nodes	Enhanced resiliency of the system
[102]	Security issues for the smart city	Blockchain utility in smart communities	An in-depth survey covering various perspectives of blockchain in smart cities
[103]	Social issues	Solving social solutions through blockchain application	Applications and research opportunities in the paradigm of a smart city using BC
[104]	Supply chain data management	Implementation of salient features of BC, viz., immutability, transparency, decentralization, etc., to improve the efficacy of supply chain management in the industry	BC chain-based food traceability system as a case study with the deployment of BC in order to enhance the efficacy of supply chain management in the industry

Table 6. *Cont.*

Reference	Objectives	Solutions/Results	Advantages/Opportunities
[105]	Models and applications with secure transactions	Through surveys, they identified research opportunities	Creating new applications and interoperability between models
[106]	Carbon emissions monitoring	A three-step blockchain that uses smart contracting	Enhanced security with advanced features
[107]	Efficient urban mobility	Traffic decongestion	Data transparency, immutability for enhanced resilient traffic management
[108]	Augmented democracy	Involvement of citizens in decision making	Determination in real-time of the persons participating in the elections of the citizens in a decentralized and confidential way
[109]	Synergy of IoT and BC	Use of multilayer blockchain	The technology used ensures the competitive efficiency of cryptographic security and confidentiality
[110]	BC for industrial IoT	Improving the performance of industrial IoT devices by minimizing unfair, permissioned BC	Development of novel algorithms considering waiting time for packing of permissioned BC data

3.6.2. Architecture of Blockchain for SCs

The prospects of the IoT determine the smart city architecture, the multitude of sensors and smart objects that help collect data collected from public infrastructure, public access to data, increasing the quality of services and costs of environmental protection, and economic development. The general architecture of blockchain for SCs is shown in Figure 18.

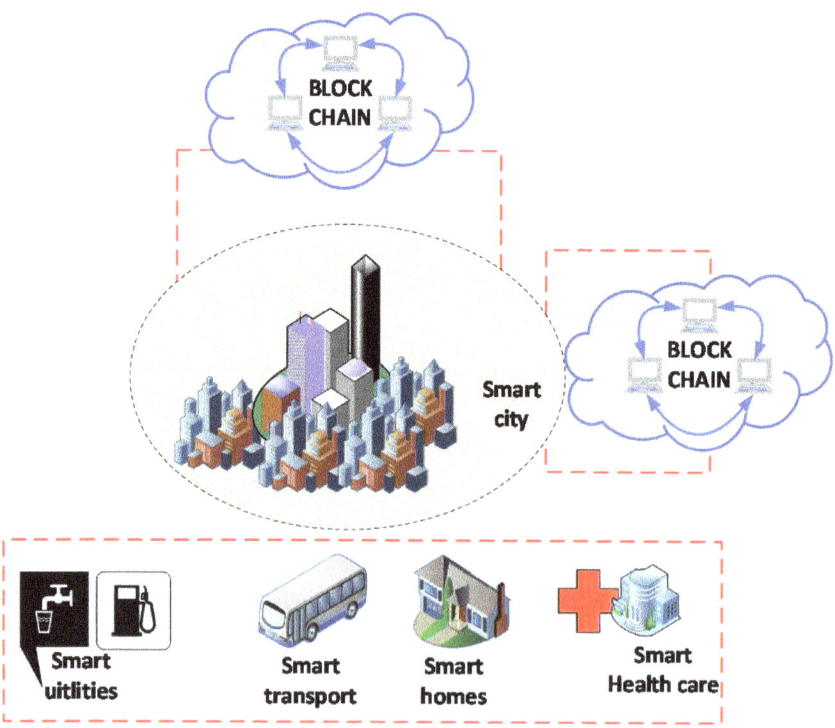

Figure 18. Blockchain architecture for SC applications.

Using the same blockchain network for all the services in the smart city is not feasible. Therefore, multiple blockchain networks have to be used depending on the size of the city and the nature of smart services provided. Each blockchain network may cater to the requirements of a single service. Smart devices, i.e., smart transport vehicles, smart sensors in homes, smart monitoring devices in hospitals, etc., generate data that is put into a blockchain related to its service. Proper protocols and blockchain frameworks will be needed to ensure the smooth operation of the services.

3.6.3. Challenges for Blockchain for SCs

SC has many different entities. The blockchain network used by the SCs' various entities varies with the application type. These applications have diverse requirements. For example, in the case of smart transportation, the devices are changing their locations, and in the case of smart lighting, the devices are static. The blockchain architecture must be planned according to the nature of the application. Additionally, the entities are spread over a large geographical area, and to meet the criterion for real-time analysis, the blockchain must be fast and secure. Additionally, interoperability between different blockchain networks in the SC is a challenge.

3.7. Blockchain for Energy Management System

Developing and implementing the distributed system, both in production and consumption and energy marketing, brought new benefits to producers and consumers. Moreover, the increasing energy use from wind turbines and photovoltaic panels necessitated changing the energy market's architecture and secure energy transactions. Blockchain technology can be used for this purpose.

3.7.1. Architecture of Blockchain for Energy Management System

The blockchain has enormous potential in the transaction related to energy marketing. EMS aims to ensure reliable energy trading in real-time, including all energy market entities such as generation systems (both renewable energy sources and non-renewable energy sources), customer systems, grid operators, etc. The blockchain architecture for EMS is shown in Figure 19.

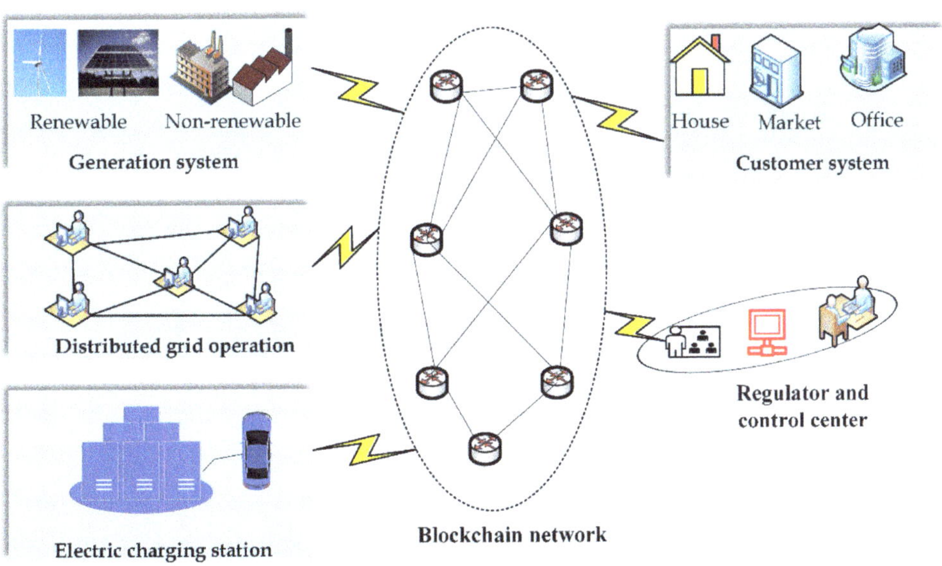

Figure 19. The architecture of blockchain-based EMS.

The SG envisioned integrating renewable energy sources with conventional energy sources as generation sources. On the consumer side, there are individual homes, residential buildings, offices, market complexes, etc. In addition to these, the EV charging stations also fall in the consumer domain of the SG. However, the consumer domain entities act not only as the electricity consumer but also as the electricity producer. Such consumers can be referred to as prosumers. When surplus electricity is available at prosumers, it is contributed to reducing the burden on the generation system.

On the one hand, it reduces the burden of the generation system, but on the other hand, it becomes vital to monitor the energy trading between users. Additionally, security and privacy in the energy trading market are equally important. To achieve this objective, blockchain can be integrated into the EMS.

The blockchain aims to integrate all domains of the SG, such as the generation system, operation system, the consumer system, regulator, and control center, using the blockchain network as shown in Figure 19. The blockchain-based EMS ensures the security and privacy of energy transactions through its distributed approach, interoperability, and smart contracts. The private blockchain can implement data permissions and selective consortium access to ensure security and privacy in energy trading. Due to distributed approach, blockchain-based EMS augments the transparency without compromising privacy in peer-to-peer energy trading.

3.7.2. A Panoramic Overview of Blockchain for EMS

The research on blockchain for EMS is gaining momentum and has been discussed in many recent works. As the amount of energy increases in trading, the greater the difficulties. So, this trading system needs to be controlled very carefully. An online energy transaction management model was proposed in ref. [111] where users can obtain information on their own trading and consumption through energetic transaction. For the transaction to be secure and fast, a payment plan was proposed based on the loan's value. Jiawei Yang et al., in ref. [112], propose a public pricing scheme based on the blockchain. The price is influenced by the share borne by the miners who are taxed with a part of the income for the power losses. The smart contract was created, although the testing was conducted with only 27 prosumers. The biggest problem when we talk about the price of energy in the trading process by using the blockchain is the high energy consumption used by this technology, which was resolved in ref. [113].

The security challenge was dealt with by Yi Zhang et al. in ref. [114] for users and energy flow. In ref. [115], S.N.G. Gourisetti et al. proposed an energy market framework using the online double auction. The authors explain the benefits and usefulness of blockchain technology and its use for transactional energy. The prognosis is that this technology and the implementation of smart contracts in stages can minimize and eliminate the challenge elements in key and certificate management. The authors stated that they expect that lower energy consumption will be achieved if users are more receptive.

Blockchain technology has allowed smart meters with enhanced security and privacy features. Further, a platform to monitor the energy generated from renewable sources by storing and trading energy between residents and network services of users was proposed in ref. [116]. The possibility of trading renewable energy generated by private producers using blockchain technology was shown in ref. [117]. The authors offer a high scalability solution based on smart contracts, which will not harm the decentralized system and data security. The costs of transactions made in this way will be lower compared with current blockchain costs. A cloud services platform for energy trading was proposed in ref. [118] by Lei Wang et al. Both users and suppliers participated in the platform, and the intelligent contract for trading between the parties was created. Antchain is used to make smart contracts, trade, and use the services offered by the cloud. The evolution over time of blockchain technology in the energy trading sector and the issues that stop the application of this technology were presented in ref. [119]. In ref. [120], the authors present the blockchain used by customers to pay for energy consumption. Some seminal work in this direction is comprehensively analyzed in Table 7.

Table 7. Summary of works on blockchain for EMS applications.

Reference	Subdomain	Objectives	Solutions/Results	Technologies	Advantages/Opportunities	Challenges
[111]	Secure energy transaction	Securing and controlling risks in energy transactions	Online transaction management is followed; Trading model; Smart contracts; Calculation of payment rates.	Blockchain	Real-time verification of individual consumer transactions.	System security with: proxy re-encryption and homomorphic encryption; Improvement credit trading management.
[112]	Energy price	Establishing the public price of traded energy	Trading with *elecoin*	Blockchain	Supervision of transactions by network members.	The power supply system should be extended to applications.
[113]	Blockchain performance. Blockchain-based virtual electricity generation	Decreasing the cost of electricity to supply the process during the operation of the blockchain.	Reduction of energy consumption during the mining process.	Blockchain	Solutions to increase the energy efficiency of the technology	Applying and deepening the study of the reduction of energy consumption consumed by the network.
[114]	Smart contract trading	Securing energy flow and users.	Value calculation according to users and producers divided according to certain criteria (diversity).	Blockchain	Differentiated tariffs taking into account a classification of producers and consumers respectively.	Classifying users and establishing quotas according to the green energy produced and consumption; The approach methods for energy production to be planned according to the real conditions have not been studied; Improving the smart contract system.
[115]	Energy market	Energy architecture objectives.	Stage implementation of smart contracts.	Blockchain	Increased security.	Key and certificate management. Coverage in a larger area of more general market/industry.
[116]	Energy trading	Energy trading between residents	Decentralized optimization algorithm, energy distribution according to a predetermined program for energy trading to the user network.	Blockchain	Efficient trading without decentralized intermediaries.	Improving energy management. The platform will be tested on a larger community of residents, by improving the algorithm.
[117]	Renewable energy	Energy trading	A blockchain scalability solution.	Blockchain	Low transaction costs.	Development and widespread use of blockchain energy trading.
[118]	Smart contract	Cloud services platform design for energy	Realization of the trading platform with intelligent contract.	Blockchain	Trading without intermediaries	Improving cloud services, adding value green certificate, energy storage and other services, application and early service in the integrated energy market
[119]	Blockchain evolution and challenges	The widespread use of blockchain technology in the energy trading process.	Use of the decentralized system; Smart contract.	Blockchain	Trading through a secure decentralized system.	Secure, decentralized energy development
[120]	Energy management-household consumers	Energy trading management between customers.	Use of electric power inverters in the network; Energy-saving technique testing.	Blockchain	Distributes energy from one home user to another within the decentralized network; Management performed for the purpose of energy distribution planning for the client; Communication networks are independent.	Energy network development.
[121]	Energy trading in microgrids	Beneficial energy trading.	Interactive double auction and blockchain technology	Blockchain	Lower price set by consensus of both producers and buyers.	Controlling the marketing of the amount of energy produced. Study the problems that occur when a network node has problems.
[122]	Energy transaction	Shared network study	Trading platform. Encouraging the use of renewable energy	Blockchain	Blockchain with distributed trading energy storage, is efficient and reliable.	High flexibility and security of the power system and subsequent exploration to be done together.

4. Blockchain for Cybersecurity in SG

The immediate need to incorporate renewable energy sources has necessitated considering a more diversified and distributed structure for the SG. This objective was achieved through distributed generation system and DER [123]. However, this has increased the complexity of the SG. Further, the SG's complex infrastructure comprises several devices such as the PMUs, smart meters, home automation sensors, remote terminal unit, spanning generation, transmission, distribution, customer, operation, marketing, and utility domains, etc. [124]. Situational awareness is vital to ensure the resiliency of such a marvelous SG infrastructure. The communication infrastructure and the communication protocols needed to support these applications vary. The core of the communication network is a wide area network (WAN). In addition to this, there exist other types of communication networks such as local area networks (LAN), home area networks (HAN), wireless sensor networks (WSN), neighborhood area networks (NAN), etc. These communication networks mostly use TCP/IP protocol suite for data communication. TCP/IP is not a secure protocol. Hence, the communication network of the SG applications can be easily attacked by exploiting its vulnerability. Despite the basic security measures such as firewall, intrusion detection, encryption, authentication, etc., which are already implemented in the SG, though it is still vulnerable to several cyber-attacks. An excellent survey on various detection algorithms was provided on false data injection in ref. [125].

The SG is a typical cyber-physical system [126]. As a cyber-physical system, cybersecurity is a vital parameter with three features: availability, confidentiality, and integrity. Availability is characterized as the property in which all data are available promptly. The cyberattack can compromise availability by blocking, delaying, and corrupting the data or even losing the data. The impact of cyber-attack on the availability of SG applications is huge. Confidentiality is characterized as the property of the system to protect the privacy and proprietary information from unauthorized access. The cyberattack on confidentiality can compromise the privacy and proprietary information of the SG application. Such incidents can grant illegal access to the application by stealing password-related information, causing enormous loss to the operation of the application. Integrity is characterized as the application's property to protect the system from unauthorized access to avoid any modification, alternation, and destruction of the data. The cyberattacks on integrity can modify the data to configure the application, resulting in an enormous loss. For example, the modification data can lead to misconfiguration of the sensors leading to failure of the SG application.

Blockchain is a distributed ledger that is immutable and does not depend upon any third party for its execution. This makes blockchain a secure method for data transactions and thus plays a vital role in SG applications. The blockchain can explicitly be used to mitigate the cyberattacks to strengthen the SG application's security. Among the different blockchains, the public blockchain is highly secure compared with the consortium and private blockchain due to the nature of the members and the consensus mechanism. The members of the public blockchain can be anonymous, whereas only the trusted nodes can be members of the consortium and private blockchains. The consensus mechanism followed in the public blockchain is proof-of-work, whereas multi-party voting in the consortium blockchain and strictly pre-approved nodes in the private blockchain are followed as a consensus mechanism. However, computational complexity is very high in the public blockchain. Thus, when security threats are fewer, and computation complexity is low, consortium and private blockchains are preferable to the public blockchain. The architecture of the blockchain for cybersecurity in SG applications is shown in Figure 20.

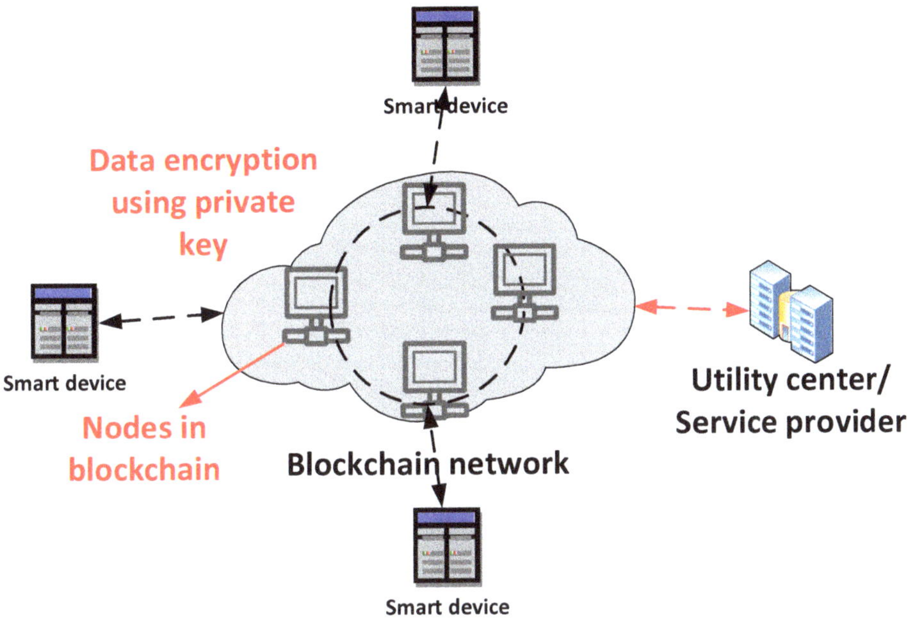

Figure 20. The architecture of SG cybersecurity using blockchain.

The smart devices generate data communicated to the blockchain network server using the TCP/IP protocol. If the devices are computationally powerful, the hashing of the data and its encryption can be performed at the device itself, thereby creating the block, which is then placed into the blockchain network. This is the most secure architecture, as any data tampering after it leaves the device results in a change in the has function, leading to the invalidation of the block. However, this puts much computational pressure on the end devices, which are already over-burdened by other tasks. The other alternative is to send the data to the servers/nodes in the blockchain using TCP/IP and generate the blocks in the blockchain. This is less secure, but it is not a computationally powerful smart devices. In the latter case, using private and public keys for extra authentication can be beneficial. This architecture envisions maximizing the security since all participants in the consortium blockchain are trusted, and the consensus mechanism is based on multi-party voting with no scope for anonymity. The administrative and management authorities select the member nodes acting as miners for the consortium and private blockchain. Next, the works related to blockchain for SG cybersecurity are comprehensively summarized in Table 8.

Table 8. Summary of works on blockchain for SG cybersecurity.

Reference	SG Application	Summary
[127]	AMI	A quantum key distribution-based secure key transmission is proposed for increasing the security of smart meters against cyber-attacks
[128]	Applicable to all	A multi-layer protocol is proposed to enhance the cyber-security of SG applications.
[81]	EMS and MG	A blockchain framework for P2P energy transactions is proposed, using a novel consensus algorithm for enhanced cyber security.
[129]	EMS and MG	A novel blockchain hyperledger is proposed for secure transactions on energy distribution.
[130]	MGs	A master-slave mechanism is proposed to protect the data against malicious attacks.
[131]	EMS	A novel rewarding scheme is presented for network security. Additionally, smart contracts are used for safe data storage.

Blockchain technology for SG applications is still in the research phase and is gradually finding practical utility. Secure mechanisms are needed that can be implemented at the device level before the data leave the device. These mechanisms should be light and can be implemented in real-time.

5. Conclusions

SG is evolving with the developments in storage and computational technologies. One such technology that can potentially transform the transactions amongst the various entities of the SG is the blockchain. The blockchain offers a decentralized and secure means of authorizing transactions, removing the need for a centralized authority. Despite its tremendous application in other domains, it has been underutilized for SG applications. This paper reviewed blockchain technology from a utility perspective for SG applications. General architectures were proposed for the important SG applications and identified challenges. The review is expected to enhance the research on developing novel technologies to meet the requirements of practical SG applications.

The blockchain-based applications are still in the nascent stage from various perspectives, which are seen as future research problems. Many SG applications operate in real-time, and the blockchain should not overburden the applications. The resource requirements for computation are a major challenge in blockchain-based systems. Blockchain must be developed to work on a lighter framework while retaining its security features. Additionally, regulatory bodies have to develop standardization procedures to make this technology interoperable and popular. Some of these research problems can be solved in the future, thoroughly revolutionizing blockchain-based applications.

Author Contributions: Conceptualization, B.A.; methodology, B.A.; software, S.K.M. (Sunil Kumar Mishra), S.K.M. (Santosh Kumar Mishra) and A.V.J.; validation, N.T., P.T. and N.B.; investigation, N.T., P.T., N.B. and B.A.; resources, S.K.M. (Sunil Kumar Mishra), S.K.M. (Santosh Kumar Mishra), A.V.J., I.S.S., F.M.E. and F.G.B.; data curation, F.M.E., N.B. and B.A.; writing—original draft preparation, S.K.M. (Sunil Kumar Mishra), S.K.M. (Santosh Kumar Mishra), A.V.J., I.S.S., F.M.E., B.A. and F.G.B.; supervision, N.B. and B.A.; project administration, N.B. and B.A.; formal analysis: N.T. and P.T.; funding acquisition: N.T. and P.T.; visualization: N.T., P.T., N.B. and B.A.; writing—review and editing: N.T., P.T., N.B. and B.A.; figures and tables: I.S.S. and A.V.J. All authors have read and agreed to the published version of the manuscript.

Funding: This work was supported in part by the Framework Agreement between University of Pitesti (Romania) and King Mongkut's University of Technology North Bangkok (Thailand), in part by an International Research Partnership "Electrical Engineering—Thai French Research Center (EE-TFRC)" under the project framework of the Lorraine Université d'Excellence (LUE) in cooperation between Université de Lorraine and King Mongkut's University of Technology North Bangkok and in part by the National Research Council of Thailand (NRCT) under Senior Research Scholar Program under Grant No. N42A640328, and in part by National Science, Research and Innovation Fund (NSRF) under King Mongkut's University of Technology North Bangkok under Grant no. KMUTNB-FF-65-20.

Institutional Review Board Statement: Not applicable.

Informed Consent Statement: Not applicable.

Conflicts of Interest: The authors declare no conflict of interest.

References

1. Jha, A.V.; Ghazali, A.N.; Appasani, B.; Ravariu, C.; Srinivasulu, A. Reliability Analysis of Smart Grid Networks Incorporating Hardware Failures and Packet Loss. *Rev. Roum. Sci. Tech. El.* **2021**, *65*, 245–252.
2. Mahmoud, M.A.; Nasir, N.R.; Gurunathan, M.; Raj, P.; Mostafa, S.A. The Current State of the Art in Research on Predictive Maintenance in Smart Grid Distribution Network: Fault's Types, Causes, and Prediction Methods—A Systematic Review. *Energies* **2021**, *14*, 5078. [CrossRef]
3. Appasani, B.; Jha, A.V.; Mishra, S.K.; Ghazali, A.N. Communication infrastructure for situational awareness enhancement in WAMS with optimal PMU placement. *Prot. Control Mod. Power Syst.* **2021**, *6*, 9. [CrossRef]

4. Yapa, C.; de Alwis, C.; Liyanage, M.; Ekanayake, J. Survey on blockchain for future smart grids: Technical aspects, applications, integration challenges and future research. *Energy Rep.* **2021**, *7*, 6530–6564. [CrossRef]
5. Kaltakis, K.; Polyzi, P.; Drosatos, G.; Rantos, K. Privacy-Preserving Solutions in Blockchain-Enabled Internet of Vehicles. *Appl. Sci.* **2021**, *11*, 9792. [CrossRef]
6. Baidya, S.; Potdar, V.; Ray, P.P.; Nandi, C. Reviewing the opportunities, challenges, and future directions for the digitalization of energy. *Energy Res. Soc. Sci.* **2021**, *81*, 102243. [CrossRef]
7. Ma, Z.; Clausen, A.; Lin, Y.; Jørgensen, B.N. An overview of digitalization for the building-to-grid ecosystem. *Energy Inform.* **2021**, *4*, 36. [CrossRef]
8. Hasankhani, A.; Hakimi, S.M.; Bisheh-Niasar, M.; Shafie-Khah, M.; Asadolahi, H. Blockchain technology in the future smart grids: A comprehensive review and frameworks. *Int. J. Electr. Power Energy Syst.* **2021**, *129*, 106811. [CrossRef]
9. Liu, C.; Zhang, X.; Chai, K.K.; Loo, J.; Chen, Y. A survey on blockchain-enabled smart grids: Advances, applications and challenges. *IET Smart Cities* **2021**, *3*, 56–78. [CrossRef]
10. Guru, D.; Perumal, S.; Varadarajan, V. Approaches towards Blockchain Innovation: A Survey and Future Directions. *Electronics* **2021**, *10*, 1219. [CrossRef]
11. Wang, Q.; Li, R.; Zhan, L. Blockchain technology in the energy sector: From basic research to real world applications. *Comput. Sci. Rev.* **2021**, *39*, 100362. [CrossRef]
12. Yagmur, A.; Dedeturk, B.A.; Soran, A.; Jung, J.; Onen, A. Blockchain-Based Energy Applications: The DSO Perspective. *IEEE Access* **2021**, *9*, 145605–145625. [CrossRef]
13. Mollah, M.B.; Zhao, J.; Niyato, D.; Lam, K.-Y.; Zhang, X.; Ghias, A.M.Y.M.; Koh, L.H.; Yang, L. Blockchain for Future Smart Grid: A Comprehensive Survey. *IEEE Internet Things J.* **2020**, *8*, 18–43. [CrossRef]
14. Kumar, N.M.; Chand, A.A.; Malvoni, M.; Prasad, K.A.; Mamun, K.A.; Islam, F.; Chopra, S.S. Distributed energy resources and the application of AI, IoT, and blockchain in smart grids. *Energies* **2020**, *13*, 5739. [CrossRef]
15. Miglani, A.; Kumar, N.; Chamola, V.; Zeadally, S. Blockchain for Internet of Energy management: Review, solutions, and challenges. *Comput. Commun.* **2020**, *151*, 395–418. [CrossRef]
16. Zia, M.F.; Benbouzid, M.; Elbouchikhi, E.; Muyeen, S.M.; Techato, K.; Guerrero, J.M. Microgrid Transactive Energy: Review, Architectures, Distributed Ledger Technologies, and Market Analysis. *IEEE Access* **2020**, *8*, 19410–19432. [CrossRef]
17. Rathore, H.; Mohamed, A.; Guizani, M. A Survey of Blockchain Enabled Cyber-Physical Systems. *Sensors* **2020**, *20*, 282. [CrossRef]
18. Khajeh, H.; Laaksonen, H.; Gazafroudi, A.S.; Shafie-Khah, M. Towards Flexibility Trading at TSO-DSO-Customer Levels: A Review. *Energies* **2019**, *13*, 165. [CrossRef]
19. Alladi, T.; Chamola, V.; Rodrigues, J.J.P.C.; Kozlov, S.A. Blockchain in Smart Grids: A Review on Different Use Cases. *Sensors* **2019**, *19*, 4862. [CrossRef]
20. Nezamabadi, H.; Vahidinasab, V. Microgrids Bidding Strategy in a Transactive Energy Market. *Sci. Iran.* **2019**, *26*, 3622–3634. [CrossRef]
21. Erturk, E.; Lopez, D.; Yu, W.Y. Benefits and Risks of Using Blockchain in Smart Energy: A Literature Review. *Contemp. Manag. Res.* **2019**, *15*, 205–225. [CrossRef]
22. Abdella, J.; Shuaib, K. Peer to Peer Distributed Energy Trading in Smart Grids: A Survey. *Energies* **2018**, *11*, 1560. [CrossRef]
23. Bodkhe, U.; Tanwar, S.; Parekh, K.; Khanpara, P.; Tyagi, S.; Kumar, N.; Alazab, M. Blockchain for Industry 4.0: A Comprehensive Review. *IEEE Access* **2020**, *8*, 79764–79800. [CrossRef]
24. Lim, M.K.; Li, Y.; Wang, C.; Tseng, M.-L. A literature review of blockchain technology applications in supply chains: A comprehensive analysis of themes, methodologies and industries. *Comput. Ind. Eng.* **2021**, *154*, 107133. [CrossRef]
25. Berdik, D.; Otoum, S.; Schmidt, N.; Porter, D.; Jararweh, Y. A Survey on Blockchain for Information Systems Management and Security. *Inf. Process. Manag.* **2020**, *58*, 102397. [CrossRef]
26. Meng, T.; Zhao, Y.; Wolter, K.; Xu, C.-Z. On Consortium Blockchain Consistency: A Queueing Network Model Approach. *IEEE Trans. Parallel Distrib. Syst.* **2021**, *32*, 1369–1382. [CrossRef]
27. Bhattacharjee, A.; Badsha, S.; Shahid, A.R.; Livani, H.; Sengupta, S. Block-Phasor: A Decentralized Blockchain Framework to Enhance Security of Synchrophasor. In Proceedings of the 2020 IEEE Kansas Power and Energy Conference (KPEC), Manhattan, KS, USA, 13–14 July 2020; pp. 1–6. [CrossRef]
28. Appasani, B.; Mohanta, D.K. A review on synchrophasor communication system: Communication technologies, standards and applications. *Prot. Control Mod. Power Syst.* **2018**, *3*, 37. [CrossRef]
29. Jha, A.; Appasani, B.; Ghazali, A.; Bizon, N. A Comprehensive Risk Assessment Framework for Synchrophasor Communication Networks in a Smart Grid Cyber Physical System with a Case Study. *Energies* **2021**, *14*, 3428. [CrossRef]
30. Jha, A.V.; Appasani, B.; Ghazali, A.N.; Pattanayak, P.; Gurjar, D.S.; Kabalci, E.; Mohanta, D.K. Smart grid cyber-physical systems: Communication technologies, standards and challenges. *Wirel. Netw.* **2021**, *27*, 2595–2613. [CrossRef]
31. Appasani, B.; Mohanta, D.K. Co-Optimal Placement of PMUs and Their Communication Infrastructure for Minimization of Propagation Delay in the WAMS. *IEEE Trans. Ind. Inform.* **2018**, *14*, 2120–2132. [CrossRef]
32. Jha, A.V.; Appasani, B.; Ghazali, A.N. Analytical Channel Modelling of Synchrophsor Communication Networks in a Smart Grid Cyber Physical System. In Proceedings of the 2021 3rd Global Power, Energy and Communication Conference (GPECOM), Antalya, Turkey, 5–8 October 2021; pp. 257–262. [CrossRef]

33. Krishna, R.R.; Priyadarshini, A.; Jha, A.V.; Appasani, B.; Srinivasulu, A.; Bizon, N. State-of-the-Art Review on IoT Threats and Attacks: Taxonomy, Challenges and Solutions. *Sustainability* **2021**, *13*, 9463. [CrossRef]
34. Jha, A.V.; Ghazali, A.N.; Appasani, B.; Mohanta, D.K. Risk Identification and Risk Assessment of Communication Net-works in Smart Grid Cyber-Physical Systems. In *Security in Cyber-Physical Systems: Foundations and Applications*; Studies in Systems, Decision and Control; Awad, A.I., Furnell, S., Paprzycki, M., Sharma, S.K., Eds.; Springer: Cham, Switzerland, 2021; Volume 339, pp. 217–253. [CrossRef]
35. Jha, A.V.; Appasani, B.; Ghazali, A.N. Performance Evaluation of Routing Protocols in Synchrophasor Communication Networks. In Proceedings of the 2019 International Conference on Information Technology (ICIT), Bhubaneswar, India, 19–21 December 2019; pp. 132–136.
36. Wazid, M.; Das, A.K.; Shetty, S.; Jo, M. A Tutorial and Future Research for Building a Blockchain-Based Secure Communication Scheme for Internet of Intelligent Things. *IEEE Access* **2020**, *8*, 88700–88716. [CrossRef]
37. Arif, S.; Khan, M.A.; Rehman, S.U.; Kabir, M.A.; Imran, M. Investigating Smart Home Security: Is Blockchain the Answer? *IEEE Access* **2020**, *8*, 117802–117816. [CrossRef]
38. Xue, J.; Xu, C.; Zhang, Y. Private Blockchain-Based Secure Access Control for Smart Home Systems. *KSII Trans. Internet Inf. Syst.* **2018**, *12*, 6057–6078. [CrossRef]
39. Tantidham, T.; Aung, Y.N. Emergency Service for Smart Home System Using Ethereum Blockchain: System and Architecture. In Proceedings of the 2019 IEEE International Conference on Pervasive Computing and Communications Workshops (PerCom Workshops), Kyoto, Japan, 11–15 March 2019; pp. 888–893. [CrossRef]
40. Dwivedi, A.D.; Srivastava, G.; Dhar, S.; Singh, R. A Decentralized Privacy-Preserving Healthcare Blockchain for IoT. *Sensors* **2019**, *19*, 326. [CrossRef]
41. Monrat, A.A.; Schelen, O.; Andersson, K. Blockchain Mobility Solution for Charging Transactions of Electrical Vehicles. In Proceedings of the 2020 IEEE/ACM 13th International Conference on Utility and Cloud Computing (UCC), Leicester, UK, 7–10 December 2020; pp. 253–348. [CrossRef]
42. Rupasinghe, T.; Burstein, F.; Rudolph, C.; Strange, S. Towards a Blockchain based Fall Prediction Model for Aged Care. In Proceedings of the PervasiveHealth: Pervasive Computing Technologies for Healthcare, Sydnet, Australia, 29–31 January 2019; pp. 1–10. [CrossRef]
43. Almadhoun, R.; Kadadha, M.; Alhemeiri, M.; Alshehhi, M.; Salah, K. A User Authentication Scheme of IoT Devices using Blockchain-Enabled Fog Nodes. In Proceedings of the 2018 IEEE/ACS 15th International Conference on Computer Systems and Applications (AICCSA), Aqaba, Jordan, 28 October–1 November 2018; pp. 1–8. [CrossRef]
44. Decker, C.; Wattenhofer, R. A Fast and Scalable Payment Network with Bitcoin Duplex Micropayment Channels. In *Lecture Notes in Computer Science (Including Subseries Lecture Notes in Artificial Intelligence and Lecture Notes in Bioinformatics)*; Springer: Cham, Switzerland, 2015; Volume 9212, pp. 3–18. [CrossRef]
45. Zhong, L.; Wu, Q.; Xie, J.; Li, J.; Qin, B. A secure versatile light payment system based on blockchain. *Futur. Gener. Comput. Syst.* **2018**, *93*, 327–337. [CrossRef]
46. Majeed, R.; Abdullah, N.A.; Ashraf, I.; Bin Zikria, Y.; Mushtaq, M.F.; Umer, M. An Intelligent, Secure, and Smart Home Automation System. *Sci. Program.* **2020**, *2020*, 4579291. [CrossRef]
47. El Houda, Z.A.; Hafid, A.; Khoukhi, L. Blockchain Meets AMI: Towards Secure Advanced Metering Infrastructures. In Proceedings of the IEEE International Conference on Communications, Dublin, Ireland, 7–11 June 2020.
48. Kamal, M.; Tariq, M. Light-Weight Security and Blockchain Based Provenance for Advanced Metering Infrastructure. *IEEE Access* **2019**, *7*, 87345–87356. [CrossRef]
49. Khalid, R.; Javaid, N.; Almogren, A.; Javed, M.U.; Javaid, S.; Zuair, M. A Blockchain-Based Load Balancing in Decentralized Hybrid P2P Energy Trading Market in Smart Grid. *IEEE Access* **2020**, *8*, 47047–47062. [CrossRef]
50. Sovacool, B.K.; Kester, J.; Noel, L.; de Rubens, G.Z. Actors, business models, and innovation activity systems for vehicle-to-grid (V2G) technology: A comprehensive review. *Renew. Sustain. Energy Rev.* **2020**, *131*, 109963. [CrossRef]
51. Islam, M.; Shahjalal; Hasan, M.K.; Jang, Y.M. Blockchain-based Energy Transaction Model for Electric Vehicles in V2G Network. In Proceedings of the 2020 International Conference on Artificial Intelligence in Information and Communication (ICAIIC), Fukuoka, Japan, 19–21 February 2020; pp. 628–630. [CrossRef]
52. Rehman, A.; Hassan, M.F.; Yew, K.H.; Paputungan, I.; Tran, D.C. State-of-the-art IoV trust management a meta-synthesis systematic literature review (SLR). *PeerJ Comput. Sci.* **2020**, *6*, e334. [CrossRef] [PubMed]
53. Pal, R.; Chavhan, S.; Gupta, D.; Khanna, A.; Padmanaban, S.; Khan, B.; Rodrigues, J.J.P.C. A comprehensive review on IoT-based infrastructure for smart grid applications. *IET Renew. Power Gener.* **2021**, *15*, 3761–3776. [CrossRef]
54. Gschwendtner, C.; Sinsel, S.R.; Stephan, A. Vehicle-to-X (V2X) implementation: An overview of predominate trial configurations and technical, social and regulatory challenges. *Renew. Sustain. Energy Rev.* **2021**, *145*, 110977. [CrossRef]
55. Khan, M.A.; Ghosh, S.; Busari, S.A.; Huq, K.M.S.; Dagiuklas, T.; Mumtaz, S.; Iqbal, M.; Rodriguez, J. Robust, Resilient and Reliable Architecture for V2X Communications. *IEEE Trans. Intell. Transp. Syst.* **2021**, *22*, 4414–4430. [CrossRef]
56. Nguyen, D.C.; Pathirana, P.N.; Ding, M.; Seneviratne, A. Blockchain for 5G and beyond networks: A state of the art survey. *J. Netw. Comput. Appl.* **2020**, *166*, 102693. [CrossRef]
57. Xu, C.; Wu, H.; Liu, H.; Li, X.; Liu, L.; Wang, P. An Intelligent Scheduling Access Privacy Protection Model of Electric Vehicle Based on 5G-V2X. *Sci. Program.* **2021**, *2021*, 1198794. [CrossRef]

58. Bhattacharya, P.; Tanwar, S.; Bodkhe, U.; Kumar, A.; Kumar, N. EVBlocks: A Blockchain-Based Secure Energy Trading Scheme for Electric Vehicles underlying 5G-V2X Ecosystems. *Wirel. Pers. Commun.* **2021**, 1–41. [CrossRef]
59. Jameel, F.; Javed, M.A.; Zeadally, S.; Jantti, R. Efficient Mining Cluster Selection for Blockchain-Based Cellular V2X Communications. *IEEE Trans. Intell. Transp. Syst.* **2020**, *22*, 4064–4072. [CrossRef]
60. Rawat, D.B.; Doku, R.; Adebayo, A.; Bajracharya, C.; Kamhoua, C. Blockchain Enabled Named Data Networking for Secure Vehicle-to-Everything Communications. *IEEE Netw.* **2020**, *34*, 185–189. [CrossRef]
61. Rasheed, M.B.; Javaid, N.; Ahmad, A.; Awais, M.; Khan, Z.A.; Qasim, U.; Alrajeh, N. Priority and delay constrained demand side management in real-time price environment with renewable energy source. *Int. J. Energy Res.* **2016**, *40*, 2002–2021. [CrossRef]
62. Khan, P.; Byun, Y.-C. Blockchain-Based Peer-to-Peer Energy Trading and Charging Payment System for Electric Vehicles. *Sustainability* **2021**, *13*, 7962. [CrossRef]
63. Sadiq, A.; Javed, M.U.; Khalid, R.; Almogren, A.; Shafiq, M.; Javaid, N. Blockchain Based Data and Energy Trading in Internet of Electric Vehicles. *IEEE Access* **2020**, *9*, 7000–7020. [CrossRef]
64. Musleh, A.S.; Yao, G.; Muyeen, S.M. Blockchain Applications in Smart Grid–Review and Frameworks. *IEEE Access* **2019**, *7*, 86746–86757. [CrossRef]
65. Kim, M.; Park, K.; Yu, S.; Lee, J.; Park, Y.; Lee, S.-W.; Chung, B. A Secure Charging System for Electric Vehicles Based on Blockchain. *Sensors* **2019**, *19*, 3028. [CrossRef] [PubMed]
66. Dorokhova, M.; Vianin, J.; Alder, J.-M.; Ballif, C.; Wyrsch, N.; Wannier, D. A Blockchain-Supported Framework for Charging Management of Electric Vehicles. *Energies* **2021**, *14*, 7144. [CrossRef]
67. Buterin, V. Ethereum Platform Review—Opportunities and Challenges for Private and Consortium Blockchains. Available online: http://www.smallake.kr/wp-content/uploads/2016/06/314477721-Ethereum-Platform-Review-Opportunities-and-Challenges-for-Private-and-Consortium-Blockchains.pdf (accessed on 10 January 2022).
68. Saxena, S.; Farag, H.E.Z.; Brookson, A.; Turesson, H.; Kim, H. A Permissioned Blockchain System to Reduce Peak Demand in Residential Communities via Energy Trading: A Real-World Case Study. *IEEE Access* **2020**, *9*, 5517–5530. [CrossRef]
69. Huang, Z.; Li, Z.; Lai, C.S.; Zhao, Z.; Wu, X.; Li, X.; Tong, N.; Lai, L.L. A Novel Power Market Mechanism Based on Blockchain for Electric Vehicle Charging Stations. *Electronics* **2021**, *10*, 307. [CrossRef]
70. Birleanu, G.F.; Bizon, N. Control and Protection of the Smart Microgrids Using Internet of Things: Technologies, Architecture and Applications. In *Microgrid Architectures, Control and Protection Methods*, 1st ed.; Tabatabaei, N.M., Kabalci, E., Bizon, N., Eds.; Springer: Cham, Switzerland, 2020; pp. 749–770. [CrossRef]
71. Birleanu, G.F.; Anghelescu, P.; Bizon, N.; Pricop, E. Cyber Security Objectives and Requirements for Smart Grid. In *Smart Grids and Their Communication Systems*, 1st ed.; Kabalci, E., Kabalci, Y., Eds.; Springer: Singapore, 2019; pp. 607–634. [CrossRef]
72. Birleanu, G.F.; Anghelescu, P.; Bizon, N. Malicious and Deliberate Attacks and Power System Resiliency. In *Power Systems Resiliency: Modeling, Analysis and Practice*, 1st ed.; Tabatabaei, N.M., Ravadanegh, S.N., Bizon, N., Eds.; Springer: Cham, Switzerland, 2018; pp. 223–246. [CrossRef]
73. Ahmethodzic, L.; Music, M. Comprehensive review of trends in microgrid control. *Renew. Energy Focus* **2021**, *38*, 84–96. [CrossRef]
74. Foti, M.; Vavalis, M. What blockchain can do for power grids? *Blockchain Res. Appl.* **2021**, *2*, 100008. [CrossRef]
75. Valdivia, A.D.; Balcell, M.P. Connecting the grids: A review of blockchain governance in distributed energy transitions. *Energy Res. Soc. Sci.* **2021**, *84*, 102383. [CrossRef]
76. Aybar-Mejía, M.; Rosario-Weeks, D.; Mariano-Hernández, D.; Domínguez-Garabitos, M. An approach for applying blockchain technology in centralized electricity markets. *Electr. J.* **2021**, *34*, 106918. [CrossRef]
77. Yahaya, A.S.; Javaid, N.; Alzahrani, F.A.; Rehman, A.; Ullah, I.; Shahid, A.; Shafiq, M. Blockchain Based Sustainable Local Energy Trading Considering Home Energy Management and Demurrage Mechanism. *Sustainability* **2020**, *12*, 3385. [CrossRef]
78. Mengelkamp, E.; Notheisen, B.; Beer, C.; Dauer, D.; Weinhardt, C. A blockchain-based smart grid: Towards sustainable local energy markets. *Comput. Sci. Res. Dev.* **2017**, *33*, 207–214. [CrossRef]
79. Antal, C.; Cioara, T.; Antal, M.; Mihailescu, V.; Mitrea, D.; Anghel, I.; Salomie, I.; Raveduto, G.; Bertoncini, M.; Croce, V.; et al. Blockchain based decentralized local energy flexibility market. *Energy Rep.* **2021**, *7*, 5269–5288. [CrossRef]
80. Vieira, G.; Zhang, J. Peer-to-peer energy trading in a microgrid leveraged by smart contracts. *Renew. Sustain. Energy Rev.* **2021**, *143*, 110900. [CrossRef]
81. Kavousi-Fard, A.; Almutairi, A.; Al-Sumaiti, A.; Farughian, A.; Alyami, S. An effective secured peer-to-peer energy market based on blockchain architecture for the interconnected microgrid and smart grid. *Int. J. Electr. Power Energy Syst.* **2021**, *132*, 107171. [CrossRef]
82. van Leeuwen, G.; AlSkaif, T.; Gibescu, M.; van Sark, W. An integrated blockchain-based energy management platform with bilateral trading for microgrid communities. *Appl. Energy* **2020**, *263*, 114613. [CrossRef]
83. Yildizbasi, A. Blockchain and renewable energy: Integration challenges in circular economy era. *Renew. Energy* **2021**, *176*, 183–197. [CrossRef]
84. Tsao, Y.-C.; Thanh, V.-V. Toward sustainable microgrids with blockchain technology-based peer-to-peer energy trading mechanism: A fuzzy meta-heuristic approach. *Renew. Sustain. Energy Rev.* **2020**, *136*, 110452. [CrossRef]
85. Wang, X.; Liu, P.; Ji, Z. Trading platform for cooperation and sharing based on blockchain within multi-agent energy internet. *Glob. Energy Interconnect.* **2021**, *4*, 384–393. [CrossRef]

86. Li, Q.; Li, A.; Wang, T.; Cai, Y. Interconnected hybrid AC-DC microgrids security enhancement using blockchain technology considering uncertainty. *Int. J. Electr. Power Energy Syst.* **2021**, *133*, 107324. [CrossRef]
87. Mahesh, G.S.; Babu, G.D.; Rakesh, V.; Mohan, S.; Ranjit, P. Energy management with blockchain technology in DC microgrids. *Mater. Today Proc.* **2021**, *47*, 2232–2236. [CrossRef]
88. Wang, S.; Xu, Z.; Ha, J. Secure and decentralized framework for energy management of hybrid AC/DC microgrids using blockchain for randomized data. *Sustain. Cities Soc.* **2021**, *76*, 103419. [CrossRef]
89. Fineberg, S.J.; Nandyala, S.V.; Marquez-Lara, A.; Oglesby, M.; Patel, A.A.; Singh, K. Incidence and risk factors for postoperative delirium after lumbar spine surgery (Phila Pa 1976). *Spine* **2013**, *38*, 1790–1796. [CrossRef]
90. Wang, T.; Hua, H.; Wei, Z.; Cao, J. Challenges of blockchain in new generation energy systems and future outlooks. *Int. J. Electr. Power Energy Syst.* **2021**, *135*, 107499. [CrossRef]
91. Ahl, A.; Yarime, M.; Goto, M.; Chopra, S.S.; Kumar, N.M.; Tanaka, K.; Sagawa, D. Exploring blockchain for the energy transition: Opportunities and challenges based on a case study in Japan. *Renew. Sustain. Energy Rev.* **2019**, *117*, 109488. [CrossRef]
92. Perez-DeLaMora, D.; Quiroz-Ibarra, J.E.; Fernandez-Anaya, G.; Hernandez-Martinez, E. Roadmap on community-based microgrids deployment: An extensive review. *Energy Rep.* **2021**, *7*, 2883–2898. [CrossRef]
93. Enescu, F.M.; Bizon, N.; Ionescu, V.M. Blockchain—A new tehnology for the smart village. In Proceedings of the 2021 13th International Conference on Electronics, Computers and Artificial Intelligence (ECAI), Pitesti, Romania, 1–3 July 2021; pp. 1–6. [CrossRef]
94. Enescu, F.M.; Bizon, N.; Cirstea, A.; Stirbu, C. Blockchain Technology Applied in Health the Study of Blockchain Application in the Health System (I). In Proceedings of the 2018 10th International Conference on Electronics, Computers and Artificial Intelligence (ECAI), Iasi, Romania, 28–30 June 2018; pp. 1–4. [CrossRef]
95. Cirstea, A.; Enescu, F.M.; Bizon, N.; Stirbu, C.; Ionescu, V.M. Blockchain Technology Applied in Health the Study of Blockchain Application in the Health System (II). In Proceedings of the 2018 10th International Conference on Electronics, Computers and Artificial Intelligence (ECAI), Iasi, Romania, 28–30 June 2018; pp. 1–4. [CrossRef]
96. Gul, M.J.; Subramanian, B.; Paul, A.; Kim, J. Blockchain for public health care in smart society. *Microprocess. Microsyst.* **2020**, *80*, 103524. [CrossRef]
97. Jiang, S.; Cao, J.; Wu, H.; Yang, Y.; Ma, M.D.; He, J. BlocHIE: A BLOCkchain-Based Platform for Healthcare Information Exchange. In Proceedings of the 2018 IEEE International Conference on Smart Computing (SMARTCOMP), Taormina, Italy, 18–20 June 2018; pp. 49–56. [CrossRef]
98. Enescu, F.M.; Bizon, N.; Onu, A.; Răboacă, M.S.; Thounthong, P.; Mazare, A.G.; Șerban, G. Implementing Blockchain Technology in Irrigation Systems That Integrate Photovoltaic Energy Generation Systems. *Sustainability* **2020**, *12*, 1540. [CrossRef]
99. Raboaca, M.S.; Bizon, N.; Trufin, C.; Enescu, F.M. Efficient and Secure Strategy for Energy Systems of Interconnected Farmers' Associations to Meet Variable Energy Demand. *Mathematics* **2020**, *8*, 2182. [CrossRef]
100. Enescu, F.M.; Bizon, N.; Ionescu, V.M. Use of Blockchain Technology in Irrigation Systems of small farmers' association. In Proceedings of the 2019 11th International Conference on Electronics, Computers and Artificial Intelligence (ECAI), Pitesti, Romania, 27–29 June 2019; pp. 1–6. [CrossRef]
101. Sharma, P.K.; Park, J.H. Blockchain based hybrid network architecture for the smart city. *Futur. Gener. Comput. Syst.* **2018**, *86*, 650–655. [CrossRef]
102. Bhushan, B.; Khamparia, A.; Sagayam, K.M.; Sharma, S.K.; Ahad, M.A.; Debnath, N.C. Blockchain for smart cities: A review of architectures, integration trends and future research directions. *Sustain. Cities Soc.* **2020**, *61*, 102360. [CrossRef]
103. Mora, H.; Mendoza-Tello, J.C.; Varela-Guzmán, E.G.; Szymanski, J. Blockchain technologies to address smart city and society challenges. *Comput. Hum. Behav.* **2021**, *122*, 106854. [CrossRef]
104. Wu, H.; Cao, J.; Yang, Y.; Tung, C.L.; Jiang, S.; Tang, B.; Liu, Y.; Wang, X.; Deng, Y. Data Management in Supply Chain Using Blockchain: Challenges and a Case Study. In Proceedings of the 2019 28th International Conference on Computer Communication and Networks (ICCCN), Valencia, Spain, 29 July–1 August 2019; pp. 1–8. [CrossRef]
105. Aggarwal, S.; Chaudhary, R.; Aujla, G.S.; Kumar, N.; Choo, K.-K.R.; Zomaya, A.Y. Blockchain for smart communities: Applications, challenges and opportunities. *J. Netw. Comput. Appl.* **2019**, *144*, 13–48. [CrossRef]
106. Al Sadawi, A.; Madani, B.; Saboor, S.; Ndiaye, M.; Abu-Lebdeh, G. A comprehensive hierarchical blockchain system for carbon emission trading utilizing blockchain of things and smart contract. *Technol. Forecast. Soc. Chang.* **2021**, *173*, 121124. [CrossRef]
107. Zia, M. B-DRIVE: A blockchain based distributed IOT network for smart urban transportation. *Blockchain Res. Appl.* **2021**, *2*, 100033. [CrossRef]
108. Pournaras, E. Proof of witness presence: Blockchain consensus for augmented democracy in smart cities. *J. Parallel Distrib. Comput.* **2020**, *145*, 160–175. [CrossRef]
109. Paul, R.; Ghosh, N.; Sau, S.; Chakrabarti, A.; Mohapatra, P. Blockchain based secure smart city architecture using low resource IoTs. *Comput. Netw.* **2021**, *196*, 108234. [CrossRef]
110. Jiang, S.; Cao, J.; Wu, H.; Yang, Y. Fairness-Based Packing of Industrial IoT Data in Permissioned Blockchains. *IEEE Trans. Ind. Inform.* **2020**, *17*, 7639–7649. [CrossRef]
111. Hu, W.; Li, H. A blockchain-based secure transaction model for distributed energy in Industrial Internet of Things. *Alex. Eng. J.* **2020**, *60*, 491–500. [CrossRef]

112. Yang, J.; Paudel, A.; Gooi, H.B.; Nguyen, H.D. A Proof-of-Stake public blockchain based pricing scheme for peer-to-peer energy trading. *Appl. Energy* **2021**, *298*, 117154. [CrossRef]
113. Nair, R.; Gupta, S.; Soni, M.; Shukla, P.K.; Dhiman, G. An approach to minimize the energy consumption during blockchain transaction. *Mater. Today Proc.* **2020**. [CrossRef]
114. Zhang, Y.; Shi, Q. An intelligent transaction model for energy blockchain based on diversity of subjects. *Alex. Eng. J.* **2020**, *60*, 749–756. [CrossRef]
115. Gourisetti, S.N.G.; Sebastian-Cardenas, D.J.; Bhattarai, B.; Wang, P.; Widergren, S.; Borkum, M.; Randall, A. Blockchain smart contract reference framework and program logic architecture for transactive energy systems. *Appl. Energy* **2021**, *304*, 117860. [CrossRef]
116. Yang, Q.; Wang, H.; Wang, T.; Zhang, S.; Wu, X.; Wang, H. Blockchain-based decentralized energy management platform for residential distributed energy resources in a virtual power plant. *Appl. Energy* **2021**, *294*, 117026. [CrossRef]
117. Wongthongtham, P.; Marrable, D.; Abu-Salih, B.; Liu, X.; Morrison, G. Blockchain-enabled Peer-to-Peer energy trading. *Comput. Electr. Eng.* **2021**, *94*, 107299. [CrossRef]
118. Wang, L.; Ma, Y.; Zhu, L.; Wang, X.; Cong, H.; Shi, T. Design of integrated energy market cloud service platform based on blockchain smart contract. *Int. J. Electr. Power Energy Syst.* **2021**, *135*, 107515. [CrossRef]
119. Choobineh, M.; Arab, A.; Khodaei, A.; Paaso, A. Energy innovations through blockchain: Challenges, opportunities, and the road ahead. *Electr. J.* **2021**, *35*, 107059. [CrossRef]
120. Mathew, R.; Mehbodniya, A.; Ambalgi, A.P.; Murali, M.; Sahay, K.B.; Babu, D.V. RETRACTED: In a virtual power plant, a blockchain-based decentralized power management solution for home distributed generation. *Sustain. Energy Technol. Assess.* **2021**, *49*, 101731. [CrossRef]
121. Zhang, C.; Yang, T.; Wang, Y. Peer-to-Peer energy trading in a microgrid based on iterative double auction and blockchain. *Sustain. Energy Grids Netw.* **2021**, *27*, 100524. [CrossRef]
122. Xie, Y.-S.; Lee, Y.; Chang, X.-Q.; Yin, X.; Zheng, H. Research on the transaction mode and mechanism of grid-side shared energy storage market based on blockchain. *Energy Rep.* **2021**, *8*, 224–229. [CrossRef]
123. Rafique, Z.; Khalid, H.M.; Muyeen, S.M. Communication Systems in Distributed Generation: A Bibliographical Review and Frameworks. *IEEE Access* **2020**, *8*, 207226–207239. [CrossRef]
124. Jha, A.V.; Appasani, B.; Ghazali, A.N. A Comprehensive Framework for the Assessment of Synchrophasor Communication Networks from the Perspective of Situational Awareness in a Smart Grid Cyber Physical System. *Technol. Econ. Smart Grids Sustain. Energy* **2022**, *7*, 20. [CrossRef]
125. Musleh, A.S.; Chen, G.; Dong, Z.Y. A Survey on the Detection Algorithms for False Data Injection Attacks in Smart Grids. *IEEE Trans. Smart Grid* **2019**, *11*, 2218–2234. [CrossRef]
126. Mahmoud, M.S.; Khalid, H.M.; Hamdan, M.M. *Cyberphysical Infrastructures in Power Systems: Architectures and Vulnerabilities*; Elsevier: Amsterdam, The Netherlands, 2021. [CrossRef]
127. Alkhiari, A.M.; Mishra, S.; AlShehri, M. Blockchain-Based SQKD and IDS in Edge Enabled Smart Grid Network. *Comput. Mater. Contin.* **2022**, *70*, 2149–2169. [CrossRef]
128. Chen, J.; Mohamed, M.A.; Dampage, U.; Rezaei, M.; Salmen, S.H.; Al Obaid, S.; Annuk, A. A Multi-Layer Security Scheme for Mitigating Smart Grid Vulnerability against Faults and Cyber-Attacks. *Appl. Sci.* **2021**, *11*, 9972. [CrossRef]
129. Khan, A.A.; Laghari, A.A.; Liu, D.-S.; Shaikh, A.A.; Ma, D.-D.; Wang, C.-Y.; Wagan, A.A. EPS-Ledger: Blockchain Hyperledger Sawtooth-Enabled Distributed Power Systems Chain of Operation and Control Node Privacy and Security. *Electronics* **2021**, *10*, 2395. [CrossRef]
130. Xu, W.; Li, J.; Dehghani, M.; GhasemiGarpachi, M. Blockchain-based secure energy policy and management of renewable-based smart microgrids. *Sustain. Cities Soc.* **2021**, *72*, 103010. [CrossRef]
131. Vasukidevi, G.; Sethukarasi, T. BBSSE: Blockchain-Based Safe Storage, Secure Sharing and Energy Scheme for Smart Grid Network. In *Wireless Personal Communications*; Springer: Berlin/Heidelberg, Germany, 2021; pp. 1–12. [CrossRef]

sustainability

Article

Developing a Generalized Multi-Level Inverter with Reduced Number of Power Electronics Components

Hossein Shayeghi [1,2,*], Ali Seifi [2,3], Majid Hosseinpour [1,2] and Nicu Bizon [4,5,6,*]

1 Energy Management Research Center, University of Mohaghegh Ardabili, Ardabil 56199-11367, Iran; hoseinpour.majid@uma.ac.ir
2 Department of Electrical Engineering, University of Mohaghegh Ardabili, Ardabil 56199-11367, Iran; ali_seifi@student.uma.ac.ir
3 Faculty of Electrical and Computer Engineering, University of Tabriz, Tabriz 51666-16471, Iran
4 Faculty of Electronics, Communication and Computers, University of Pitesti, 110040 Pitesti, Romania
5 ICSI Energy, National Research and Development Institute for Cryogenic and Isotopic Technologies, 240050 Ramnicu Valcea, Romania
6 Doctoral School, University Politehnica of Bucharest, Splaiul Independentei Street No. 313, 060042 Bucharest, Romania
* Correspondence: hshayeghi@gmail.com (H.S.); nicu.bizon@upit.ro (N.B.)

Abstract: Reducing the number of components of power electronic converters has been an important research topic over the past few decades. This paper introduces a new structure for a multi-level inverter based on reduced switch basic modules. The proposed basic module requires fewer switches and auxiliary devices. In addition, a lesser number of on-state switches for the synthesis of each voltage level results in less conduction losses, which enhances the converter efficiency. The proposed structure is capable of being implemented in both symmetrical and asymmetrical topologies. This is a merit feature for the proposed topology, which produces high voltage levels with a limited number of elements. The proposed structure is controlled using the fundamental frequency control scheme. The proposed basic module consists of six unidirectional switches and five DC voltage sources, generating five positive voltage levels. The performance of the recommended topology is analyzed from the various circuitry parameters, and a comprehensive comparison carried out with similar recent structures. The presented comparison reveals the advantage of the recommended inverter from different aspects of the circuitry parameters. The suggested structure is simulated using Matlab/Simulink software, and its performance is validated using a laboratory prototype. The results are reported for various steady-state and dynamic conditions.

Keywords: multi-level inverter; reduced switch basic modules; efficiency

Citation: Shayeghi, H.; Seifi, A.; Hosseinpour, M.; Bizon, N. Developing a Generalized Multi-Level Inverter with Reduced Number of Power Electronics Components. *Sustainability* **2022**, *14*, 5545. https://doi.org/10.3390/su14095545

Academic Editor: Pablo García Triviño

Received: 16 March 2022
Accepted: 27 April 2022
Published: 5 May 2022

1. Introduction

Multi-level inverters are widely used in various applications due to their various features, such as low dv/dt stress, modularity, and high-power quality. In high voltage applications, equipment with low and medium voltage levels is often used [1]. Multi-level inverters are usually utilized in high power quality [2], FACTS devices [3], electric vehicles [4], variable speed drives [5], smart grids [6], high voltage applications [7], etc. The traditional multi-level inverters are segmented into three primary categories, which include Diode Clamped Multi-Level Inverter (DC-MLI) or Neutral Point Clamped Multi-Level Inverters (NPC-MLI), Flying Capacitor Multi-Level Inverter (FC-MLI), and Cascade H-Bridge Multi-Level Inverters (CHB-MLI) [8,9]. From the number of circuitry components view, the DC-MLI structure requires multiple diodes if the number of levels increases, making circuit control complex and tedious [1]. In the FC-MLI structure, the voltage balance problem of the capacitors can be solved by using additional switching modes. However, the number of passive components will increase which will be a threat to the reliability of the converter [10]. The CHB-MLI structure is modular and relatively simple compared to the other two topologies. Besides, it does not need additional circuits to match the voltage, but instead uses several isolated DC sources [11]. Despite the simplicity and modularity of CHB-MLIs, adding one module increases four switches and related devices. The need for more circuit devices in the classic topologies of multi-level inverters is a research motivation for new topologies by reducing the number of devices [12].

An advanced configuration was represented in [13] for a symmetrical voltage source multi-level inverter that the high voltage levels are synthesized by extending basic cells. This structure has fewer switches compared to the cascaded H-bridge inverter. However, the number of switches in this topology is still high. In [14], a sub-multi-level structure is introduced, and its cascaded structure was investigated, which requires many switches and drivers. In [14,15], a six-switch H-bridge configuration was utilized for an incremental combination of DC voltage sources. To generalize the voltage levels of this structure, bidirectional switches are considered on both sides of the basic structure. These structures also require a large number of switches. In [16], an extended cascaded structure is introduced, which was based on a basic module and level booster circuit. The number of power electronic devices is high in this structure. An improved symmetric generalized topology was introduced in [17]. This topology consists of two parts, the Level Generation Unit (LGU) and the Polarity Generation Unit (PGU), which uses bidirectional switches. This structure requires a high number of switches, and also the total blocking voltage is significant. In [18], a multi-level cascaded inverter structure was introduced with a new design in symmetrical and asymmetrical topologies, where the number of switches is relatively high. In [19], a bidirectional cascaded multi-level inverter topology was reported, where can operate in symmetrical and asymmetrical modes. This structure was designed by improving the other two structures, which have fewer switches. However, in this inverter, the number of switches and the blocking voltage are also high. In [20], a multi-level inverter for dynamic loads was suggested, which provides two separate structures for symmetrical and asymmetrical topologies. Its symmetrical structure consists of an advanced cascaded H-bridge with many semiconductors. The reference [21] introduces a symmetrical modular multilevel topology with a relatively low number of switches. However, the voltage sources number in this structure is a high value. The reference [22] introduced a switch-ladder structure for a new multi-level converter. This structure was implemented symmetrically and asymmetrically using unidirectional and bidirectional switches. A diode-containing bidirectional structure for a multi-level inverter was reported in [23], which solves the voltage spike problem in such diode-based structures. However, this structure uses a current sensor to realize the current polarity, complicating the control scheme. Also, a generalized diode-containing bidirectional structure for multi-level inverters was represented in [24]. This structure solved the current sensor problem in [23] yet has many switches and considerable blocking voltage. In [25], a modified K-type multi-level inverter structure was reported, utilizing the cascading method to generalize to higher voltage levels. This structure requires eight unidirectional switches to produce seven levels, while the number of switches is not small. A Cross-Switched T-Type (CT-Type) was presented in [26], which used a T-type inverter embedded on either side of the structure to yield more voltage levels. This inverter can operate symmetrically or asymmetrically, while the number of switches was not reduced.

This paper develops a novel configuration for multi-level inverters with symmetrical and asymmetrical topologies. Fewer power electronic devices and fewer switches in the current path of different voltage levels are the main design goals of the recommended topology. These goals make the suggested inverter useful for different applications. The main features of the proposed structure are as follows.

1. The proposed basic module has a lesser number of switches, which by generalizing the basic module, the proposed extended structure is realized.

2. As the number of switches decreases, the axillary circuit number of devices, including the number of gate drivers, snubber circuits, heat sinks, etc., decreases as well, which reduces the cost and volume of the suggested inverter.
3. The voltage stress of the recommended basic module switches is low.
4. The maximum number of conducting switches in the current path of each voltage level is low for the proposed structure. Thus, the switches total conduction losses are reduced and the efficiency increased.

The rest of this paper is organized as follows. Section 2 presents the configuration of the suggested MLI converter and its functionality. Section 3 compares the recommended structure with other topologies in terms of circuitry parameters, including switches and gate driver's number, and total standing voltage (TSV) for different voltage levels. Section 4 presents the mathematical equations and simulations related to losses and efficiency. The analysis of the simulation and laboratory results is reported in Section 5, and finally, the conclusions are presented in Section 6.

2. Suggested Structure

2.1. The Reduced Switch Basic Module

The proposed reduced switch basic module (RSBM) is depicted in Figure 1. It consists of five DC sources and six semiconductor switches along with their anti-parallel diodes. The proposed reduced switch basic module is responsible for producing positive levels. The switches (S_1, $\overline{S_1}$), (S_2, $\overline{S_2}$), (S_3, $\overline{S_3}$) of the proposed RSBM operate complementarily. In other words, the conducting switches in the current path of any voltage level are always half of the total switches of the suggested RSBM. The recommended RSBM can produce five positive voltage levels.

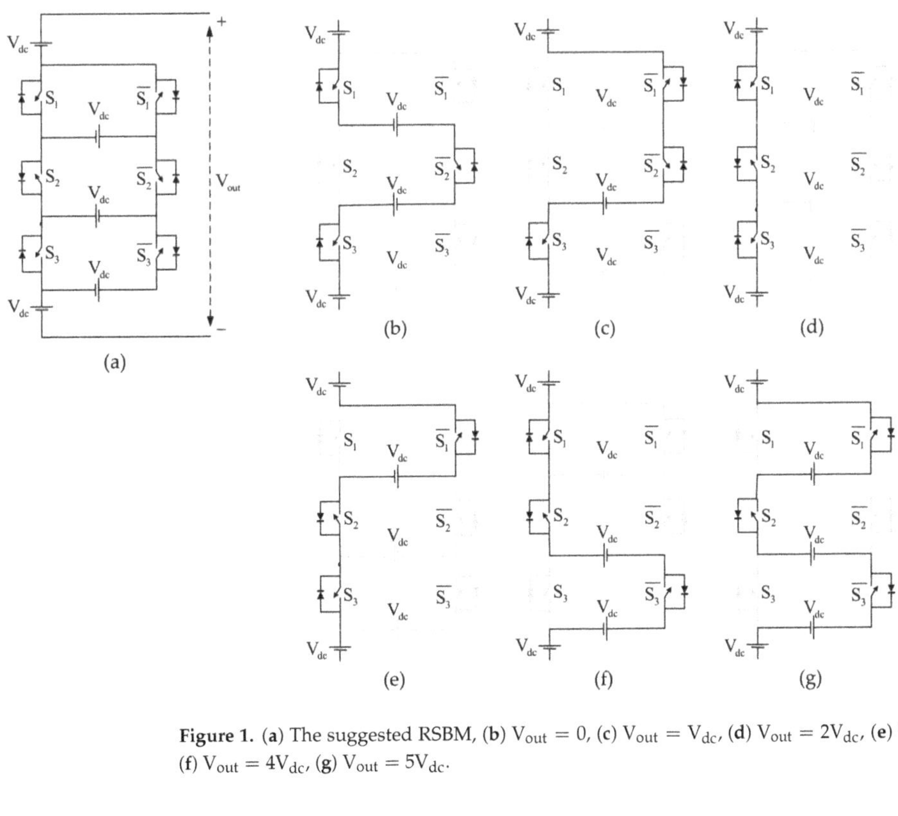

Figure 1. (**a**) The suggested RSBM, (**b**) $V_{out} = 0$, (**c**) $V_{out} = V_{dc}$, (**d**) $V_{out} = 2V_{dc}$, (**e**) $V_{out} = 3V_{dc}$, (**f**) $V_{out} = 4V_{dc}$, (**g**) $V_{out} = 5V_{dc}$.

The reduced switch basic module consists of three switches S_1, S_2, S_3, and their complementary pairs. According to Figure 1, if the switches S_1, S_3 are turned on, the output voltage 0 is generated, and if the switch S_3 is turned on, the first voltage level is produced. To generate the second voltage level, the switches S_1, S_2, S_3 are turned on, and switches S_2, S_3 are turned on to generate the third voltage level. The RSBM also generates the fourth voltage level when the switches S_1, S_2 are turned on, and the fifth voltage level is generated when the switch S_2 is turned on. Table 1 demonstrates the switching pattern of the suggested basic module to produce voltage levels. In this table, 1 means on-state, and 0 means off-state for switches. Naturally, complementary switches behave inversely with the main switches.

Table 1. The switching logic of the RSBM.

S_1	S2	S_3	V_{out}
1	0	1	0
0	0	1	V_{dc}
1	1	1	$2V_{dc}$
0	1	1	$3V_{dc}$
1	1	0	$4V_{dc}$
0	1	0	$5V_{dc}$

2.2. Blocking Voltage of the Proposed Reduced Switch Basic Module

Maximum blocking voltage (MBV) indicates the peak voltage across the off-state switch. Considering the MBV of all switches of the converter, the total blocking voltage (TBV) for the converter is obtained, which can be expressed as follows:

$$\text{TBV} = \sum_{\text{Switches}} \text{MBV} \tag{1}$$

The MBV and TBV parameters are an important challenge for multi-level inverters. The parameter MBV is the most important in selecting the voltage rating of switches. The price of the switches is proportional to their allowable voltage rating. Therefore, the lower the MBV value of the converter switches, the lower the total cost. The MBV value for each switch of the proposed basic module is given in the following relations.

$$S_1 = \overline{S_1} = V_{dc} \tag{2}$$

$$S_2 = \overline{S_2} = S_3 = \overline{S_3} = 2\,V_{dc} \tag{3}$$

The MBV relations of the proposed RSBM reveal that this basic unit can generate five voltage levels using low-voltage rating switches.

2.3. The Proposed Generalized Inverter Structure

The proposed reduced switch basic module can generate only positive voltage levels. So, a structure must generate positive and negative voltage polarity. The H-bridge module can be used for this purpose. Figure 2 displays the proposed multi-level inverter structure. According to this figure, the proposed structure consists of two parts. The first part is related to the level generator, and the second part is related to the polarity generator. In the proposed multi-level inverter, the reduced switch basic modules are connected in series and form the level generator. All switches of the proposed structure operate complementarily. This means that only half of the switches are in on-state at any output voltage level. In other words, the low number of on-state switches reduces the conduction losses of the switches. The size of voltage sources in the RSBMs can be designed and adjusted both symmetrically and asymmetrically.

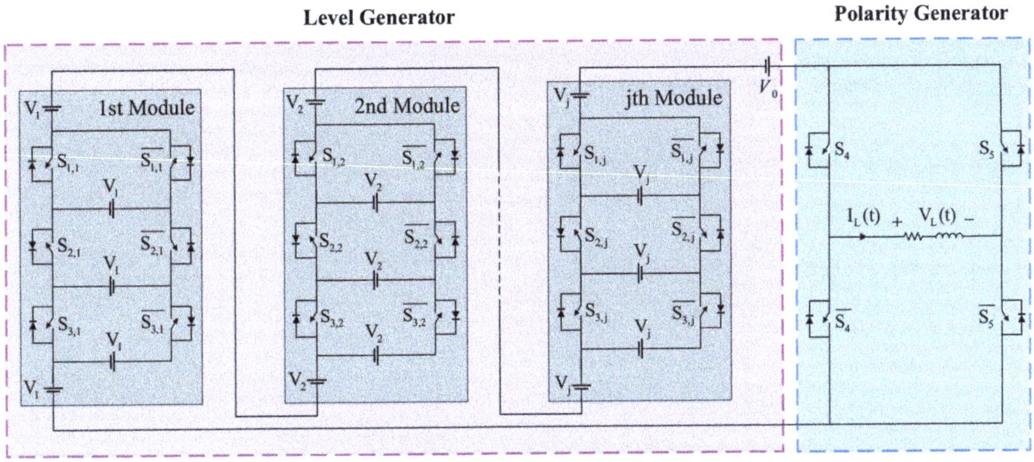

Figure 2. The proposed generalized inverter structure.

In a symmetric topology, the size of voltage sources of the RSBMs is equal. If the voltage source size of the RSBMs is assumed to be V_{dc}, the following relations provide the various parameters of the proposed multi-level inverter in the symmetric topology.

$$V_0 = V_1 = V_2 = \ldots = V_j = V_{dc} \tag{4}$$

$$N_L = 11j + 1 \tag{5}$$

$$N_{IGBT} = N_S = 6j + 4 \tag{6}$$

$$N_{GD} = 6j + 4 \tag{7}$$

$$TBV = 30j \times V_{dc} \tag{8}$$

In the above relations, j shows number of RSBMs, V_j shows the size of the voltage sources of the jth basic module, N_L shows the number of output voltage levels that can be synthesized by the topology, N_{IGBT}, and N_S show the number of IGBT and switches, respectively. Since the bidirectional switch is not utilized in the structure, the number of switches and IGBTs are equal. N_{GD} shows the number of gate drivers, and the TBV shows the total blocking voltage by the switches of the structure.

In an asymmetric topology, voltage source sizes can have different values using various algorithms. Table 2 presents the allowable algorithms for the size of voltage sources in asymmetric topology, where j represents the jth RSBM.

Table 2. Proposed algorithms for asymmetric topology voltage source size.

Proposed Algorithm	Magnitude of dc Voltage Sources	N_{IGBT}	N_L	N_D
1st proposed algorithm	$V_0 = V_1 = V,\ V_2 = \ldots = V_j = 2V_{dc}$	$6j + 4$	$20j + 11$	$6j + 4$
2nd proposed algorithm	$V_0 = V_1 = V_{dc},\ V_2 = 2V_{dc},\ V_j = 2V_{j-1}$	$6j + 4$	$10V_j + 10V_{j-1} + \ldots + 11V_1 + 1$	$6j + 4$
3rd proposed algorithm	$V_0 = V_1 = V_{dc},\ V_2 = 3V_{dc},\ V_j = 3V_{j-1}$	$6j + 4$	$10V_j + 10V_{j-1} + \ldots + 11V_1 + 1$	$6j + 4$
4th proposed algorithm	$V_0 = V_1 = V_{dc},\ V_2 = 6V_{dc},\ V_j = 6V_{j-1}$	$6j + 4$	$10V_j + 10V_{j-1} + \ldots + 11V_1 + 1$	$6j + 4$

2.4. The Proposed Multi-Level Inverter Structure

The structure shown in Figure 2 with a reduced switch basic module is assumed to be the basic cell. By connecting these cells in series, the cascaded structure of the proposed topology is realized. Figure 3 shows the proposed cascaded topology with n basic cells. The cascading method is also a method for generalizing the proposed structure to achieve high voltage levels by using a small number of sources. In addition, in the cascaded topology, the voltage range of H-bridge switches is limited, and it is possible to achieve medium and even high voltage and power levels using limited Maximum Voltage Blocking (MBV) switches.

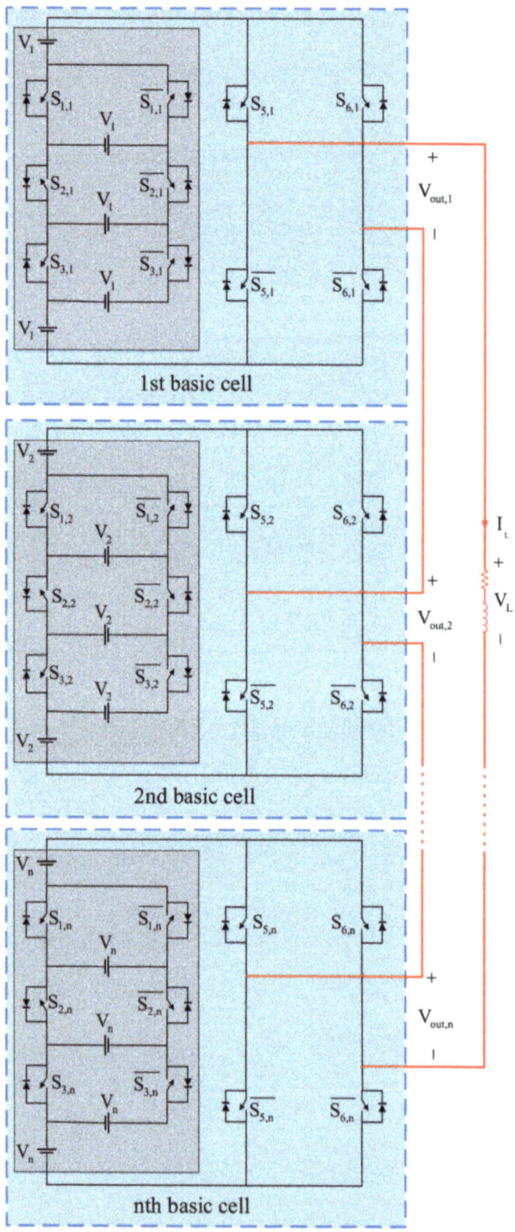

Figure 3. The proposed cascaded topology.

In the cascaded topology, the output voltage (V_L) is obtained from the output voltage of the multiple basic cells:

$$V_L = V_{out,1} + V_{out,2} + \ldots + V_{out,n} \tag{9}$$

In the cascaded topology, different algorithms can be used to generate voltage levels. The cascaded topology can be implemented with the following algorithm.

In this algorithm, the voltage value of independent voltage sources is determined as follows:

$$V_1 = V_{dc},\ V_2 = 11V_{dc},\ \ V_n = 5V_{n-1} + 5V_{n-2} + \ldots + 5V_1 + V_{dc} \tag{10}$$

In the main structure shown in Figure 2, the voltage sources simply have an incremental combination. However, the voltage sources in the cascaded topology, in addition to incremental combination, can also have a decreasing combination, which leads to a significant increase in the number of output voltage levels. The number of voltage levels, the number of IGBTs, and the number of drivers for the cascaded topology are presented in the following equations.

$$N_L = 2(5V_1 + 5V_2 + \ldots + 5V_n) + 1 \tag{11}$$

$$N_{IGBT} = N_S = 10n \tag{12}$$

$$N_D = 10n \tag{13}$$

where n represents the number of basic cells.

One of the advantages of cascaded topology is the replacing capability for the basic cells. If one of the basic cells is damaged, it can be taken out of the circuit, and the structure will continue to work with fewer voltage levels.

3. Efficiency Calculation

In this section, the calculation and simulation of the conduction losses and switching losses are presented to estimate the efficiency of an 11-level basic cell. The conduction losses are divided into two parts: switch conductance losses and anti-parallel diode conduction losses. A detailed analysis of the power losses of a power electronic converter is investigated in the following.

3.1. Conduction Losses

Power electronic switches have conduction losses when conducting the current in ON-state. The conduction losses of the switch and its anti-parallel diode is calculated by the following equations:

$$P_{c,S}(t) = [V_{S,ON} + R_S i^{\alpha}(t)]i(t) \tag{14}$$

$$P_{c,D}(t) = [V_{D,ON} + R_D i(t)]i(t) \tag{15}$$

where S indicates the switch and D indicates the diode. The voltages vs. and V_D are the voltage drop across the switch and the anti-parallel diode in their conduction interval. The resistors R_S and R_D represent the equivalent resistance of the switch and its anti-parallel diode, i(t) is the current flowing through the switch and the anti-parallel diode at the conduction moments. The parameter α is a switch constant that depends on the switch specifications, which is introduced by the manufacturer in the switch datasheet. The conduction losses are calculated from the sum of the conduction losses presented in Equations (14) and (15). The amount of conduction losses of a multi-level inverter depends on the number of switches conducted at different voltage levels. Considering N_S as the conducting number of switches and N_D as the conducting anti-parallel diodes in a time interval, the average conduction losses of the converter in an output voltage period can be represented by (16):

$$P_c = \frac{1}{2\pi}\int_0^{2\pi}\left[N_S(t)P_{c,S}(t) + N_D(t)P_{c,D}(t)\right]d(t) \tag{16}$$

3.2. Switching Losses

The switching losses are based on the energy losses due to the non-ideal switch performance. The energy losses include switch ON and OFF losses calculated by Equations (17) and (18):

$$\begin{aligned} E_{ON,j} &= \int_0^{t_{ON}} [v(t)\ i(t)]\, d(t) \\ &= \int_0^{t_{ON}} \left[\left(\frac{V_{S,j}}{t_{ON}}\right)\left(-\frac{I'}{t_{ON}}(t - t_{ON})\right)\right] d(t) = \frac{1}{6}V_{S,j}\, I'\, t_{ON} \end{aligned} \tag{17}$$

$$\begin{aligned} E_{OFF,j} &= \int_0^{t_{OFF}} [v(t)\ i(t)]\, d(t) \\ &= \int_0^{t_{OFF}} \left[\left(\frac{V_{S,j}}{t_{OFF}}\right)\left(-\frac{I}{t_{OFF}}(t - t_{OFF})\right)\right] d(t) = \frac{1}{6}V_{S,j} I t_{OFF} \end{aligned} \tag{18}$$

which $E_{ON,j}$ and $E_{OFF,j}$ are the energy dissipation of the switch j at the moments of turning on and off, t_{ON} and t_{OFF} are the time intervals required to turn a switch on and off, respectively. The parameters I and I' are the current that passes through the switch before turning it off, and after turning it on. $V_{S,j}$ is the reverse voltage across the switch after it is turned off. The switching power losses of switches in an output voltage period can be written as follows:

$$P_s = \sum_{j=1}^{N_S}\left[\sum_{i=0}^{N_{ON,j}} E_{ON,ji} + \sum_{i=0}^{N_{OFF,j}} E_{OFF,ji}\right] f \tag{19}$$

which $N_{ON,j}$ and $N_{OEE,j}$ are the number of times that switch turns on and off in a cycle, and f is the output voltage frequency. Finally, the total losses are calculated by Equation (20):

$$P_{Total} = P_c + P_s \tag{20}$$

To investigate the efficiency of the proposed 11-level basic cell, power losses on proposed MLI at the output loads Z_1 = 50 Ω, Z_2 = 50 + j12.56 Ω, and Z_3 = 50 + j25.12 Ω are simulated with output voltage steps of 50 V. Figure 4a, b show the conduction and switching power losses for all three types of the loads, respectively. Besides, the efficiency and total losses are also displayed in Figure 4c. To compare the efficiency of the proposed 11-level structure, the efficiency curve for different output power is shown in Figure 5.

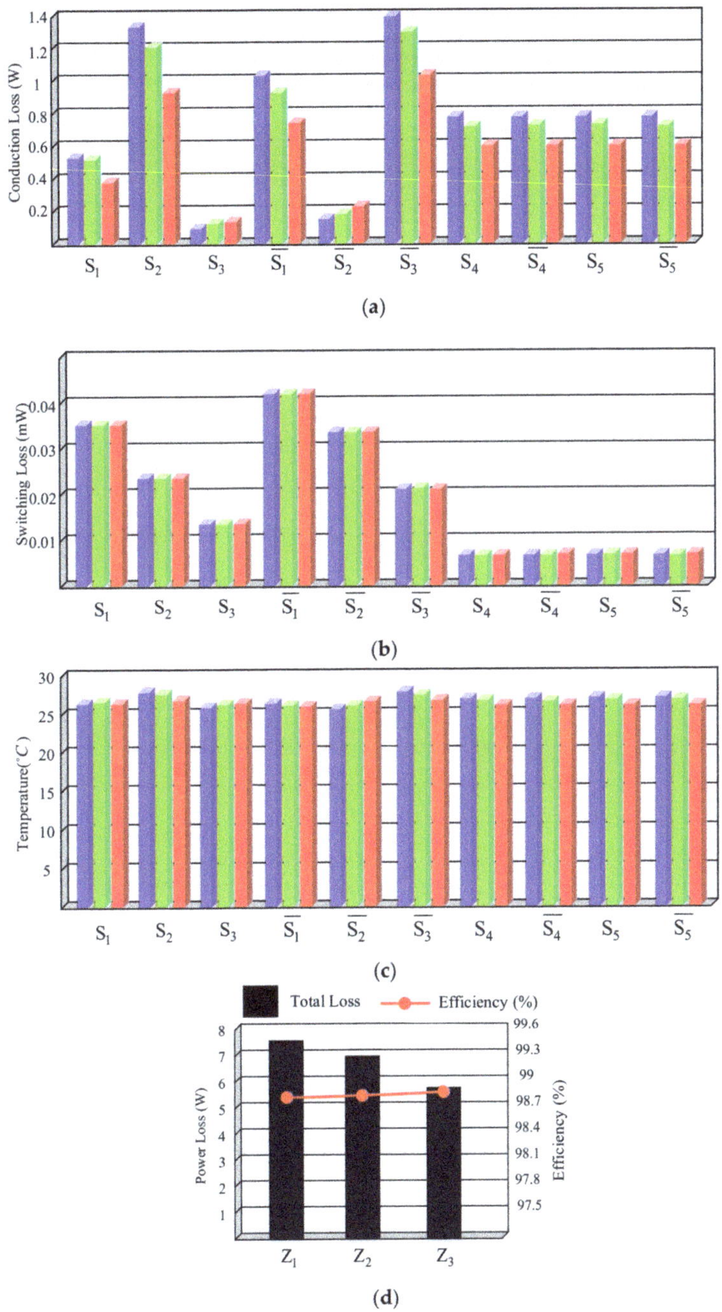

Figure 4. (**a**) Power losses and efficiency curves for the proposed 11-level basic cell topology at three types of loads, (**a**) Conduction losses, (**b**) Switching losses, (**c**) Temperature of switches, (**d**) Efficiency and total losses.

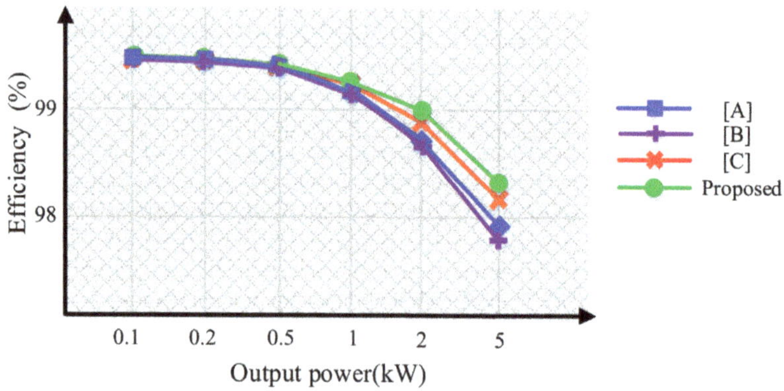

Figure 5. Efficiency comparison of the proposed 11-level basic cell with other topologies. [A] Alishah et al., 2021, [B] Jayabalan et al., 2017, [C] Ponraj et al., 2021.

4. Comparative Study

In this section, the proposed topology is compared with other recent topologies presented in [13–31] to evaluate the validity and capability of the proposed MLI. For a fair comparison of the structures, a graph of the number of switches to different voltage levels is presented, and in addition, the N_{IGBT}/N_L ratio is calculated and presented in Table 3 for the proposed basic module and the comparative structures. Moreover, Table 3 lists other comparative parameters, including N_{IGBT} (number of IGBTs), N_{GD} (number of drivers), N_L (number of voltage levels synthesized by structure), N_{DC} (number of DC voltage source), TBV (total blocking voltage), N_D (number of Diodes) and $N_{IGBT, ON}$ (number of conducting IGBTs in each voltage level) for the proposed topology and other symmetric topologies. According to this Table, the presented basic module utilizes fewer switches for various voltage levels.

To fairly compare the number of IGBTs in the proposed structure with other structures, a graph of the number of IGBTs relative to the number of voltage levels (N_{IGBT}/N_L) is evaluated. This ratio also provides cost-effectiveness of structures. The larger this ratio, the steeper the slope of the comparison curve, and the more IGBTs are used to achieve higher output voltage levels. Furthermore, the smaller this ratio, the lower the slope of the comparison curve, and the fewer IGBTs are required.

As shown in Figure 6a, using the proposed structure, a large number of voltage levels can be achieved with a smaller number of IGBTs, which produces high-quality output voltage with a smaller number of switches. Some structures use bidirectional common-emitter switches, which makes the number of gate drivers different from the number of IGBTs. For a fair comparison of the number of gate drivers of the proposed structure with other structures, the ratio of the number of gate drivers to the number of voltage levels (N_{GD}/N_L) is presented. Figure 6 demonstrates the (N_{IGBT}/N_L) and (N_{GD}/N_L) diagrams for the proposed and other structures. As shown in Figure 6a, the proposed topology has the lowest slope for (N_{IGBT}/N_L) diagram, which means the proposed structure utilizes the lowest number of switches to generate different voltage levels. Figure 6b also shows a comparison of the number of gate drivers, in which the proposed structure does not have the lowest curve slope regarding the number of gate drivers since it has not utilized a bidirectional switch. Nevertheless, the proposed structure still has a relatively good condition regarding the number of gate drivers compared to most comparative structures.

Table 3. Comparing the parameters of the proposed topology with other basic topologies.

	N_{IGBT}	N_{GD}	N_L	N_{DC}	TBV(* V_{dc})	N_D	$N_{IGBT,ON}$	N_{IGBT}/N_L
[13]	10	10	9	4	22	0	5	1.11
[14]	12	10	9	4	24	0	7	1.33
[15]	12	9	7	3	18	0	7	1.71
[16]	10	10	7	3	20	0	5	1.42
[17]	10	7	7	3	21	1	4	1.42
[18]	8	7	7	3	14	0	4	1.14
[19]	8	7	7	3	18	0	4	1.14
[20]	5	5	3	2	9	4	2	1.66
[21]	6	6	7	4	12	0	3	0.85
[22]	12	10	9	4	24	0	7	1.33
[23]	11	10	11	5	31	1	5	1
[24]	11	10	11	5	31	2	4	1
[25]	8	8	7	3	12	0	3	1.14
[26]	10	8	9	4	20	0	4	1.11
[27]	10	10	9	4	20	0	5	1.11
[28]	10	7	7	3	20	0	3	1.42
[29]	10	9	7	3	14	0	4	1.42
[30]	10	9	9	4	18	0	4	1.11
[31]	9	9	9	4	22	1	5	1
Proposed	10	10	13	5	30	0	5	0.77

* Indicates product symbol.

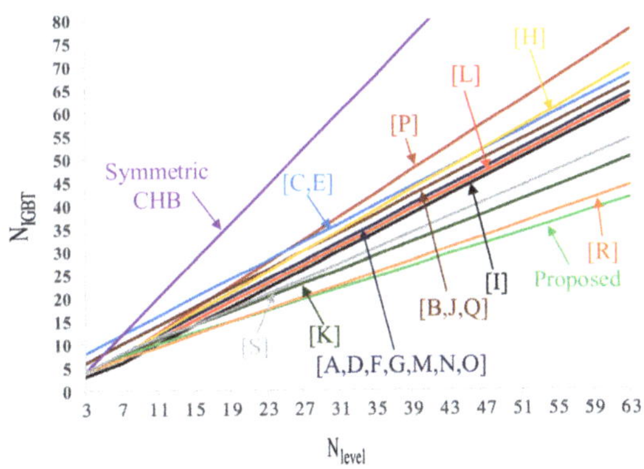

(**a**)

Figure 6. *Cont.*

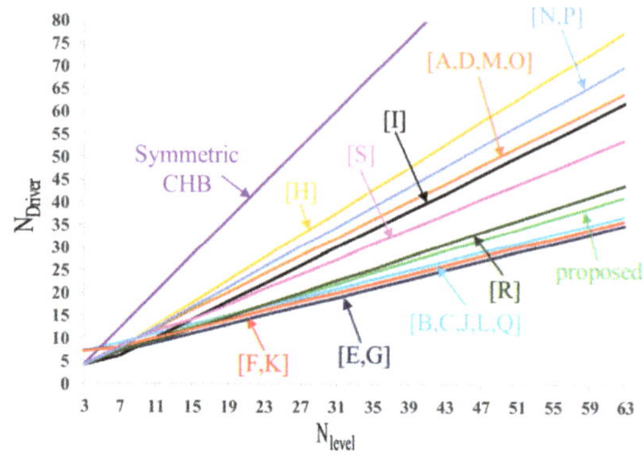

(b)

Figure 6. Comparative diagrams including: (**a**) N_{IGBT}/N_L, (**b**) N_{GD}/N_L. [A] Oskuee et al., 2015, [B] Alishah et al., 2021, [C] Jayabalan et al., 2017, [D] Ponraj et al., 2021, [E] Peddapati 2020, [F] Siddique et al., 2019, [G] Samsami et al., 2017, [H] Dhanamjayulu et al., 2017, [I] Gohari et al., 2019, [J] Alishah et al., 2016, [K] Hosseinpour et al., 2020, [L] Hosseini Montazer et al., 2021, [M] Selvaraj et al., 2020, [N] Meraj et al., 2019, [O] Ponraj et al., 2021, [P] Lee et al., 2017, [Q] Siddique et al., 2019, [R] Ali 2018, [S] Seifi et al 2020.

The total blocking voltage (TBV), the sum of the maximum blocking voltage (MBV) of the converter switches, is an essential parameter in comparing and evaluating structures because the voltage rating of required IGBTs for the structure is determined based on the MBV parameter, and the TBV parameter is directly related to the cost of the structure switches. Here, for a fair comparison of the TBV value, a graph of the ratio of this parameter to the number of output voltage levels (TBV/N_L) is used, and this graph is plotted for different topologies, as shown in Figure 7. As Figure 7 displays, the proposed structure provides a relatively good TBV compared to other structures.

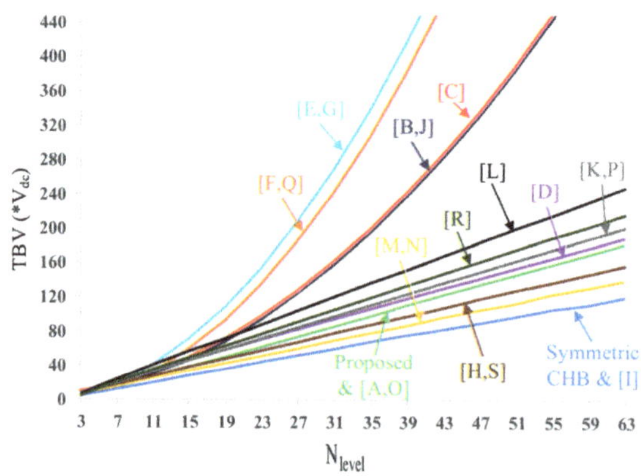

Figure 7. TBV/N_L diagram for the proposed structure and other structures.* Indicates product symbol. [A] Oskuee et al., 2015, [B] Alishah et al., 2021, [C] Jayabalan et al., 2017, [D] Ponraj et al., 2021, [E] Peddapati 2020, [F] Siddique et al., 2019, [G] Samsami et al., 2017, [H] Dhanamjayulu et al., 2017, [I] Gohari et al., 2019, [J] Alishah et al., 2016, [K] Hosseinpour et al., 2020, [L] Hosseini Montazer et al., 2021, [M] Selvaraj et al., 2020, [N] Meraj et al., 2019, [O] Ponraj et al., 2021, [P] Lee et al., 2017, [Q] Siddique et al., 2019, [R] Ali 2018, [S] Seifi et al., 2020.

5. Simulation and Laboratory Results of the Proposed Structure

To investigate the performance of the proposed structure, a prototype containing two RSBM is simulated, and a laboratory prototype is implemented. The proposed structure containing two RSBM is investigated for the symmetric and asymmetric topologies. The laboratory sample hardware includes MOSFET IRFP460 power switches. The ARDUINO MEGA2560 microcontroller is used to generate gate pulses, which are isolated and amplified to drive the switches using the TLP250 optocoupler. The laboratory prototype and related instruments are shown in Figure 8.

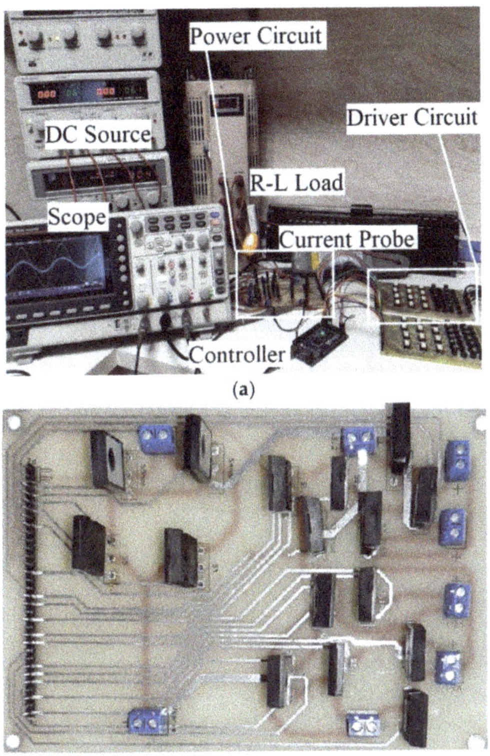

(a)

(b)

Figure 8. (**a**) A laboratory prototype of the proposed structure, (**b**) The power circuit of proposed structure.

As it is known, multi-level inverter switching methods are classified into two categories according to the switching frequency, which can be low frequency or high frequency, respectively. From the first class it can be mentioned the staircase switching, fundamental frequency switching, active harmonic elimination, nearest level modulation (NLM) and selective harmonic elimination (SHE) technique. The second class includes sinusoidal PWM and space vector modulation (SVM) techniques. The modulation methods in both classes can be easily adapted and implemented to the topology proposed in this paper. The NLM method has been used to generate the switching pulses (see Figure 9a), using an integer that is close to the nearest voltage level as the reference signal. For example, if the voltage is in the range of 1.5 to 2.5, then the reference of 2Vdc will be generated. Figure 9b presents the NLM operation. The switching frequency is not well-known in NLM method. But it is low and is relatively near to output voltage frequency. The output voltage frequency is considered 50 Hz.

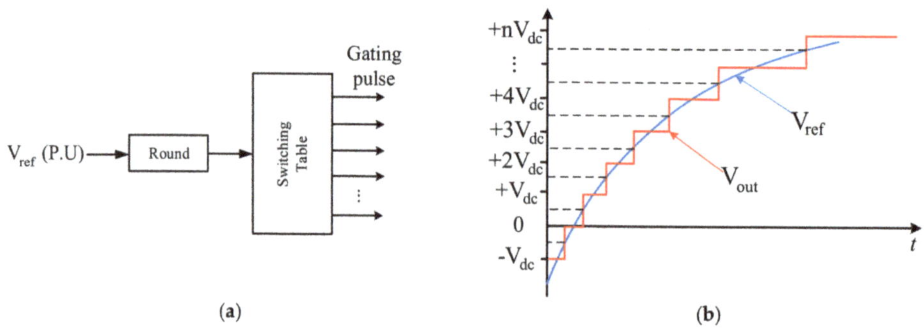

Figure 9. (**a**) NLM method, (**b**) The NLM operation.

In symmetric topology, the value of DC voltage sources is $V_0 = V_1 = V_2 = 6$ V. First, the inverter results for a purely resistive load $Z_{load1} = 6.6\ \Omega$ are presented in Figure 10. The peak output voltage of the inverter with 11 levels of 6 V steps results in 66 volts. The peak load current, in this case, is 10A. The harmonic spectrum of the proposed topology is presented in Figure 10c, highlighting that the total harmonic distortion (THD) is 3.25%, with advantage in size and cost of the output filter.

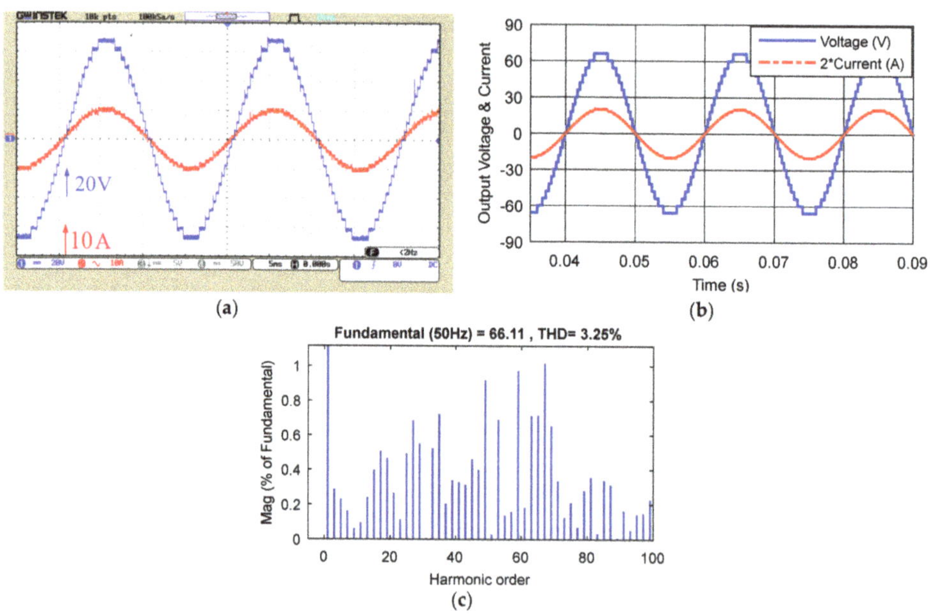

Figure 10. The waveforms for resistive output load in symmetric topology, (**a**) laboratory sample voltage and current, (**b**) simulation voltage and current, (**c**) output voltage THD. * Indicates product symbol.

Additionally, Figure 11 displays the results of the proposed structure in the case of symmetric sources for a resistive-inductive load $Z_{load2} = 6.6 + j4.71\ \Omega$. The peak load current in the case of R-L output load is 8.1 A, in which the inductance of the load filters, the current, and the load current is obtained similar to a sine wave.

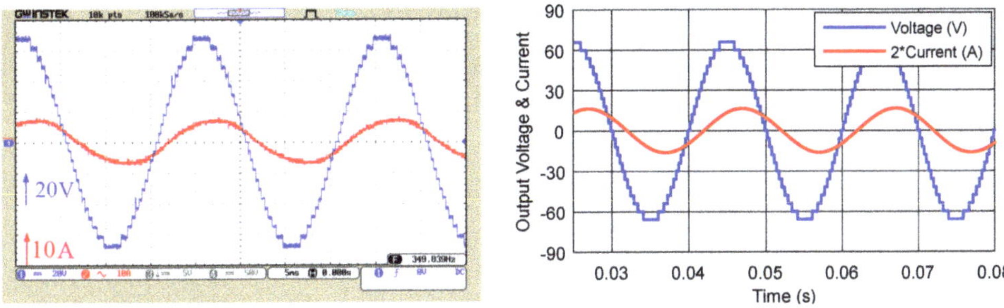

(a) (b)

Figure 11. Resistive-inductive load waveforms in symmetric topology, (**a**) laboratory sample voltage and current, (**b**) simulation voltage and current. * Indicates product symbol.

To analyze and introduce the response of the proposed MLI in a dynamic situation, the modulation index changes abruptly, and the system response is evaluated. It should be noted that, this test is only open loop dynamic responses. As shown in Figure 12, for this purpose, the modulation index is once changed from 1 to 0.65 according to Figure 12a, and again according to Figure 12b, the modulation index is changed from 0.65 to 0.3. The proper performance of the proposed multi-level inverter in dynamic conditions is visible for changing the modulation index in Figure 12.

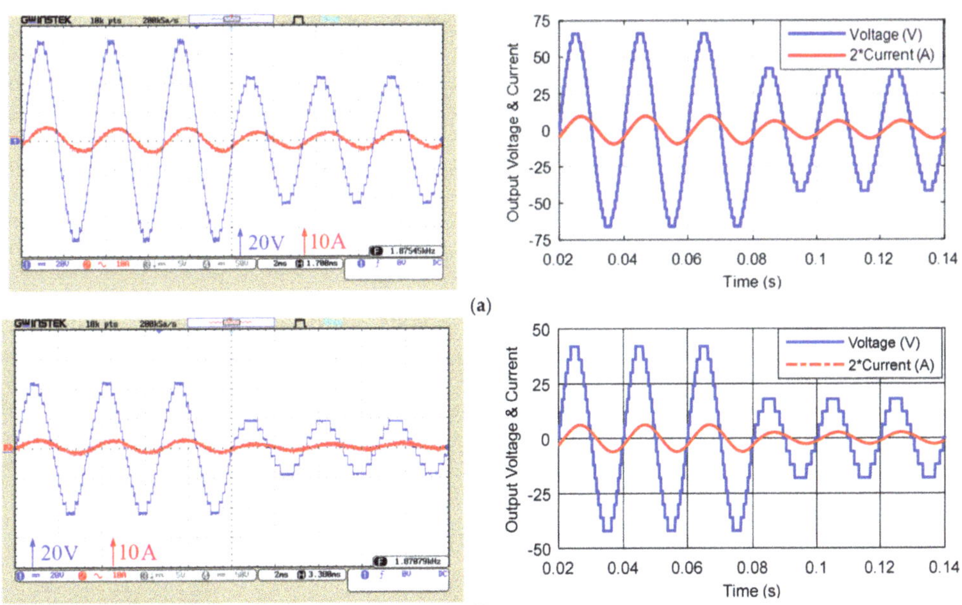

(b)

Figure 12. Voltage and current waveforms for changes of the modulation index: (**a**) from 1 to 0.65 and (**b**) from 0.65 to 0.3. * Indicates product symbol.

The behavior of the proposed structure for a sudden change in load is evaluated as well. For this purpose, the load changes abruptly from resistive-inductive to pure resistive. Figure 13 illustrates inverter output voltage and current for a sudden change of the resistive-inductive load $Z_{load2} = 6.6 + j4.71\ \Omega$ to a purely resistive load $Z_{load3} = 4\ \Omega$. The proposed structure performs well under dynamic load change conditions as well.

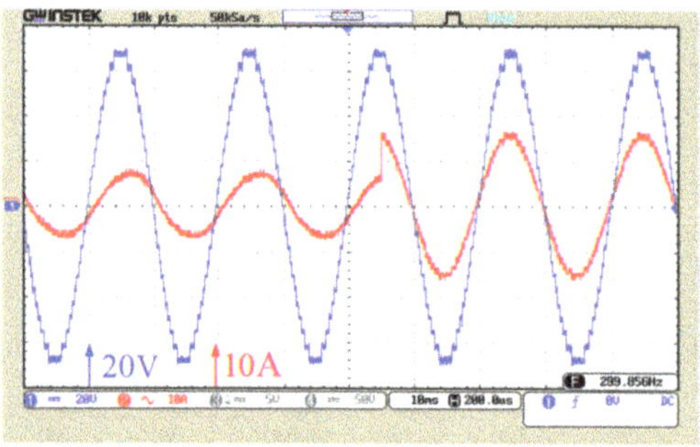

(a)

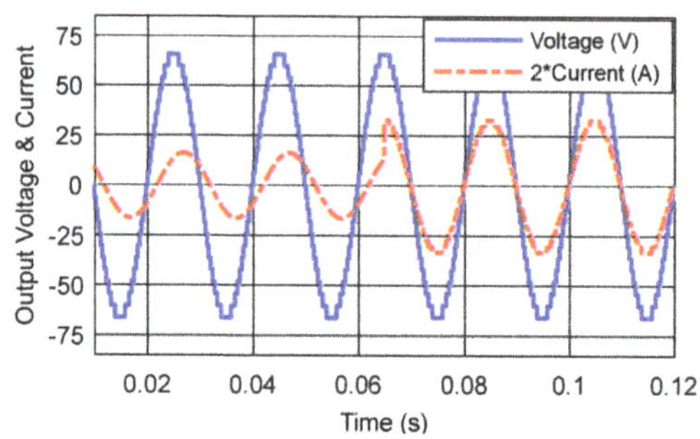

(b)

Figure 13. Voltage and current waveforms related to load's dynamic change, (**a**) laboratory result, (**b**) simulation result. * Indicates product symbol.

Figure 14 displays the blocking voltage for switches of the proposed structure in the symmetric topology. Based on these curves, the maximum voltage across each switch is determined. These diagrams confirm the correct operation of the proposed structure.

To evaluate the performance of multi-level inverter in asymmetric topology, the value of DC voltage sources is selected based on the proposed fourth algorithm according to Table 2 and assumed as $V_0 = V_1 = 2$ V, $V_2 = 12$ V. The performance results of the proposed multi-level inverter with asymmetric topology for a resistive-inductive load are presented in Figure 15. The number of voltage steps increases from 11 levels in the symmetric topology to 36 levels of 2 V steps in the asymmetric topology. In this case, the peak output voltage results in 72 V. The peak load current equals 11.4 A in this condition. Figure 15b displays the zoomed staircase waveform of the output voltage. The proper performance of the proposed structure is clearly shown in this figure in the production of voltage steps. In addition, the THD value is decreased from 3.26% in the symmetric topology to 1.01% in the asymmetric topology. This THD value demonstrates that the asymmetric topology of the suggested MLI can operate properly without any output filter, which results in more reduction in overall size and cost.

Dynamic load testing is necessary to evaluate the capability of the proposed structure in the application of electric motor drives. Based on the simulation results and laboratory results of the prototype, it can be concluded that the performance of the proposed structure is satisfactory, and the proposed topology provides a desirable performance.

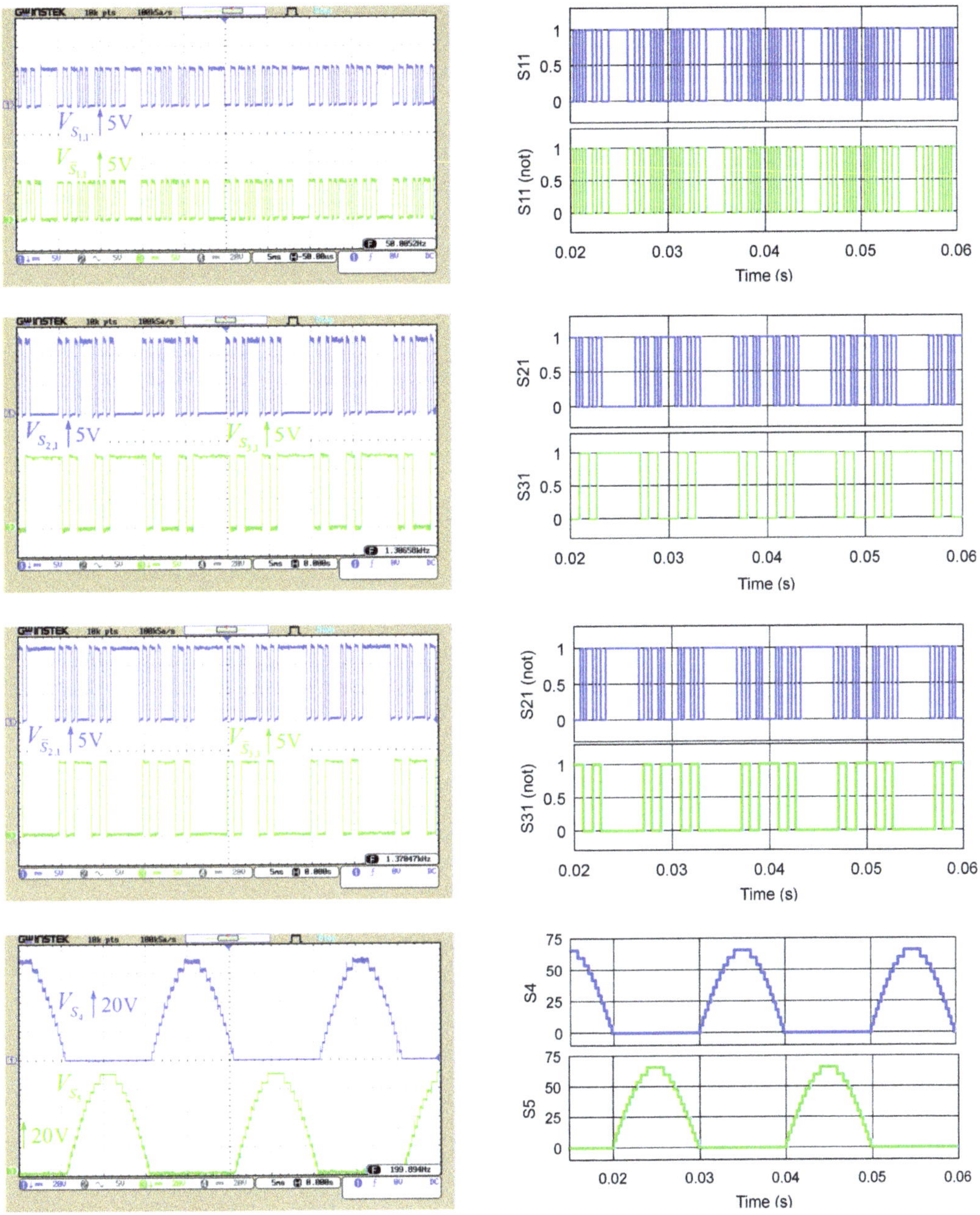

Figure 14. Blocking voltage of the proposed structure switches for symmetric topology.

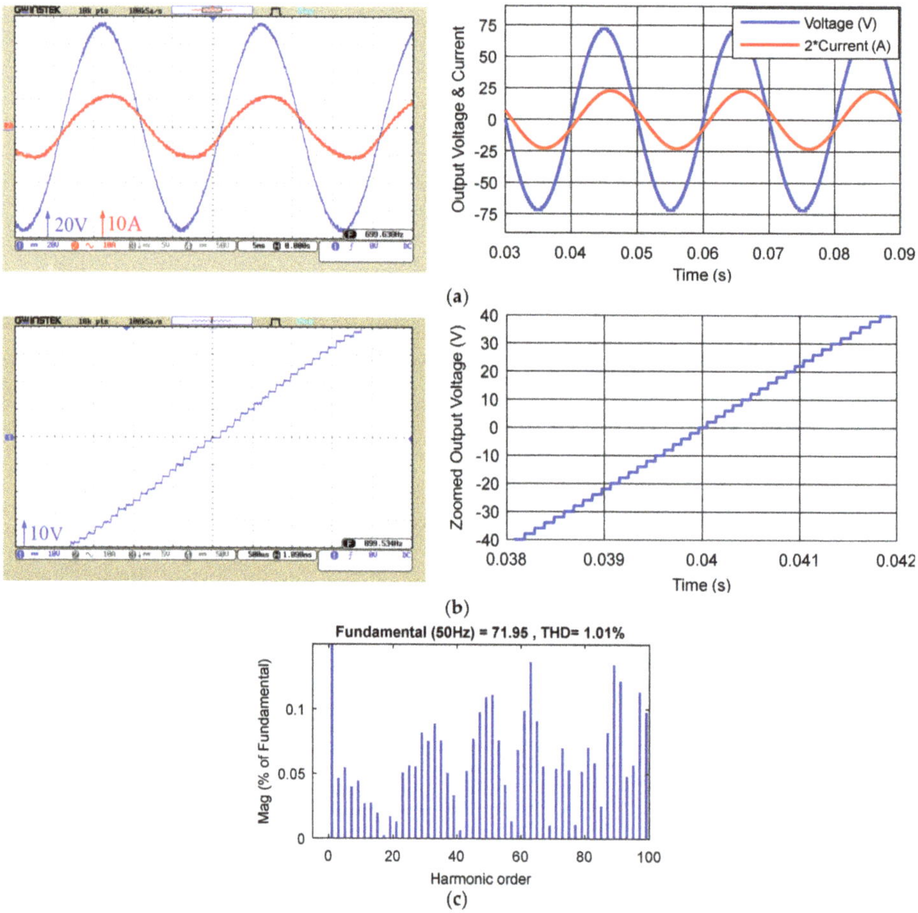

Figure 15. The waveforms of Proposed MLI in asymmetric topology (**a**) voltage and current, (**b**) zoomed voltage, (**c**) output voltage THD. * Indicates product symbol.

6. Conclusions

In this paper, a novel configuration is proposed for multi-level voltage source inverters to reduce the number of switches. The proposed structure uses a reduced switch basic module to generate more output voltage levels. This structure is investigated for symmetric and asymmetric topologies. The efficiency of the structure is presented for different loads and compared with several other structures. The result is shown that the suggested inverter has higher efficiency. Besides, the comparison in terms of circuit devices reveals that the number of switches of the proposed structure is less than other reported structures. This difference in the number of switches is clearly evident at higher levels, and the graph slope of the number of switches to the number of levels is 0.6, which is less than other structures. Also, the number of gate driver, as well as the TBV parameter are in good situation and superior in comparison with most of similar structures. These results make that the proposed structure has a low cost and size. The performance of the proposed structure is confirmed by simulation and laboratory results. In the laboratory prototype, different loading conditions, including resistive load, resistive-inductive load as well as dynamic load, have been used to validate the proposed structure. Modulation index change is also performed for the proposed structure. Results evaluation of simulations and laboratory samples, as well as the results of comparisons, indicate the appropriate performance of the proposed structure.

Author Contributions: Conceptualization, A.S., M.H. and H.S.; methodology, A.S., M.H. and H.S.; software, A.S. and M.H.; validation, M.H. and H.S.; investigation, M.H. and H.S.; resources, N.B.; data curation, N.B.; writing—original draft preparation, A.S., M.H. and H.S.; supervision, H.S. and N.B.; Funding acquisition: N.B.; Visualization: H.S. and N.B.; writing—review and editing: M.H., H.S. and N.B.; project administration, N.B.; Formal analysis: M.H., N.B. and H.S. All authors have read and agreed to the published version of the manuscript.

Funding: This research received no external funding.

Institutional Review Board Statement: Not applicable.

Informed Consent Statement: Not applicable.

Data Availability Statement: Not applicable.

Conflicts of Interest: The authors declare no conflict of interest.

References

1. Sunddararaj, S.P.; Srinivasarangan Rangarajan, S. An extensive review of multilevel inverters based on their multifaceted structural configuration, triggering methods and applications. *Electronics* **2020**, *9*, 433. [CrossRef]
2. Ahmed, S.A.M.; Abd El Sattar, M. Dynamic Performance and Effectiveness of Voltage Disturbances on the Improvement of Power Quality for Grid-Connected DFIG System Based Wind Farm. *J. Electr. Eng. Electron. Control. Comput. Sci.* **2019**, *5*, 25–30. Available online: https://jeeeccs.net/index.php/journal/article/view/126 (accessed on 3 September 2019).
3. Ahmed, S.A.M. Performance Analysis of Power Quality Improvement for Standard IEEE 14-Bus Power System based on FACTS Controller. *J. Electr. Eng. Electron. Control Comput. Sci.* **2019**, *5*, 11–18. Available online: https://jeeeccs.net/index.php/journal/article/view/134 (accessed on 3 December 2019).
4. Sorlei, I.S.; Bizon, N.; Thounthong, P.; Varlam, M.; Carcadea, E.; Culcer, M.; Iliescu, M.; Raceanu, M. Fuel Cell Electric Vehicles—A Brief Review of Current Topologies and Energy Management Strategies. *Energies* **2021**, *14*, 252. [CrossRef]
5. Benbouhenni, H.; Bizon, N. Improved Rotor Flux and Torque Control Based on the Third-Order Sliding Mode Scheme Applied to the Asynchronous Generator for the Single-Rotor Wind Turbine. *Mathematics* **2021**, *9*, 2297. [CrossRef]
6. Shahzad, U. Significance of Smart Grids in Electric Power Systems: A Brief Overview. *J. Electr. Eng. Electron. Control. Comput. Sci.* **2020**, *6*, 7–12. Available online: https://jeeeccs.net/index.php/journal/article/view/141 (accessed on 3 February 2020).
7. Onyishi, D.U.; Ofualagba, G. Analysis of the Electricity Distribution Supply in Eastern Nigeria: Current Challenges and Possible Solutions. *J. Electr. Eng. Electron. Control Comput. Sci.* **2021**, *7*, 1–8. Available online: https://jeeeccs.net/index.php/journal/article/view/209 (accessed on 3 September 2021).
8. Seifi, A.; Hosseinpour, M.; Dejamkhooy, A. A switch-source cell-based cascaded multilevel inverter topology with minimum number of power electronics components. *Trans. Inst. Meas. Control* **2021**, *43*, 1212–1225. [CrossRef]
9. Sarebanzadeh, M.; Hosseinzadeh, M.A.; Garcia, C.; Babaei, E.; Hosseinpour, M.; Seifi, A.; Rodriguez, J. A 15-Level Switched-Capacitor Multilevel Inverter Structure with Self-Balancing Capacitor. *IEEE Trans. Circuits Syst. II Express Briefs* **2022**, *69*, 1477–1481. [CrossRef]
10. Hosseinpour, M.; Seifi, A.; Dejamkhooy, A.; Sedaghati, F. Switch count reduced structure for symmetric bi-directional multilevel inverter based on switch-diode-source cells. *IET Power Electron.* **2020**, *13*, 1675–1686. [CrossRef]
11. Esmaeilzadeh, R.; Ajami, A.; Banaei, M.R. Two-Stage Inverter Based on Combination of High Gain DC-DC Converter and Five-Level Inverter for PV-Battery Energy Conversion. *J. Oper. Autom. Power Eng.* **2018**, *6*, 101–110. Available online: http://joape.uma.ac.ir/article_641.html (accessed on 29 February 2018).
12. Siddique, M.D.; Mekhilef, S.; Shah, N.M.; Sarwar, A.; Memon, M.A. A new single-phase cascaded multilevel inverter topology with reduced number of switches and voltage stress. *Int. Trans. Electr. Energy Syst.* **2020**, *30*, 12191. [CrossRef]
13. Oskuee, M.R.; Karimi, M.; Ravadanegh, S.N.; Gharehpetian, G.B. An innovative scheme of symmetric multilevel voltage source inverter with lower number of circuit devices. *IEEE Trans. Ind. Electron.* **2015**, *62*, 6965–6973. [CrossRef]
14. Alishah, R.S.; Bertilsson, K.; Vosoughi Kurdkandi, N.; Hosseini, S.H.; Gharehkoushan, A.Z.; Ali, J.S. A New Switched-Ladder Multilevel Converter Structure with Reduced Power Electronic Components. *J. Circuits Syst. Comput.* **2021**, *30*, 2150217. [CrossRef]
15. Jayabalan, M.; Jeevarathinam, B.; Sandirasegarane, T. Reduced switch count pulse width modulated multilevel inverter. *IET Power Electron.* **2017**, *10*, 10–17. [CrossRef]
16. Ponraj, R.P.; Sigamani, T.; Subramanian, V. A developed H-bridge cascaded multilevel inverter with reduced switch count. *J. Electr. Eng. Technol.* **2021**, *16*, 1445–1455. [CrossRef]
17. Peddapati, S. A generalized symmetrical MLI topology with improved commutation. *Electr. Eng.* **2020**, *102*, 2617–2635. [CrossRef]
18. Siddique, M.D.; Mekhilef, S.; Shah, N.M.; Memon, M.A. Optimal design of a new cascaded multilevel inverter topology with reduced switch count. *IEEE Access* **2019**, *7*, 24498–24510. [CrossRef]

19. Samsami, H.; Taheri, A.; Samanbakhsh, R. New bidirectional multilevel inverter topology with staircase cascading for symmetric and asymmetric structures. *IET Power Electron.* **2017**, *10*, 1315–1323. [CrossRef]
20. Dhanamjayulu, C.; Meikandasivam, S. Implementation and comparison of symmetric and asymmetric multilevel inverters for dynamic loads. *IEEE Access* **2017**, *6*, 738–746. [CrossRef]
21. Gohari, A.; Afjei, E.S.; Torkaman, H. Novel symmetric modular hybrid multilevel inverter with reduced number of semiconductors and low-voltage stress across switches. *IEEE J. Emerg. Sel. Top. Power Electron.* **2019**, *8*, 4297–4305. [CrossRef]
22. Alishah, R.S.; Hosseini, S.H.; Babaei, E.; Sabahi, M. A new general multilevel converter topology based on cascaded connection of submultilevel units with reduced switching components, DC sources, and blocked voltage by switches. *IEEE Trans. Ind. Electron.* **2016**, *63*, 7157–7164. [CrossRef]
23. Hosseinpour, M.; Seifi, A.; Rahimian, M.M. A bidirectional diode containing multilevel inverter topology with reduced switch count and driver. *Int. J. Circuit Theory Appl.* **2020**, *48*, 1766–1785. [CrossRef]
24. Hosseini Montazer, B.; Olamaei, J.; Hosseinpour, M.; Mozafari, B. A generalized diode containing bidirectional topology for multilevel inverter with reduced switches and power loss. *Int. J. Circuit Theory Appl.* **2021**, *49*, 2959–2978. [CrossRef]
25. Selvaraj, S.; Kumaresan, G.; Sathik, M.A. Modified "K"-type multilevel inverter topology with reduced switches, DC sources, and power loss. *Int. Trans. Electr. Energy Syst.* **2020**, *30*, 12345. [CrossRef]
26. Meraj, S.T.; Hasan, K.; Masaoud, A. A novel configuration of cross-switched T-type (CT-type) multilevel inverter. *IEEE Trans. Power Electron.* **2019**, *35*, 3688–3696. [CrossRef]
27. Ponraj, R.P.; Sigamani, T. A novel design and performance improvement of symmetric multilevel inverter with reduced switches using genetic algorithm. *Soft Comput.* **2021**, *25*, 4597–4607. [CrossRef]
28. Lee, S.S.; Sidorov, M.; Idris, N.R.; Heng, Y.E. A symmetrical cascaded compact-module multilevel inverter (CCM-MLI) with pulsewidth modulation. *IEEE Trans. Ind. Electron.* **2017**, *65*, 4631–4639. [CrossRef]
29. Siddique, M.D.; Mekhilef, S.; Shah, N.M.; Sarwar, A.; Iqbal, A.; Memon, M.A. A new multilevel inverter topology with reduce switch count. *IEEE Access* **2019**, *7*, 58584–58594. [CrossRef]
30. Ali, J.S.; Alishah, R.S.; Krishnasamy, V. A new generalized multilevel converter topology with reduced voltage on switches, power losses, and components. *IEEE J. Emerg. Sel. Top. Power Electron.* **2018**, *7*, 1094–1106. [CrossRef]
31. Seifi, A.; Hosseinpour, M.; Dejamkhooy, A.; Sedaghati, F. Novel reduced switch-count structure for symmetric/asymmetric cascaded multilevel inverter. *Arab. J. Sci. Eng.* **2020**, *45*, 6687–6700. [CrossRef]

sustainability

MDPI

Article

Design, Modeling, and Model-Free Control of Permanent Magnet-Assisted Synchronous Reluctance Motor for e-Vehicle Applications

Songklod Sriprang [1,2], Nitchamon Poonnoy [2,*], Babak Nahid-Mobarakeh [3], Noureddine Takorabet [1], Nicu Bizon [4], Pongsiri Mungporn [5] and Phatiphat Thounthong [2,*]

1 Groupe de Recherche en Energie Electrique de Nancy (GREEN), Université de Lorraine, GREEN, F-54000 Nancy, France; songklod.sriprang@univ-lorraine.fr (S.S.); noureddine.takorabet@univ-lorraine.fr (N.T.)
2 Renewable Energy Research Centre (RERC), Department of Teacher Training in Electrical Engineering, Faculty of Technical Education, King Mongkut's University of Technology North Bangkok, 1518, Pracharat 1 Road, Bangsue, Bangkok 10800, Thailand
3 Department of Electrical and Computer Engineering, McMaster University, Hamilton, ON L8S 4L8, Canada; babak.nahid@mcmaster.ca
4 Faculty of Electronics, Communications and Computers, University of Pitesti, Arges, 110040 Pitesti, Romania; nicu.bizon@upit.ro
5 Thai-French Innovation Institute (TFII), King Mongkut's University of Technology North Bangkok, 1518, Pracharat 1 Road, Bangsue, Bangkok 10800, Thailand; pongsiri.m@tfii.kmutnb.ac.th
* Correspondence: nitchamon.p@fte.kmutnb.ac.th (N.P.); phatiphat.t@fte.kmutnb.ac.th (P.T.)

Citation: Sriprang, S.; Poonnoy, N.; Nahid-Mobarakeh, B.; Takorabet, N.; Bizon, N.; Mungporn, P.; Thounthong, P. Design, Modeling, and Model-Free Control of Permanent Magnet-Assisted Synchronous Reluctance Motor for e-Vehicle Applications. *Sustainability* **2022**, *14*, 5423. https://doi.org/10.3390/su14095423

Academic Editor: Mouloud Denai

Received: 17 February 2022
Accepted: 27 April 2022
Published: 30 April 2022

Publisher's Note: MDPI stays neutral with regard to jurisdictional claims in published maps and institutional affiliations.

Abstract: This paper describes the model-free control approaches for permanent magnet-assisted (PMa) synchronous reluctance motors (SynRMs) drive. The important improvement of the proposed control technique is the ability to determine the behavior of the state-variable system during both fixed-point and transient operations. The mathematical models of PMa-SynRM were firstly written in a straightforward linear model form to show the known and unknown parts. Before, the proposed controller, named here the intelligent proportional-integral (*i*PI), was applied as a control law to fix some unavoidable modeling errors and uncertainties of the motor. Lastly, a dSPACE control platform was used to realize the proposed control algorithm. A prototype 1-kW test bench based on a PMa-SynRM machine was designed and realized in the laboratory to test the studied control approach. The simulation using MATLAB/Simulink and experimental results revealed that the proposed control achieved excellent results under transient operating conditions for the motor drive's cascaded control compared to traditional PI and model-based controls.

Keywords: electric vehicle; inverter; permanent magnet-assisted synchronous reluctance motor; PMa-SynRM; model-free control; traction drive

1. Introduction

By the end of 2021, the demand for electrical traction machines, including battery electric vehicles and hybrid electric vehicles (HEVs), surpassed two million units [1–4]. Electrical traction machines are also required to further develop more electric aircrafts (MEAs) [5–7]. For these reasons, several state-of-the-art machines have been developed in the last few years, such as synchronous reluctance motors (SynRMs) and especially permanent magnet-assisted synchronous reluctance motors (PMa-SynRMs). PMa-SynRMs can produce 75% of the torque of an interior permanent magnet synchronous motor (IPMSM) for the same size and liquid cooling technology [8,9]. In addition, state-of-the-art modern motors provide more desired characteristics for electric vehicle (EV) applications, in particular, high efficiency at low and high speeds. Therefore, PMa-SynRMs constitute a promising choice for these applications. However, PMa-SynRMs have a much more complicated structure, which affects the control system, and its model is strongly nonlinear. Therefore, traditional control, such as field-oriented control (FOC) based on a proportional-integral (PI) controller, cannot accomplish high performance for all operating conditions of these modern machines.

Furthermore, in EV applications, safety, energy saving, and soft driving are mandatory and require improvement of the control performance of the motor drive system. Many studies have been conducted in the last few years regarding SynRMs and PMa-SynRMs, with special attention to machine design and optimization aspects. Multiple-flux barrier rotors and transversely laminated rotors were reported. Rotor laminations are made by traditional punching or wire cutting, resulting in easy and cheap construction [10,11]. Control characteristics have also been investigated [9,12,13]. In this regard, in the current control of PMa-SynRM drive systems, the essential objective is to ensure that the stator currents track the reference values with minimum errors in both transient and steady-state conditions. To design a robust controller with acceptable tracking performance, all the model-based control (MBC) approaches mentioned in the literature applied to PMa-SynRM require extensive knowledge about the dynamics and the model of PMa-SynRM systems. In addition, the MBC performance can be affected by unexpected dynamic variations of the system and parametric uncertainty, which are very common phenomena in industrial applications. To overcome the limitations of MBC approaches, some studies proposed model predictive control (MPC) as an appropriate current control scheme for electric motors, which ensures a fast dynamic and a remarkable safety factor [14–16]. This method's concept is based on predicting controlled variables in the next calculation step using the measured variables and a mathematical model of the controlled system. Then, the predicted results are analyzed using a cost function in terms of the difference between the desired trajectories and real outputs of the system. Compared to the previously mentioned control techniques, safety and fast dynamics are two remarkable features of the MPC method. Despite these advantages, the performance of MPC highly depends on the correctness of the model, given that a mathematical model is used in the prediction section [17]. When using the prediction at each sampling time of the MPC algorithm, some additional mathematical calculations are imported into the control algorithm.

Therefore, a control principle called model-free control (MFC) has been proposed to address the limitations of the abovementioned MPC and MBC techniques. MFC, also referred to as model-free tuning in the literature, uses a local linear approximation of the process model, which is valid for a small time window, and a fast estimator, which is employed to update the approximation [18,19]. The main advantage of MFC is that it does not require the process model in the controller tuning. Few experiments have been conducted on real-world control system structures concerning the tuning process. This paper introduces MFC development to control both torque and speed control of PMa-SynRMs. To verify the advantages of MFC, both simulations and experiments were carried out under several conditions.

This paper is organized as follows. A model-free control and control law are briefly introduced in Section 2. The main issues regarding the control of PMa-SynRMs, related state-of-the-art studies, as well as mathematical models are reviewed in Section 3, with a focus on MFC applied to PMa-SynRM drive systems. In Section 4, simulation and experimental results are provided to demonstrate the advantages of the proposed MFC. Sections 5 and 6 summarize and conclude the paper. A small-scale 1-kW test bench based on a PMa-SynRM with ferrite magnets was implemented to confirm the high performance of the designed control scheme in the laboratory [13].

2. Model-Free Control and Control Law (Brief Introduction)

2.1. Model-Free Control

The idea of model-free control accomplished for control system applications was originally proposed by Fliess et al. [20,21]. Many industrial applications have significantly changed with technology development and have become more complex. Accordingly, modeling the dynamic and process of these applications using mathematical models becomes very difficult or at least time-consuming. In this case, using the MBC methods for these kinds of applications will be impossible. Conversely, almost all industrial applications generate and save a large number of process data that contain all the necessary information related to the system's operation. In this case, it is important to use these generated data, obtained online/offline, directly for designing the controller or other purposes. In this way, the model-free control (MFC) foundation is essential in controlling industrial applications. So far, the types of the modern control system can be roughly categorized by MBC and MFC, as in Figure 1.

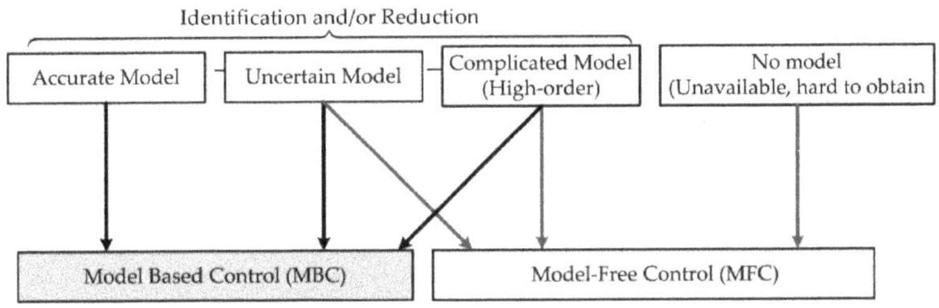

Figure 1. Control law's block diagram.

MFC is a control method that uses only the online data obtained from the controlled system to design the controller, without the additional need for information about the mathematical model or parameters of the studied system. Therefore, the MFC can be applicable for all nonlinear systems with complex or unknown structures.

The principle of model-free control is briefly introduced next. A nonlinear system can be described by a state-variable written as follows:

$$\begin{aligned} \dot{x} &= f(x,u) \\ y &= h(x,u) \end{aligned} \tag{1}$$

where

$$\begin{aligned} x &= [x_1, x_2, \ldots, x_{\mathrm{n}}]^T; x \in \mathbb{R}^n \\ u &= [u_1, u_2, \ldots, u_{\mathrm{m}}]^T; u \in \mathbb{R}^m \\ y &= [y_1, y_2, \ldots, y_{\mathrm{m}}]^T; y \in \mathbb{R}^m \end{aligned} \tag{2}$$

where x is the state variable, u is the control variable, y is the output variable, and $n, m \in \mathrm{N}$.

According to Equation (2), the system described by Equation (1) is flat. A control law of variable u can be expressed as follows [13]:

$$u = u_{\mathrm{ref}} + u_{\mathrm{feedback}}(\varepsilon) \tag{3}$$

with $\varepsilon = y_{\mathrm{ref}} - y$.

This control law is suitable for all systems with known parameters. However, if only some system parameters can be identified or the system described by Equation (1) cannot be identified, the controller needs to be modified as a partially-known model, replaced by a model-free control as follows:

$$u = \frac{\hat{\alpha}(y, \dot{y}, \ddot{y}, \ldots, y^{(n)})}{b} + \frac{F}{b} \tag{4}$$

where $\hat{\alpha}(\dot{y})/b$ is a known system, and F denotes an unknown part of the system.

The difference between $\hat{\alpha}(y, \dot{y}, \ddot{y}, \ldots, y^{(n)})$, $\hat{\alpha}(\dot{y})$, and $\dot{y}$ is that the $\hat{\alpha}(y, \dot{y}, \ddot{y}, \ldots, y^{(n)})$ is the known part of the $\alpha(y, \dot{y}, \ddot{y}, \ldots, y^{(k)})$, the $\hat{\alpha}(\dot{y})$ is the only known part of the studied system, and the $\dot{y}$ is the differential of the known part, respectively.

Alternatively, it can be rewritten and rearranged as a straightforward linear model as follows:

$$\dot{y} = -F + b \cdot u \tag{5}$$

2.2. Control Law

Figure 2 represents the control law block diagram for the model-free control technique. The control law is defined as follows:

$$u = u_{\text{ref}} + u_{\text{feedback}}(\varepsilon) + \frac{\widehat{F}}{b} \tag{6}$$

where

$$u_{\text{ref}} = \frac{\hat{\alpha}\left(y_{\text{ref}}, \dot{y}_{\text{ref}}, \ddot{y}_{\text{ref}}, \ldots y_{\text{ref}}^{(\beta+1)}\right)}{b} \tag{7}$$

and $\widehat{F}$ is the estimated value of F, which is expressed as follows:

$$\widehat{F} = b \cdot u - \dot{y} \tag{8}$$

The function $\hat{\alpha}$ is a regular function [22,23].

The feedback term u_{feedback} can be described by applying the PI controller as follows:

$$u_{\text{feedback}} = K_{\text{p}} \cdot \varepsilon + K_{\text{i}} \cdot \int \varepsilon dt \tag{9}$$

Substituting Equation (6) into Equation (5), and rearranging the expressions, Equation (5) can be expressed as follows:

$$\dot{y} = -F + b \cdot u_{\text{ref}} + b \cdot u_{\text{feedback}}(\varepsilon) + \widehat{F} \tag{10}$$

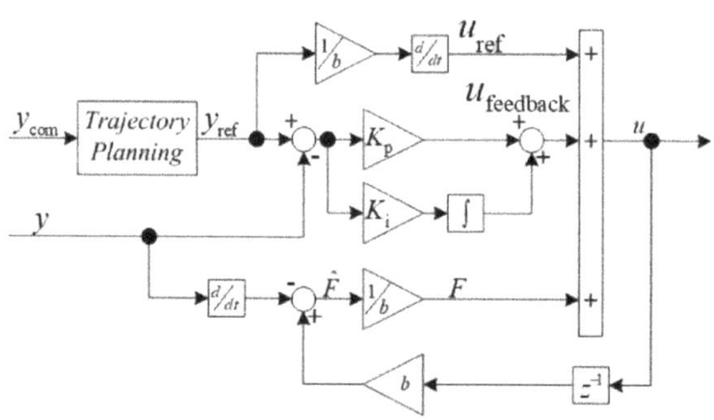

Figure 2. Control law's block diagram.

2.3. Controller Design

The observation term purposes to afford an estimated signal $\widehat{F}$ so that $\widehat{F} \to F$ as $t \to \infty$ (under global convergence assumption for the estimation). Consequently, Equation (10) can be rewritten as follows:

$$\dot{y} = b \cdot u_{\text{ref}} + b \cdot u_{\text{feedback}}(\varepsilon) \tag{11}$$

Consequently, Equation (11) describes the dynamic of the closed-loop control system. By substituting Equation (9) into Equation (11) and rearranging, Equation (11) can be expressed as follows:

$$\frac{d(y_{ref} - y)}{dt} + b \cdot K_p \cdot \varepsilon + b \cdot K_i \int \varepsilon dt = 0 \tag{12}$$

Referring to the control law displayed in Figure 2, the controller coefficients can be determined using the following expression obtained by taking time derivation in Equation (12):

$$\ddot{\varepsilon} + b \cdot K_p \cdot \dot{\varepsilon} + b \cdot K_i \cdot \varepsilon = 0 \tag{13}$$

Comparing Equation (13) to the 2nd order standard equation stated as follows:

$$\ddot{q} + 2 \cdot \zeta \cdot \omega_n \cdot \dot{q} + \omega_n^2 \cdot q = 0 \tag{14}$$

the controller coefficients become:

$$K_p = \frac{2 \cdot \zeta \cdot \omega_n}{b} \tag{15}$$

and

$$K_i = \frac{\omega_n^2}{b} \tag{16}$$

where ζ and ω_n are the tuning dominant damping ratio and natural frequency, respectively.

The gain $b \in \mathbb{R}$ is a non-physical constant parameter. Instead of α, the b is present in this paper, as shown in (4). It was chosen by the practitioner or obtained by trials and errors. F, which is continuously updated, subsumes the poor parts of the plant and the various possible disturbances without distinguishing between them [24,25]

3. Applying Model-Free Control to PMa-SynRM Drive

3.1. Mathematic Model of PMa-SynRM/Inverter

A variable speed drive (VSD), which powers the PMa-SynRM under study, is shown in Figure 3. Owing to the rotor geometries of the PMa-SynRM discussed in [26], the current control strategies in the literature differ from those applied to PMSM. The rotor geometries of PMa-SynRMs are given by the salient-pole, in which $L_d > L_q$. Its torque expression was given by Equation (17). In this case, the i_d component should not be equal to zero to take advantage of the reluctance torque produced by the high saliency ratio. Therefore, the maximum torque per ampere (MTPA) control strategy was recommended for PMa-SynRMs. The main idea of this control was to develop the requested torque using the minimum value of the stator current magnitude:

$$T_e = n_p\{\Psi_m - (L_d - L_q)i_q\} \cdot i_d \tag{17}$$

The equations of a PMa-SynRM in the rotating d_q reference frame and a mechanical equation are expressed by a state-space representation as follows:

$$\underbrace{\begin{bmatrix} \frac{di_d}{dt} \\ \frac{di_q}{dt} \\ \frac{d\omega_m}{dt} \end{bmatrix}}_{\dot{x}} = \underbrace{\begin{bmatrix} \{-R_s i_d + \omega_e(L_q i_q - \Psi_m)\}/L_d \\ (-R_s i_q - \omega_e L_d i_d)/L_q \\ [n_p\{\Psi_m i_d + (L_d - L_q)i_q i_d\} - B_f\omega_m]/J \end{bmatrix}}_{f(x)} + \underbrace{\begin{bmatrix} \frac{1}{L_d} & 0 & 0 \\ 0 & \frac{1}{L_q} & 0 \\ 0 & 0 & -\frac{1}{J} \end{bmatrix}}_{\mathbf{B}} \underbrace{\begin{bmatrix} v_d \\ v_q \\ T_L \end{bmatrix}}_{u}$$
$$y = \underbrace{\begin{bmatrix} 1 & 0 & 0 \\ 0 & 1 & 0 \\ 0 & 0 & 1 \end{bmatrix}}_{\mathbf{C}} \begin{bmatrix} i_d \\ i_q \\ \omega_m \end{bmatrix} \tag{18}$$

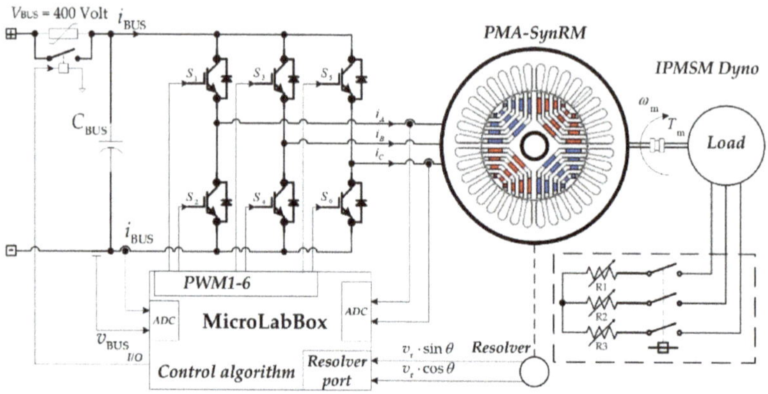

Figure 3. A three-phase inverter to control a PMa-SynRM prototype.

3.2. Model-Free of Current and Speed Control Development

The control system of PMa-SynRMs proposed in this paper (Figure 4) had a case cascade construction consisting of two loops (i.e., inner current control loop and outer speed control loop). The inner current loop was much faster than the outer speed control loop, such that the model-free control for the current control was developed first. By defining $u = [u_1\ u_2]^T = [v_d\ v_q]^T$, $y = [y_1\ y_2]^T = [i_d\ i_q]^T$, and rearranging the first and second rows in Equation (18) in the form of Equation (5), the PMa-SynRM model is expressed as follows:

$$\begin{aligned} \frac{di_d}{dt} &= -\frac{R_s i_d}{L_d} + \frac{\omega_e(L_q i_q - \Psi_m)}{L_d} + v_d \cdot \frac{1}{L_d} \\ \frac{di_q}{dt} &= -\frac{R_s i_q}{L_q} - \frac{\omega_e L_d i_d}{L_q} + v_q \cdot \frac{1}{L_q} \end{aligned} \tag{19}$$

According to the principle of the model-free as in [20,21], Equation (19) can be separated to identify the known and unknown terms as follows. The known terms are

$$\begin{aligned} \hat{\alpha}_1 &= \frac{\dot{y}_1}{b_1} = L_d \frac{di_d}{dt} \\ \hat{\alpha}_1 &= \frac{\dot{y}_2}{b_2} = L_q \frac{di_q}{dt} \end{aligned} \tag{20}$$

and the unknown terms are

$$\begin{aligned} F_1 &= \{-R_s i_d + \omega_e(L_q i_q - \Psi_m)\} \cdot \frac{1}{L_d} \\ F_2 &= (-R_s i_q - \omega_e L_d i_d) \cdot \frac{1}{L_q} \end{aligned} \tag{21}$$

According to the control law (Figure 2), the first term of the model-free control for inner current loop control is determined as follows:

$$\begin{aligned} u_{1ref} &= \frac{\dot{y}_{1ref}}{b_1} = L_d \frac{di_d}{dt} \\ u_{2ref} &= \frac{\dot{y}_{2ref}}{b_2} = L_q \frac{di_q}{dt} \end{aligned} \tag{22}$$

The estimation of unknown terms is expressed as follows:

$$\begin{aligned} \widehat{F}_1 &= \frac{1}{L_d} u_1 - \dot{y}_1 = \frac{1}{L_d} v_d - \frac{di_d}{dt} \\ \widehat{F}_2 &= \frac{1}{L_q} u_1 - \dot{y}_2 = \frac{1}{L_q} v_q - \frac{di_q}{dt} \end{aligned} \tag{23}$$

The feedback terms of d- and q-axis current control are obtained as follows:

$$b_1 \cdot u_{1\text{feedback}} = b_1\left(K_{\text{pd}} \cdot \varepsilon_{\text{d}} + K_{\text{id}} \int \varepsilon_{\text{d}} dt\right)$$
$$b_2 \cdot u_{2\text{feedback}} = b_2\left(K_{\text{pq}} \cdot \varepsilon_{\text{q}} + K_{\text{iq}} \int \varepsilon_{\text{q}} dt\right) \tag{24}$$

Concerning the design procedure in the controller design, Equation (24) can be rewritten as follows:

$$\ddot{\varepsilon}_{\text{d}} + b_1 \cdot K_{\text{pd}} \cdot \dot{\varepsilon}_{\text{d}} + b_1 \cdot K_{\text{id}} \cdot \varepsilon_{\text{d}} = 0$$
$$\ddot{\varepsilon}_{\text{q}} + b_2 \cdot K_{\text{pq}} \cdot \dot{\varepsilon}_{\text{q}} + b_2 \cdot K_{\text{iq}} \cdot \varepsilon_{\text{d}} = 0 \tag{25}$$

The controller coefficients K_{pd}, K_{id}, K_{pq}, and K_{iq} are determined as follows:

$$K_{\text{pd}} = \frac{2\zeta_1\omega_{\text{n1}}}{b_1}, K_{\text{id}} = \frac{\omega_{\text{n1}}^2}{b_1}$$
$$K_{\text{pq}} = \frac{2\zeta_1\omega_{\text{n1}}}{b_2}, K_{\text{iq}} = \frac{\omega_{\text{n1}}^2}{b_2} \tag{26}$$

The second model-free control for the outer speed control loop is developed here. The output of the speed control loop provides the torque reference of the MTPA algorithm, generating optimized d- and q-axis current references. Therefore, T_{e} was chosen as a control variable of the outer speed control loop, such that $u_3 = T_{\text{eREF}}$. Then, rewriting the mechanical equation of the PMa-SynRM represented by the third row in Equation (18) in the form of Equation (5) yields:

$$\frac{d\omega_{\text{m}}}{dt} = \left(-B_f \cdot \omega_{\text{m}} - T_{\text{L}}\right) \cdot \frac{1}{J} + T_{\text{e}} \cdot \frac{1}{J} \tag{27}$$

Separating this equation into the known and unknown terms, the known term is expressed as follows:

$$\hat{\alpha}_3 = \frac{\dot{y}_3}{b_3} = J\frac{d\omega_{\text{m}}}{dt} \tag{28}$$

The unknown term is expressed as follows:

$$F_3 = \left(-B_f\omega_m - T_L\right) \cdot \frac{K_t}{J} \tag{29}$$

Each part of the model-free control for the outer speed control loop is defined according to the following expression:

$$u_{3\text{ref}} = \frac{\dot{y}_{3\text{ref}}}{b_3} = J\frac{d\omega_{\text{m}}}{dt} \tag{30}$$

The estimation of the unknown term is expressed as follows:

$$\widehat{F}_3 = \frac{K_t}{J}u_3 - \dot{y}_3 = \frac{1}{J} \cdot T_{\text{e}} - \frac{d\omega_{\text{m}}}{dt} \tag{31}$$

$$b_3 u_{3\text{feedback}} = b_3\left(K_{\text{p}\omega} \cdot \varepsilon_\omega + K_{\text{i}\omega} \int \varepsilon_\omega dt\right) \tag{32}$$

Regarding the controller design procedure, Equation (32) can be rewritten as follows:

$$\ddot{\varepsilon}_\omega + b_1 \cdot K_{\text{p}\omega} \cdot \dot{\varepsilon}_\omega + b_1 \cdot K_{\text{i}\omega} \cdot \varepsilon_\omega = 0 \tag{33}$$

The controller coefficients $K_{\text{p}\omega}$ and $K_{\text{i}\omega}$ are determined as follows:

$$K_{\text{p}\omega} = \frac{2\zeta_2\omega_{\text{n2}}}{b_3}, K_{\text{i}\omega} = \frac{\omega_{\text{n2}}^2}{b_3} \tag{34}$$

where ζ_2 and ω_{n2} are the desired dominant damping ratio and natural frequency of the outer speed control loop, respectively.

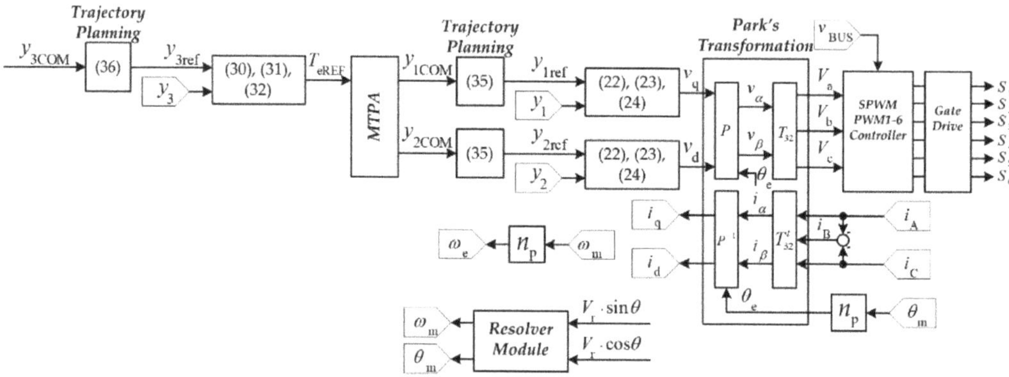

Figure 4. Control system of PMa-SynRM based on a model-free control diagram.

3.3. Trajectory Planning

Finally, as presented in Figure 2, desired trajectory planning must be implemented to generate the input set-point y_{REF}. A second order filter is often implemented to plan the desired trajectory for the controlled output. It permits limiting the derivative terms in the control law. The proposed trajectory planning for the two inner current control loops is expressed as follows:

$$\frac{y_{1REF}}{y_{1COM}} = 1 \Big/ \left\{ \left(\frac{s}{\omega_{n3}} \right)^2 + \frac{2\zeta_3}{\omega_{n3}} s + 1 \right\} \tag{35}$$

$$\frac{y_{2REF}}{y_{2COM}} = 1 \Big/ \left\{ \left(\frac{s}{\omega_{n3}} \right)^2 + \frac{2\zeta_3}{\omega_{n3}} s + 1 \right\} \tag{36}$$

where ζ_3 and ω_{n3} are the tuning dominant damping ratio and natural frequency, respectively.

The trajectory planning of the outer speed loop is expressed as follows:

$$\frac{y_{3REF}}{y_{3COM}} = 1 \Big/ \left\{ \left(\frac{s}{\omega_{n4}} \right)^2 + \frac{2\zeta_4}{\omega_{n4}} s + 1 \right\} \tag{37}$$

where ζ_4 and ω_{n4} are the desired dominant damping ratio and natural frequency of the speed loop trajectory planning, respectively.

4. Simulation and Experimental Validation of the Model-Free Control Applied to PMa-SynRM

4.1. Experimental Setup

A small-scale test bench 1-KW relying on the prototype PMa-SynRM was conceived in the laboratory, as shown in Figure 5. The prototype PMa-SynRM was supplied by a 3-kW 3-phase inverter (DC/AC) operating at a switching frequency of 16 kHz. Besides, the input DC grid voltage of the inverter was fed by a three-phase variable power supply combined with a three-phase diode rectifier. The PMa-SynRM was mechanically coupled with an IPMSM (interior permanent magnet synchronous motor) feeding a resistive load (see Figure 3). The measurements for the speed and rotor angle were acquired by a resolver placed on the rotor shaft. The developed control scheme relying on the model-free control was modeled in the Matlab/Simulink software, and then it was incorporated in the dSPACE 1202 MicroLabBox real-time interface to generate the gate control signals applied to the VSI.

The main PMa-SynRM parameters are listed in Table 1, whereas the model-free controller parameters are listed in Table 2.

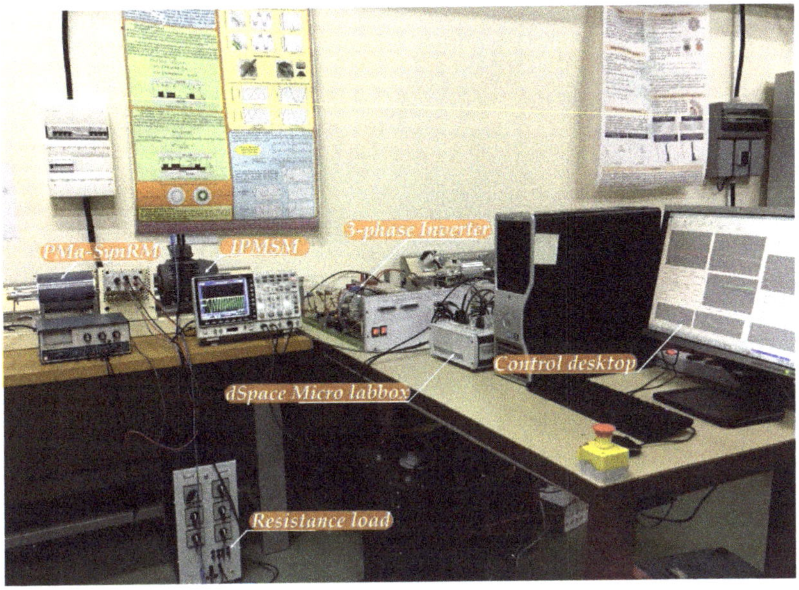

Figure 5. Experimental setup.

Table 1. Specifications and parameters of the motor/inverter.

Symbol	Quantity	Value
P_{rated}	Rated power	1 kW
n_{rated}	Rated speed	1350 rpm
T_{rated}	Rated torque	7.07 Nm
n_{p}	Number of pole pairs	2
P.F.	Power factor	0.80
R_{s}	Resistance (motor + inverter)	3.2 Ω
L_{d}	Nominal *d*-axis inductance	288 mH
L_{q}	Nominal *q*-axis inductance	38 mH
J	Equivalent inertia	0.017 kg m^2
B_{f}	Viscous friction coefficient	0.008 Nm s/rad
Ψ_{m}	PMs flux linkage	0.138 Wb
f_{s}	Switching frequency	16 kHz
V_{dc}	DC bus voltage	400 V

Table 2. Current/torque and speed regulation parameters.

Symbol	Quantity	Value
$\zeta_{1\text{d}}$	Damping ratio 1	0.7
ω_{n1d}	Natural frequency 1	3000 Rad s^{-1}
$\zeta_{1\text{q}}$	Damping ratio 1	0.7 pu.
ω_{n1q}	Natural frequency 1	2000 Rad s^{-1}
ζ_2	Damping ratio 2	0.7

Table 2. *Cont.*

Symbol	Quantity	Value
ω_{n2}	Natural frequency 2	107.1419 Rad s^{-1}
ζ_{3d}	Damping ratio 3	1
ω_{n3d}	Natural frequency 3	300 Rad s^{-1}
ζ_{3q}	Damping ratio 3	1
ω_{n3q}	Natural frequency 3	200 Rad s^{-1}
ζ_4	Damping ratio 4	1
ω_{n4}	Natural frequency 4	150 Rad s^{-1}
T_{emax}	Maximum torque reference	+6 Nm
T_{emin}	Minimum torque reference	−6 Nm
Vdc	DC bus voltage	400 V
f_s	Switching frequency	16 kHz

4.2. Simulations

The developed MFC algorithm for the PMa-SynRM drive was simulated under different operation conditions before its implementation. Figure 6 shows the simulation results of the set-point tracking d-axis inner loop current control response using the model-free control. Interestingly, note that, during the transient response, the d-axis current tracked the reference very well, and there was no steady-state error. The simulation conditions were set as follows: for d-axis testing, the q-axis current command i_{qCOM} was set to zero.

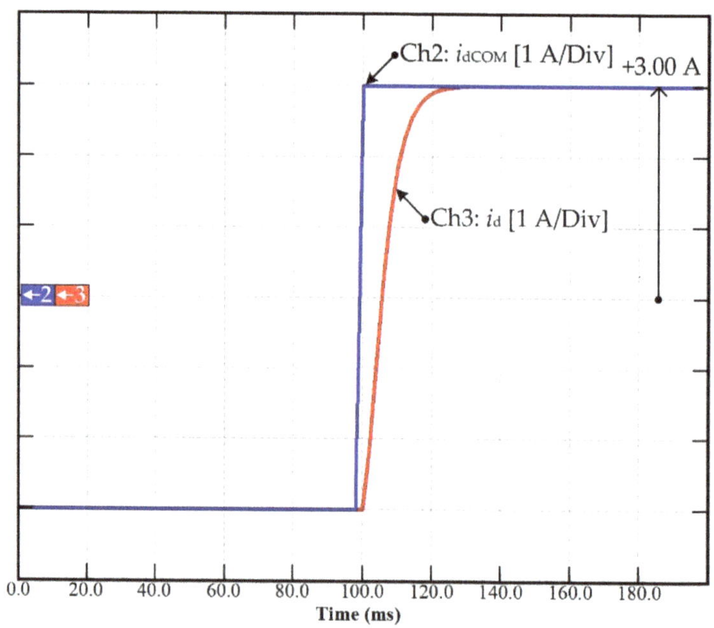

Figure 6. Simulation results: Dynamic response of the set-point tracking d-axis current control with the MFC.

Figure 7 shows the set-point tracking q-axis current control simulation results using the model-free control. Note that the control performance was satisfactory, with good set-point tracking and zero steady-state error. The simulation conditions were set as follows: during q-axis testing, the d-axis command i_{dCOM} was set to zero, and the load was the rated one.

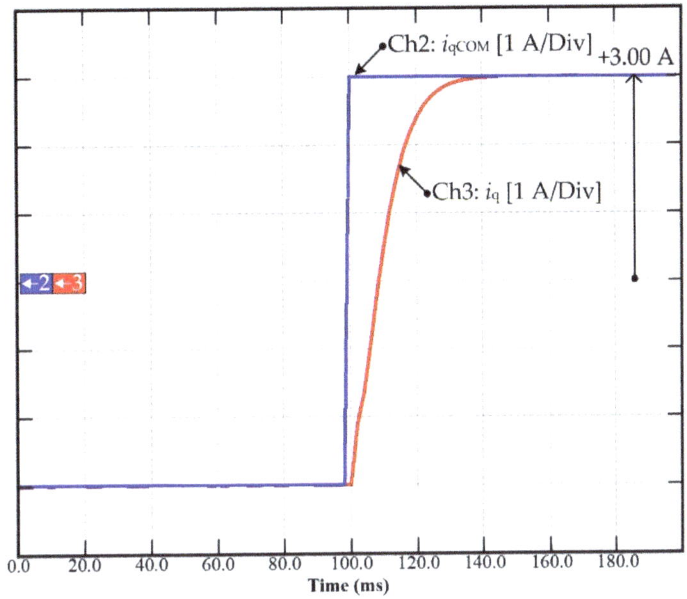

Figure 7. Simulation results: Dynamic response of q-axis currents with the MFC applied to the PMa-SynRM drive.

Another simulation result is depicted in Figure 8. It shows the drive response to a step change on the speed reference from 0 to 1000 rpm. In this figure, Chs 1, 2, and 4 represent the speed command n_{COM}, speed reference n_{REF}, and measured speed n, respectively. Chs 3, 5, and 6 represent the torque reference T_{eREF} and the d- and q-axis currents i_d and i_q, and Chs 7 and 8 represent the d-axis voltage and q-axis voltage, respectively. The parameters of the simulated drive are those of the test bench that will be later used for experimental validation. They are reported in Section 4.1. The MFC was designed to keep the torque within the range ±6 Nm. Note that the speed response was satisfactory with small overshoot and without steady-state error.

Although no torque sensor was employed in the experimental setup, the torque seemed to be limited to the allowed range. Moreover, i_q and i_d were generated on the basis of the MTPA algorithm discussed in [22].

Figure 9 shows the simulation of the disturbance rejection ability of the MFC applied to the PMa-SynRM drive. In this figure, Ch 4 represents the measured speed n, Ch2 represents the d-axis current i_d, Ch3 represents the q-axis current i_q, and Ch1 represents the torque reference T_{eREF}. The simulation conditions were as follows: n = 1000 rpm; sudden increase of 3.7 Nm on the load torque T_L at 0.3 s; and subsequent clearance of the load torque at 0.7 s. Note that, under the action of the proposed model-free control, when the load changed suddenly, the motor speed deviated slightly from its set-point, but it recovered very quickly. Figure 9 also shows the disturbance rejection capability of the MFC. As a result, the speed control performance was significantly improved, confirming the feasibility of the proposed MFC for this application.

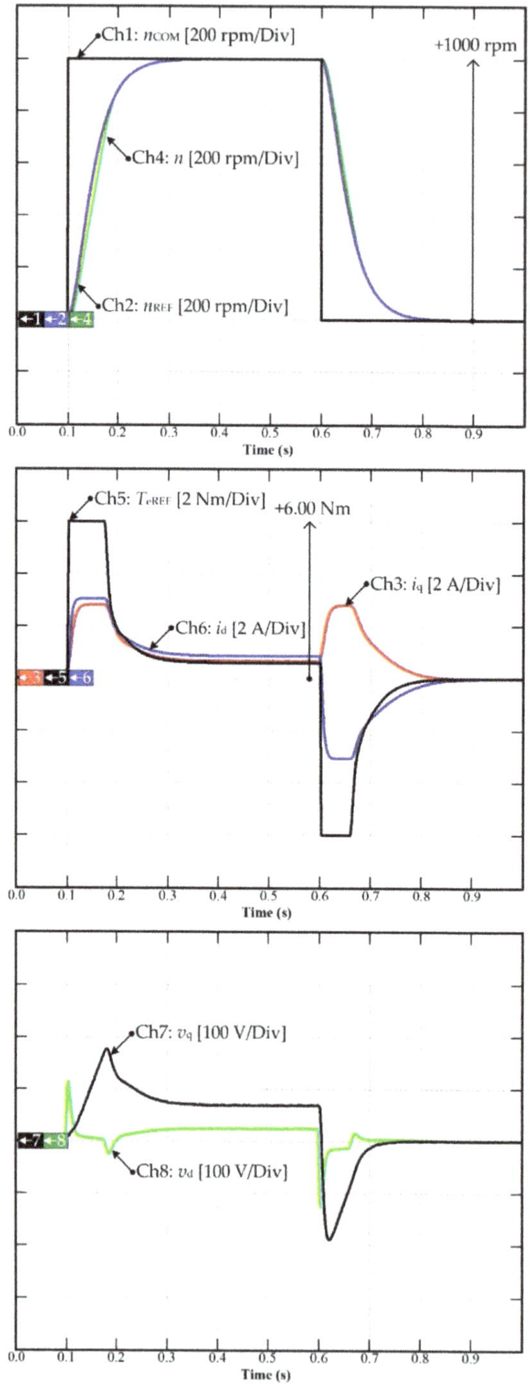

Figure 8. Simulation results: Simulated drive response to a 0–1000 rpm reference speed pulse. From top to bottom: speed reference and response, *d*- and *q*-currents, *d*- and *q*-voltages, and active region number of the MFC controller.

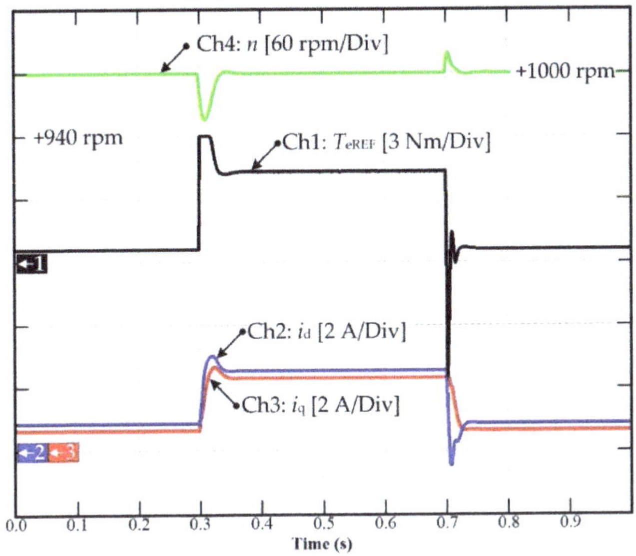

Figure 9. Simulation results: Disturbance rejection of MFC applied to the PMa-SynRM drive.

4.3. Experimental Validation of PMa-SynRM Drive Based on Model-Free Control

The designed MFC for the PMa-SynRM drive was experimentally validated on a laboratory test bench. The experimental setup is depicted in Figure 5. The entire controller parameters of the current/torque and speed are presented in Table 2. The model-free control stability and response were easy to set compared to the FOC with PI controller. Thus, by defining and selecting the governing damping and natural frequency as mentioned in the literature [19], the controller coefficients of the PI controller for both the current and speed loops control may be calculated by (26) and (34). The PI controller was provided to deal with inevitable modeling errors and uncertainties. Therefore, the PI controller guaranteed the stability of the model-free control the ensure that the current and speed control achieved the steady-state error.

Figure 10 shows the current control test of the set-point tracking d-axis inner loop. In this figure, d-axis command i_{dCOM}, d-axis reference i_{dREF}, which is provided by the d-axis trajectory planning, and the actual d-axis current are represented. Ch5, Ch6, and Ch7 represent the measured stator phase currents A, B, and C, respectively. These results are similar to those obtained by simulation and confirm that the current control performance was satisfactory.

The same test was conducted with the q-axis current while the d-axis current was regulated to zero. In this case, the motor was at a stand-still. Figure 11 depicts the experimental data, where Ch1 represents the q-axis current command i_{qCOM}, Ch2 represents the q-axis current reference i_{qREF}, and Ch3 represents the q-axis current measurement i_q. Ch5, Ch6, and Ch7 represent the measured stator phase currents A, B, and C, respectively. Overall, the current control performance seemed satisfactory.

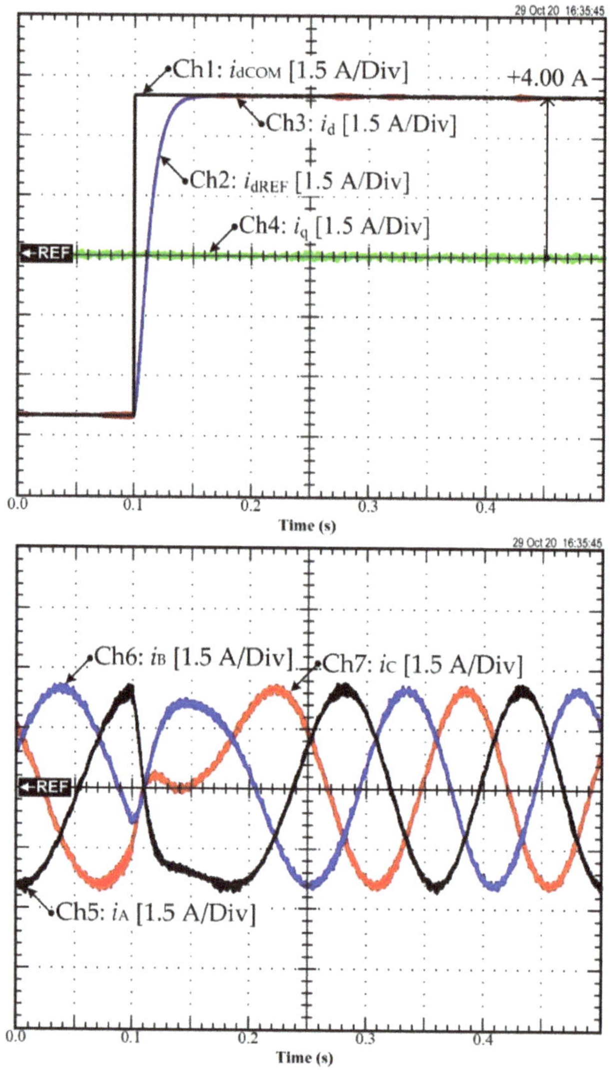

Figure 10. Experimental result: Set-point tracking *d*-axis current control response curve based on MFC.

Figure 12 depicts the speed startup of the PMa-SynRM drive using the MFC. In this figure, Chs 1, 2, 3, and 4 represent the torque reference T_{eREF}, *d*-axis current i_d, *q*-axis current i_q, and measured speed *n*, respectively. Chs 5 and 6 represent the output v_q and v_d, chosen as the output of the MFC. Moreover, Chs 7 and 8 represent the estimated unknown terms of the *d*- and *q*-axis models. As expected, the torque was limited, and the speed response showed neither overshoot nor steady-state error. It is worth recalling that the torque reference generated i_q and i_d command references according to the MTPA algorithm.

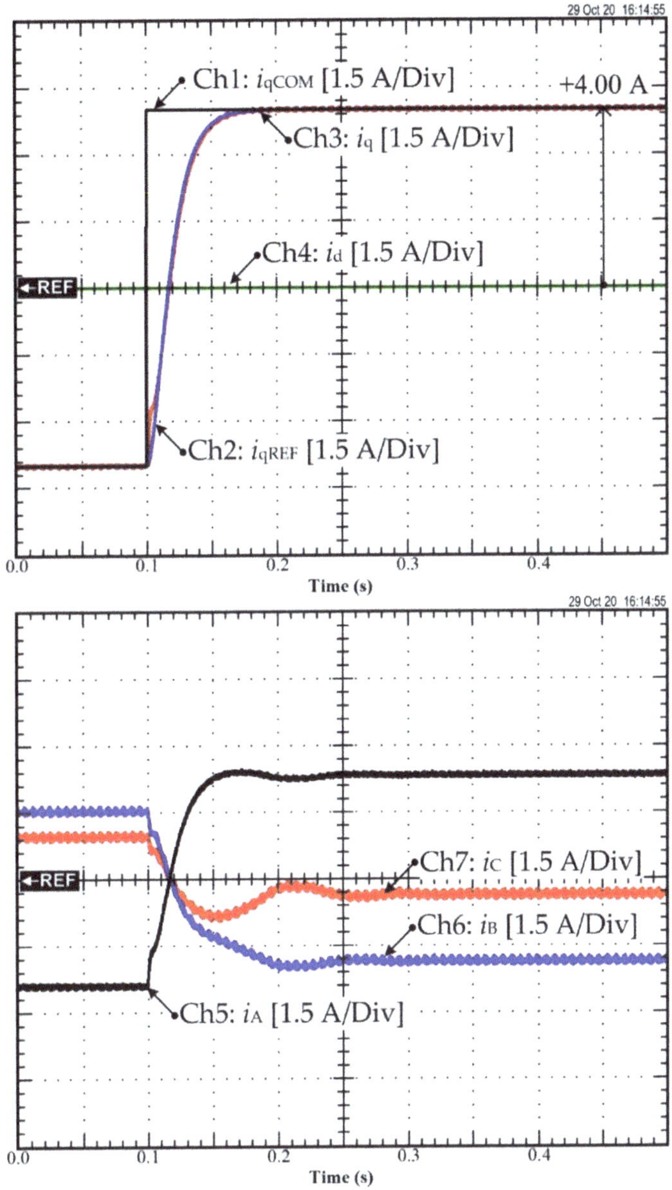

Figure 11. Experimental result: Set-point tracking q-axis current control response curve based on MFC.

Figure 13 shows the experimental validation of the disturbance rejection ability of the proposed MFC applied to the PMa-SynRM drive. In this figure, Chs 1, 2, 3, and 4 represent the torque reference T_{eREF}, d-axis current i_d, q-axis current i_q, and measured speed n, respectively. Chs 5 and 6 represent the output v_q and v_d, chosen as the output of the MFC. Moreover, Chs 7 and 8 represent the estimated unknown terms of the d- and q-axis current models. The experimental conditions were set as follows: n_{REF} = 1000 rpm, and sudden increase of the load torque (T_L) to 3.7 Nm at 0.2 s. Note that the proposed model-free control compensated for the load torque variation and rejected its effect on the motor speed in a short time. This figure shows the effectiveness of the MFC in rejecting load torque disturbance and maintaining zero steady-state speed error. As a result, the speed loop control performance was good. This result confirms the feasibility of the proposed MFC for speed control of PMa-SynRM.

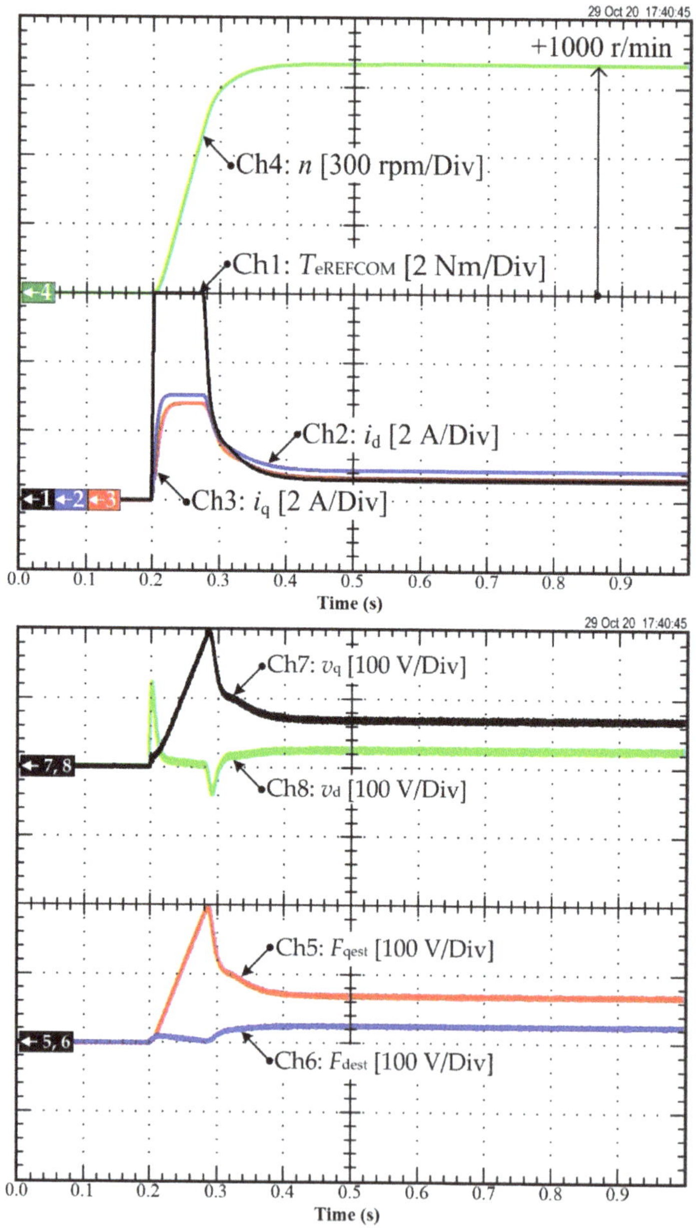

Figure 12. Experimental result: Speed acceleration curve based on MFC.

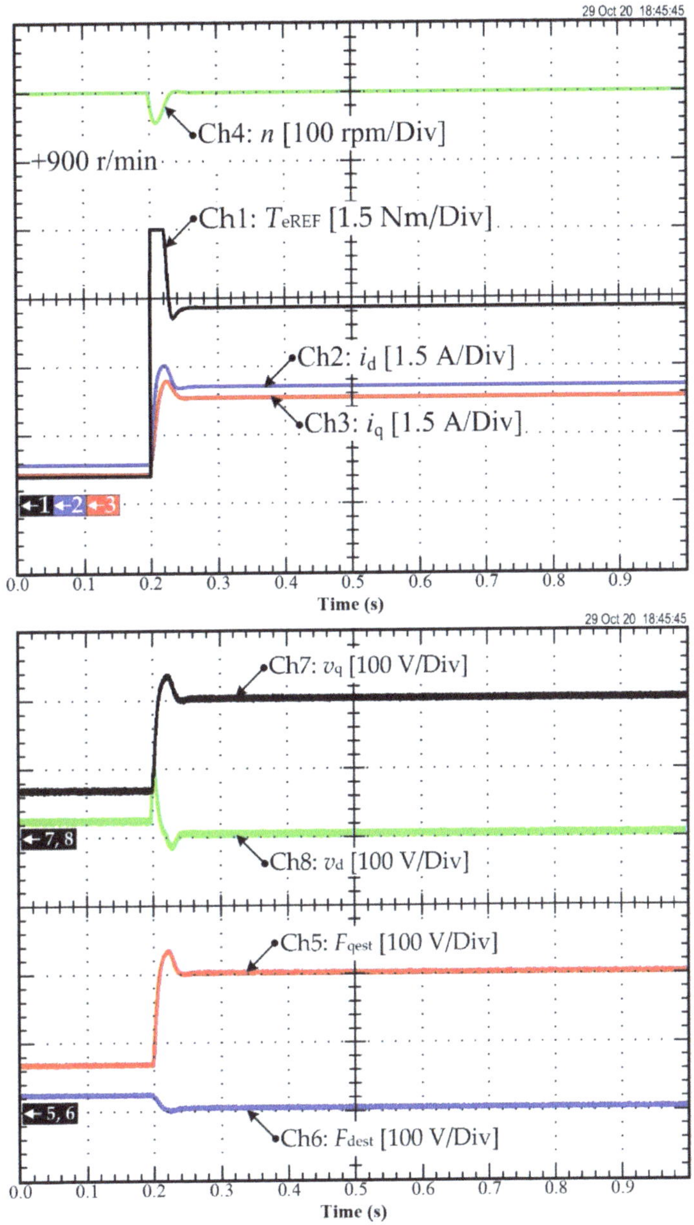

Figure 13. Experimental result: Disturbance rejection ability based on MFC.

5. Comparison of Traditional FOC with PI Controller, MBC, and Model-Free Control

Traditional FOC based on the PI controller applied to PMa-SynRM was introduced in a previous study [24]. In addition, the MBC based on differential flatness-based control applied to PMa-SynRM was proposed in [13]. Thus, the comparison of the experimental results between the FOC with the PI controller and the MBC (the differential flatness-based control) is expressed below.

Figure 14a shows the current control test of the set-point tracking *d*-axis inner loop of the FOC with the PI controller, and Figure 14b illustrates the current control test of the set-point tracking *d*-axis inner loop of the differential flatness-based control applied to the PMa-SynRM drive system. In Figure 14a,b Ch1 is the current i_{dCOM}, Ch3 is the measured current i_d, Ch4 is the measured current i_q, and Ch5 is the measured speed n. As shown in Figure 14a,b, in a transitory operation, the i_d of the FOC with the PI controller exhibits a small overshoot compared to the differential flatness-based controller, and the i_q of the FOC with PI controller shows oscillations.

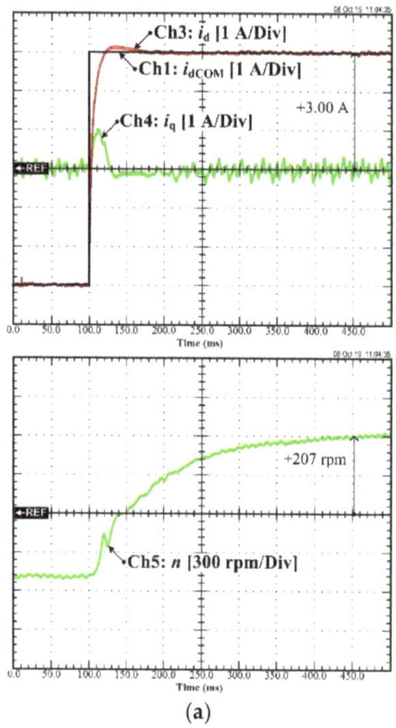

Figure 14. Experimental result: Comparison of the set-point tracking between (**a**) the FOC with the PI controller and (**b**) the MBC.

However, the differential flatness-based control was the model-based control (MBC), as mentioned in the introduction. Its performance depends on the system model. More clearly, the control laws of the model-free control and the differential flatness-based control are shown in Figure 15.

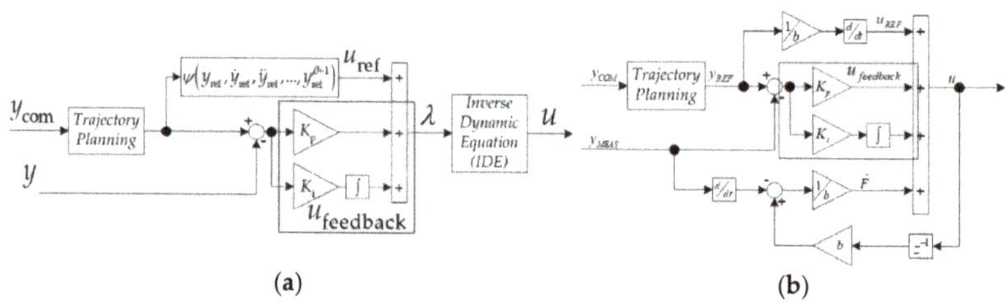

Figure 15. The difference between (**a**) the differential flatness-based control law and (**b**) the model-free control law.

The control law of the differential flatness-based control (See Figure 15a) has the inverse dynamic equation, which contains the system models including R_s, L_d, L_q, and Ψ_m. In contrast, the control law of the model-free control (See Figure 15b) estimated all the system parameters through the unknown term, F.

As a more concise summary, Table 3 shows a comparison of the advantages of traditional FOC+PI, differential flatness-based control, and model-free control.

Table 3. Comparison of three different control techniques applied to the PMa-SynRM drive system.

FOC + PI Controller	Differential Flatness-Based Control	Model-Free Control
- *More suitable for a linear motor drive system* - *Design controller coefficient using Laplace transform* - *Control performance depending on system parameters*	- *More effective with a nonlinear motor drive system* - *Model-based control system* - *Control performance depending on system parameters* - *Performance enhancement using observer* - *Require more computation resources*	- *Tailored for the control of unknown or partially known systems* - *Partially known parameters (inductance for current control)*

6. Conclusions

In this study, we analyzed the application of an MFC for the current and speed control of motor drives. This novel control approach was applied to PMa-SynRMs for the combined control of the outer speed control loop and inner current control loop. After a brief introduction of the MFC fundamentals, the design approach was comprehensively described, providing a step-by-step procedure. Suggestions for extending the design to different drive controllers were also provided. Simulations and numerous experimental results highlighted the promising features and characteristics of MFC applied to electrical motor drives. Finally, the potential of MFC pointed out in this study should stimulate further exploration and analysis of this type of controller to achieve the expertise required to transfer the results to practical applications.

Interestingly, the proposed MFC provided high performance for the PMa-SynRM drives compared to FOC with the traditional PI controller. Besides, it had a higher dynamic performance than the PMa-SynRM drive using the differential flatness-based control.

In this study, the simulation and the experimental validation were performed by a prototype PMa-SynRM at GREEN Lab, Université de Lorraine. This machine can operate in constant torque and constant power regions if a proper field weakening control is applied. In summary, by applying MFC, the performance of the PMa-SynRM was improved not only in terms of the inner current control loop but also the outer speed control loop. Moreover, the controller coefficients of the proposed MFC are not complicated to define, and a unique design approach can be applied for the PMa-SynRM drive.

Author Contributions: Conceptualization, B.N.-M. and N.T.; methodology, S.S., N.P. and P.T.; validation, S.S., N.P. and P.T.; formal analysis, N.B. and P.M.; writing—original draft preparation, P.T.; writing—review and editing, N.B.; visualization, N.B.; supervision, B.N.-M. and N.T. All authors have read and agreed to the published version of the manuscript.

Funding: This work was supported in part by the Framework Agreement between the University of Pitesti (Romania) and King Mongkut's University of Technology North Bangkok (Thailand), in part by an International Research Partnership "Electrical Engineering–Thai French Research Center (EE-TFRC)" under the project framework Lorraine Université d'Excellence (LUE) in cooperation with Université de Lorraine and King Mongkut's University of Technology North Bangkok and in part by the National Research Council of Thailand (NRCT) under Senior Research Scholar Program, Grant No. N42A640328, and in part by King Mongkut's University of Technology North Bangkok under Grant no. KMUTNB-64-KNOW-20.

Institutional Review Board Statement: Not applicable.

Informed Consent Statement: Not applicable.

Data Availability Statement: Not applicable.

Acknowledgments: The authors would like to express their gratitude to the GREEN laboratory at the University of Lorraine and King Mongkut's University of Technology North Bangkok (KMUTNB) for their constant support in boosting collaborations between France and Thailand. Besides, the authors would like to express their gratitude to the Rajamangala University of Technology Rattanakosin Wang Klai Kangwon Campus as an original agency of the first author (A25/2021).

Conflicts of Interest: The authors declare no conflict of interest.

References

1. Sadeghi, Z.; Shahparasti, M.; Rajaei, A.; Laaksonen, H. Three-Level Reduced Switch AC/DC/AC Power Conversion System for High Voltage Electric Vehicles. *Sustainability* **2022**, *14*, 1620. [CrossRef]
2. Krings, A.; Monissen, C. Review and Trends in Electric Traction Motors for Battery Electric and Hybrid Vehicles. In Proceedings of the 2020 International Conference on Electrical Machines (ICEM), Gothenburg, Sweden, 23–26 August 2020. [CrossRef]
3. Husain, I.; Ozpineci, B.; Islam, M.; Gurpinar, E.; Su, G.; Yu, W.; Chowdhury, S.; Xue, L.; Rahman, D.; Sahu, R. Electric Drive Technology Trends, Challenges, and Opportunities for Future Electric Vehicles. *Proc. IEEE* **2021**, *109*, 1039–1059. [CrossRef]
4. Lukic, M.; Giangrande, P.; Hebala, A.; Nuzzo, S.; Galea, M. Review, Challenges, and Future Developments of Electric Taxiing Systems. *IEEE Trans. Transp. Electrif.* **2019**, *5*, 1441–1457. [CrossRef]
5. Rahrovi, B.; Ehsani, M. A Review of the More Electric Aircraft Power Electronics. In Proceedings of the 2019 IEEE Texas Power and Energy Conference (TPEC), College Station, TX, USA, 7–8 February 2019. [CrossRef]
6. Noland, J.K.; Leandro, M.; Suul, J.A.; Molinas, M. High-Power Machines and Starter-Generator Topologies for More Electric Aircraft: A Technology Outlook. *IEEE Access* **2020**, *8*, 130104–130123. [CrossRef]
7. Tang, M.; Bifaretti, S.; Pipolo, S.; Formentini, A.; Odhano, S.; Zanchetta, P. A Novel Low Computational Burden Dual-Observer Phase-Locked Loop with Strong Disturbance Rejection Capability for More Electric Aircraft. *IEEE Trans. Ind. Appl.* **2021**, *57*, 3832–3841. [CrossRef]
8. Ooi, S.; Morimoto, S.; Sanada, M.; Inoue, Y. Performance Evaluation of a High-Power-Density PMASynRM with Ferrite Magnets. *IEEE Trans. Ind. Appl.* **2013**, *49*, 1308–1315. [CrossRef]
9. Trancho, E.; Ibarra, E.; Arias, A.; Kortabarria, I.; Jurrgens, J.; Marengo, L.; Fricasse, A.; Gragger, J.V. PM-Assisted Synchronous Reluctance Machine Flux Weakening Control for EV and HEV Applications. *IEEE Trans. Ind. Electron.* **2017**, *65*, 2986–2995. [CrossRef]
10. Capecchi, E.; Guglielmi, P.; Pastorelli, M.; Vagati, A. Position-sensorless control of the transverse-laminated synchronous reluctance motor. *IEEE Trans. Ind. Appl.* **2001**, *37*, 1768–1776. [CrossRef]
11. Niazi, P. Permanent Magnet Assisted Synchronous Reluctance Motor Design and Performance Improvement. Ph.D. Thesis, Texas A & M University, College Station, TX, USA, 2005.
12. Joo, K.; Kim, I.; Lee, J.; Go, S. Robust Speed Sensorless Control to Estimated Error for PMa-SynRM. *IEEE Trans. Magn.* **2017**, *53*, 1–4. [CrossRef]
13. Sriprang, S.; Poonnoy, N.; Guilbert, D.; Nahid-Mobarakeh, B.; Takorabet, N.; Bizon, N.; Thounthong, P. Design, Modeling, and Differential Flatness Based Control of Permanent Magnet-Assisted Synchronous Reluctance Motor for e-Vehicle Applications. *Sustainability* **2021**, *13*, 9502. [CrossRef]
14. Wang, Y.; Li, H.; Liu, R.; Yang, L.; Wang, X. Modulated Model-Free Predictive Control with Minimum Switching Losses for PMSM Drive System. *IEEE Access* **2020**, *8*, 20942–20953. [CrossRef]
15. Lin, C.-K.; Agustin, C.A.; Yu, J.-T.; Cheng, Y.-S.; Chen, F.-M.; Lai, Y.-S. A Modulated Model-Free Predictive Current Control for Four-Switch Three-Phase Inverter-Fed SynRM Drive Systems. *IEEE Access* **2021**, *9*, 162984–162995. [CrossRef]
16. Lyu, Z.; Wu, X.; Gao, J.; Tan, G. An Improved Finite-Control-Set Model Predictive Current Control for IPMSM under Model Parameter Mismatches. *Energies* **2021**, *14*, 6342. [CrossRef]
17. Hashjin, S.A.; Pang, S.; Miliani, E.-H.; Ait-Abderrahim, K.; Nahid-Mobarakeh, B. Data-Driven Model-Free Adaptive Current Control of a Wound Rotor Synchronous Machine Drive System. *IEEE Trans. Transp. Electrif.* **2020**, *6*, 1146–1156. [CrossRef]
18. Precup, R.-E.; Radac, M.-B.; Roman, R.-C.; Petriu, E.M. Model-free sliding mode control of nonlinear systems: Algorithms and experiments. *Inf. Sci.* **2017**, *381*, 176–192. [CrossRef]
19. Sriprang, S.; Nahid-Mobarakeh, B.; Takorabet, N.; Pierfederici, S.; Mungporn, P.; Thounthong, P.; Bizon, N.; Kuman, P.; Shah, Z. Model Free-Based Torque Control of Permanent Magnet Synchronous Motor Drives. In Proceedings of the 2019 Research, Invention, and Innovation Congress (RI2C), Bangkok, Thailand, 11–13 December 2019. [CrossRef]
20. Fliess, M.; Join, C. Model-free control. *Int. J. Control* **2013**, *86*, 2228–2252. [CrossRef]
21. Fliess, M.; Join, C. Stability margins and model-free control: A first look. In Proceedings of the 2014 European Control Conference (ECC), Strasbourg, France, 24–27 June 2014. [CrossRef]

22. Battiston, A.; Miliani, E.-H.; Martin, J.-P.; Nahid-Mobarakeh, B.; Pierfederici, S.; Meibody-Tabar, F. A Control Strategy for Electric Traction Systems Using a PM-Motor Fed by a Bidirectional Z-Source Inverter. *IEEE Trans. Veh. Technol.* **2014**, *63*, 4178–4191. [CrossRef]
23. Thounthong, P.; Sikkabut, S.; Poonnoy, N.; Mungporn, P.; Yodwong, B.; Kumam, P.; Bizon, N.; Pierfederici, S.; Poonnoi, N.; Nahidmobarakeh, B. Nonlinear Differential Flatness-Based Speed/Torque Control with State-Observers of Permanent Magnet Synchronous Motor Drives. *IEEE Trans. Ind. Appl.* **2018**, *54*, 2874–2884. [CrossRef]
24. Fliess, M.; Sira–Ramírez, H. An algebraic framework for linear identification. *ESAIM Control Optim. Calc. Var.* **2003**, *9*, 151–168. [CrossRef]
25. Fliess, M.; Sira-Ramírez, H. Closed-loop Parametric Identification for Continuous-time Linear Systems via New Algebraic Techniques. In *Identification of Continuous-Time Models from Sampled Data*; Part of the Advances in Industrial Control Book Series (AIC); Springer: London, UK, 2008; pp. 363–391. [CrossRef]
26. Sriprang, S.; Nahid-Mobarakeh, B.; Takorabet, N.; Pierfederici, S.; Kumam, P.; Bizon, N.; Taghavi, N.; Vahedi, A.; Mungporn, P.; Thounthong, P. Design and control of permanent magnet assisted synchronous reluctance motor with copper loss minimization using MTPA. *J. Electr. Eng.* **2020**, *71*, 11–19. [CrossRef]

sustainability

MDPI

Article

Simplified Super Twisting Sliding Mode Approaches of the Double-Powered Induction Generator-Based Multi-Rotor Wind Turbine System

Habib Benbouhenni [1], Nicu Bizon [2,3,4,*], Ilhami Colak [1], Phatiphat Thounthong [5,6] and Noureddine Takorabet [6]

1 Department of Electrical & Electronics Engineering, Faculty of Engineering and Architecture, Nisantasi University, Istanbul 34398, Turkey; habib.benbouenni@nisantasi.edu.tr (H.B.); ilhcol@gmail.com (I.C.)
2 Faculty of Electronics, Communication and Computers, University of Pitesti, 110040 Pitesti, Romania
3 Doctoral School, Polytechnic University of Bucharest, 313 Splaiul Independentei, 060042 Bucharest, Romania
4 ICSI Energy Department, National Research and Development Institute for Cryogenic and Isotopic Technologies, 240050 Ramnicu Valcea, Romania
5 ICSI Renewable Energy Research Centre (RERC), Faculty of Technical Education, King Mongkut's University of Technology North Bangkok, Bangkok 10800, Thailand; phatiphat.t@fte.kmutnb.ac.th
6 Group of Research in Electrical Engineering of Nancy (GREEN), University of Lorraine-GREEN, F-54000 Nancy, France; noureddine.takorabet@univ-lorraine.fr
* Correspondence: nicu.bizon@upit.ro

Citation: Benbouhenni, H.; Bizon, N.; Colak, I.; Thounthong, P.; Takorabet, N. Simplified Super Twisting Sliding Mode Approaches of the Double-Powered Induction Generator-Based Multi-Rotor Wind Turbine System. *Sustainability* **2022**, *14*, 5014. https://doi.org/10.3390/su14095014

Academic Editor: Miguel Carrión

Received: 28 March 2022
Accepted: 18 April 2022
Published: 22 April 2022

Abstract: This work proposes a new indirect filed-oriented control (IFOC) scheme for double-powered induction generators (DPIGs) in multi-rotor wind turbine systems ($MRWT_S$). The IFOC strategy is characterized by its simplicity, ease of use, and fast dynamic speed. However, there are drawbacks to this method. Among its disadvantages is the presence of ripples in the level of torque, active power, and current. In addition, the total harmonic distortion (THD) value of the electric current is higher compared to the direct torque control method. In order to overcome these shortcomings and in terms of improving the effectiveness and performance of this method, a new algorithm is proposed for the super twisting algorithm (STA). In this work, a new STA method called simplified STA (SSTA) algorithm is proposed and applied to the traditional IFOC strategy in order to reduce the ripples of torque, current, and active power. On the other hand, the inverter of the DPIG is controlled by using a five-level fuzzy simplified space vector modulation (FSSVM) technique to obtain a signal at the inverter output of a fixed frequency. The results obtained from this proposed IFOC-SSTA method with FSSVM strategy are compared with the classical IFOC method which uses the classical controller based on a proportional-integral (PI) controller. The proposed method is achieved using the Matlab/Simulink software, where a generator with a large capacity of 1.5 megawatts is used. The generator is placed in a multi-rotor electric power generation system. On the other hand, the two methods are compared in terms of ripple ratio, dynamic response, durability, and total harmonic distortion (THD) value of the electric current. Through the results obtained from this work, the proposed method based on SSTA provided better results in terms of ripple ratio, response dynamic, and even THD value compared to the classical method, and this shows the robustness of the proposed method in improving the performance and efficiency of the generator in the multi-rotor wind system.

Keywords: double-powered induction generator; indirect filed-oriented control; five-level fuzzy simplified space vector modulation; super-twisting algorithm; simplified STA; multi-rotor wind turbine systems

1. Introduction

Traditionally, field-oriented control (FOC) is among the most widely used control methods in the field of controlling electrical machines, whether they are motors or electric generators, due to its simplicity and ease of implementation. This method is based on the use of a classic proportional-integral (PI) controller, which gives this method a fast dynamic response. On the other hand, this method is based on the use of the traditional pulse width modulation (PWM) to control the inverter of the machine. There are two types of this method, the direct FOC method and the indirect FOC method [1]. In the field of control, this method was used to control several electrical machines such as the asynchronous motor [2,3] and the synchronous motor [4]. However, this method has been used in the field of renewable energies, as according to the reference [5], this method is among the most widely used methods in the field of generating electric power from wind.

In the FOC method, internal loops are used to control electrical machines, creating ripples in both the current and the torque level of the machine. Moreover, this method gives a slow dynamic response compared to both direct torque control (DTC) and direct power control (DPC) [6]. Among the disadvantages of this method is that the total harmonic distortion (THD) value of the current is much higher compared to both DTC and DPC [7].

In the field of scientific research, there are several solutions that have been suggested in order to improve the performance and effectiveness of the FOC method. Artificial intelligence methods and nonlinear methods are among the most widely used methods. The authors of [8] implemented a super twisting algorithm (STA) on a multi-phase induction motor system using the FOC method where the intention was to minimize the current and torque ripples. Despite the advantages that are achieved using the STA algorithm, compared to the FOC strategy with PI controllers, the main drawback of this proposed method is the unstable frequency and presence of torque and current ripples due to the use of the pulse width modulation (PWM) technique. Another nonlinear FOC strategy is proposed in [9] to control the active and reactive power of the induction generator (IG). According to the study that was completed in [10] with the use of the nonlinear method based on second-order sliding mode control (SOSMC), the results were compared with the classical FOC method based on PI controller, and significant improvements were noticed in both the current and the active power compared to the classical FOC method. The problem with the proposed method is that there is a problem of chattering. However, one of the major drawbacks of employing the SOSMC technique to control electrical machines is that the mathematical form of the studied system must be precisely known. Another nonlinear method was applied to an asynchronous generator controlled by the FOC method. In this proposed method, the third-order sliding mode control (TOSMC) technique is used to compensate for the PI and improve the quality of the active power. One of the results obtained from this work is that the control by TOSMC reduces the chatter problem present in the SMC in addition to reducing the ripples at the level of torque, current, and effective power of the asynchronous generator. In [11], two different methods were combined in principle, where the first method is represented by neural networks, which is characterized by precision and speed of response, while the second method is the TOSMC technique, which is characterized by durability and unaffected by external factors such as noise or change in the parameters of the studied system. The result of combining these two methods is to obtain a more durable method and get excellent quality of the electric current of the induction generator.

In the field of electric machine control, response speed or dynamic response is very important. This response is considered a criterion among the criteria that is selected and differentiated between the control methods. In [12], a new method is proposed under the name of terminal synergetic control in order to improve the performance and effectiveness of the direct FOC strategy for an asynchronous generator integrated into a wind turbine. This proposed method is a new method based on the use of nonlinear error instead of using linear error. The obtained results showed the effectiveness of the proposed method in improving the dynamic response of the induction generator while reducing the ripples of torque and the active power of the induction generator. In [13], a method of artificial intelligence based on the neural networks and fuzzy logic was used to improve the performance and effectiveness of the IFOC strategy of the asynchronous generator, where the classic PI controller was replaced by an intelligent controller under the name of the neuro-fuzzy controller. The proposed IFOC strategy with the neuro-fuzzy controller is more robust compared to the traditional IFOC strategy. The results showed the effectiveness of the proposed method in improving the THD of stator current value compared to the classical method. Moreover, there is a fast dynamic response to the proposed method, where we find that the response time is much reduced compared to the PI controller. In [14], the author proposed a new FOC strategy based on an adaptive observer to control the induction generator-based wind turbine. The proposed FOC strategy with adaptive observer was experimentally tested on an 11 kW induction generator using a 27 kW DC motor to rotate the generator. The experimental results showed the effectiveness of the proposed FOC method in obtaining a fast dynamic response compared to the classical FOC method.

In this work, a new FOC strategy method is introduced using a new nonlinear method in order to obtain a more robust method and reduce the ripples of torque, active power, and electric current generated by the double-powered induction generator (DPIG). In addition, a new method for the STA algorithm is proposed, where this method is simplified more than the traditional STA method. Thus, a simpler method that can be easily developed and applied to any system regardless of its complexity is designed. The proposed nonlinear method is called the simplified STA (SSTA) algorithm. Among the advantages of this nonlinear method is the fact that the design of the method is not related to the studied system, regardless of what this system will be. This proposed nonlinear method is used in order to improve the performance and effectiveness of the STA method and on the other hand to improve the dynamic response speed of the generator placed in the multi-rotor wind electric power generation system.

The novelty and main contributions of this work accomplished in this paper are summarized in the following points:

- A new super twisting algorithm is proposed and confirmed.
- A new space vector modulation (SVM) scheme is proposed based on the fuzzy logic controller to control the five-level inverter of the DPIG.
- A new IFOC strategy scheme is proposed to control the DPIG-based multi-rotor wind turbine system.
- A new SSTA algorithm is designed to improve the dynamic characteristics of DPIG-based multi-rotor wind turbine systems.

This work is structured as follows. Section 2 presents the dynamic modeling of the multi-rotor wind turbine system. In Sections 3 and 4, the proposed new super twisting algorithm (STA) and the proposed multilevel fuzzy SVM technique are presented. Section 5 presents the traditional IFOC strategy with the PI controller. Section 6 presents the proposed IFOC strategy using the proposed simplified STA controller and five-level fuzzy SVM technique. Finally, Section 7 concludes the work by presenting the main findings and future directions of research, as well as some comments and recommendations.

2. Multi-Rotor Wind Turbine System

The multi-rotor wind turbine is a new technology that has appeared in recent years in order to overcome the problems and defects of the old technology (single-rotor wind turbine). In this new technology, the output torque or power gained from the wind is greater than that gained in the case of a single-rotor wind turbine. In addition, this new technology surpasses the winds generated by wind farms, and thus the yield is greater compared to the single-rotor wind turbine [15]. Figure 1 represents the electric power generation system using the DPIG placed in the electric power generation system using a multi-rotor wind turbine.

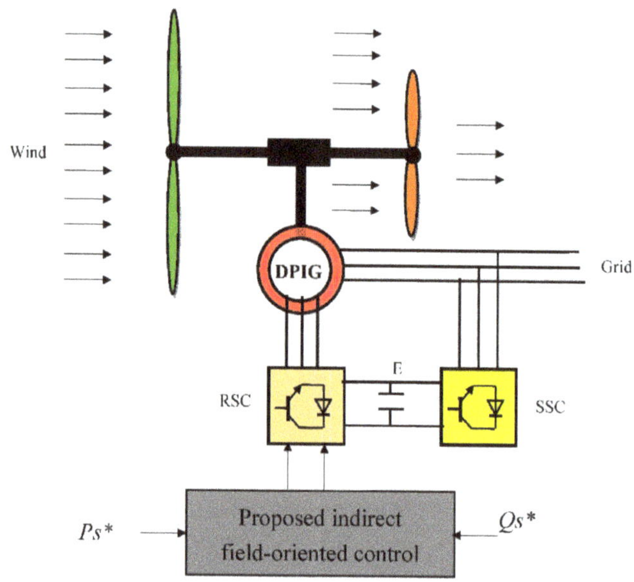

Figure 1. Structure of the multi-rotor wind turbine system.

In the studied wind system, a large-power DPIG (1.5 megawatts) is used. This generator is fed from the electrical network using two different inverters. The first inverter is on the side of the electrical network to convert alternating voltage into constant voltage, while the second inverter, which is on the side of the DPIG, aims to convert the constant voltage resulting from the first inverter into alternating voltage. To rotate the generator, a dual-rotor wind turbine is used.

The two-rotor wind turbine is used in this work in order to increase the energy gained from the wind. On the other hand, a dual-rotor wind turbine is two turbines linked together forming one turbine. The two turbines have the same axis of rotation, and the torque and energy resulting from them can be expressed by the following equations:

$$P_t = P_{ST} + P_{LT} \tag{1}$$

$$T_t = T_{ST} + T_{LT} \tag{2}$$

where T_t is the output torque of the dual-rotor wind turbine, T_{LT} and T_{ST} are the output torque of the large and small wind turbines, P_t is the output mechanical power of the dual-rotor wind turbine, P_{LT} and P_{ST} are the output mechanical power of the large and small wind turbine torque.

The term torque for both large and small turbines is shown in Equations (3) and (4), respectively [16].

$$T_{LT} = \frac{C_p}{2\lambda_{LT}^3}\rho \cdot \pi \cdot R_{LT}^5 \cdot w_{LT}^2 \tag{3}$$

$$T_{ST} = \frac{C_p}{2\lambda_{ST}^3}\rho \cdot \pi \cdot R_{ST}^5 \cdot w_{ST}^2 \tag{4}$$

From the two equations, it can be seen that the torque of the two turbines is related to each of the air density (ρ), coefficient of power (C_p), the mechanical speed of the small and large turbines (w_{ST} and w_{LT}), the blade radius of the small and large turbines (R_{ST}, R_{LT}), and the tip speed ration of the small and large turbines (λ_{ST} and λ_{LT}).

The energy gained from wind by a dual-rotor turbine is related to the power factor, the value of which is related to pitch angle (β) and tip speed ratio (λ). This parameter can be expressed by the following equation:

$$C_p(\beta,\ \lambda) = \frac{1}{0.08\beta + \lambda} + \frac{0.035}{\beta^3 + 1} \tag{5}$$

The terms of the energy produced by both the small turbine and the large turbine are shown in Equations (6) and (7), respectively [17].

$$P_{ST} = \frac{C_p(\beta,\ \lambda)}{2} \rho \cdot S_{ST} \cdot w_{ST}^3 \tag{6}$$

$$P_{LT} = \frac{C_p(\beta,\ \lambda)}{2} \rho \cdot S_{LT} \cdot w_{LT}^3 \tag{7}$$

The value of the tip speed ratios of the small turbine and the large turbine are given in Equations (8) and (9), respectively [15].

$$\lambda_{ST} = \frac{w_{ST} \cdot R_{ST}}{V_{ST}} \tag{8}$$

$$\lambda_{LT} = \frac{w_{LT} \cdot R_{LT}}{V_{LT}} \tag{9}$$

The values of the tip speed ratios for a large and a small turbine are related to the wind speed (V_{ST} and V_{LT}), The speed of the large and small turbine (w_{ST} and w_{LT}), and the blade radius of the small and large turbines (R_{ST}, R_{LT}).

The wind speed between the large and small turbines is different from the wind speed before the large turbines. The wind speed can be calculated at any point between the turbines given the following relationship [17]:

$$V_x = V_{LT}\left(1 - \frac{1 - \sqrt{(1 - C_T)}}{2}\left(1 + \frac{2x}{\sqrt{1 + 4x^2}}\right)\right) \tag{10}$$

The wind speed of the small turbine is related to the wind speed of the large turbine (V_{LT}) and a constant value (C_T = 0.9), as well as the separation distance (x) between the large and small turbine. In this case, this distance between the center of the large and small turbine is 15 m [15].

In this work, a multi-rotor turbine is used to drive an DPIG. The latter was used in this work, and this is due to the advantages that distinguish it compared to other generators. As it is known, the DPIG is characterized by durability, easy control, and low cost.

In order to give the generator a mathematical form, the park transformation is used. To give the generator the mathematical form, the equations of voltage, flux, and mechanical equation are used. In addition, the torque equation is given for the generator, since torque is related to both current and flux. Quadrature and direct rotor voltages are shown in Equation (11). The direct and quadrature rotor flux are related to the stator/rotor current and are represented in Equation (12) [18–20].

$$\begin{cases} V_{dr} = R_r I_{dr} - w_r \Psi_{qr} + \frac{d}{dt}\Psi_{dr} \\ V_{qr} = R_r I_{qr} + w_r \Psi_{dr} + \frac{d}{dt}\Psi_{qr} \end{cases} \tag{11}$$

$$\begin{cases} \Psi_{dr} = L_r I_{dr} + M I_{ds} \\ \Psi_{qr} = M I_{qs} + L_r I_{qr} \end{cases} \tag{12}$$

Equation (13) represents each of the direct and quadrature stator voltages. Through this equation, the tension is related to stator resistance (R_s), direct and quadrature stator current (I_{ds} and I_{qs}), direct and quadrature stator (Ψ_{qs} and Ψ_{ds}), and electrical pulsation of the stator (w_s). The direct and quadrature stator flux are represented in Equation (14). These two fluxes are related to both the inductance of the stator (L_s), direct and quadrature rotor current (I_{dr} and I_{qr}), and direct and quadrature rotor current (I_{ds} and I_{qs}).

$$\begin{cases} V_{qs} = R_s I_{qs} + w_s \Psi_{ds} + \frac{d}{dt}\Psi_{qs} \\ V_{ds} = R_s I_{ds} - w_s \Psi_{qs} + \frac{d}{dt}\Psi_{sd} \end{cases} \quad (13)$$

$$\begin{cases} \Psi_{qs} = M I_{qr} + L_s I_{qs} \\ \Psi_{ds} = L_s I_{ds} + M I_{dr} \end{cases} \quad (14)$$

The value of the generated torque (T_e) is related to rotor current, the number of pole pairs (p), and stator flux and its expression can be given by Equation (15) [19].

$$T_e = 1.5\, p \frac{M}{L_s} \left(-\Psi_{sd} I_{rq} + \Psi_{sq} I_{rd} \right) \quad (15)$$

The active power and the reactive power of the generator are represented in the Equation (16). The active power is related to stator current and stator voltage, and the same is related to the reactive power.

$$\begin{cases} P_s = 1.5\left(V_{qs} I_{qs} + I_{ds} V_{ds} \right) \\ Q_s = 1.5\left(-I_{qs} V_{ds} + V_{qs} I_{ds} \right) \end{cases} \quad (16)$$

The mechanical part of the DPIG is represented by Equation (17). This equation gives the relationship between torque and speed (Ω).

$$T_e = J \frac{d\Omega}{dt} + f\Omega + T_r \quad (17)$$

where T_r is the load torque, Ω is the mechanical rotor speed, J is the inertia, f is the viscous friction coefficient.

3. Simplified STA Controller

Super twisting algorithm (STA) is among the most widely used nonlinear methods in the field of electrical machine control, due to its durability and ease of implementation [21]. The use of the STA algorithm in automated systems gives great effectiveness in improving the performance and efficiency of electrical machines. On the other hand, the STA algorithm is a type of SOSMC technique. The STA algorithm reduces chattering problems compared to the traditional SMC technique [22]. Equation (18) represents the super twisting algorithm [23]. The classical STA algorithm can be illustrated in Figure 2.

$$\begin{cases} u = K_p |e|^r Sign\ (e) + u_1 \\ \frac{du_1}{dt} = K_i Sign\ (e) \end{cases} \quad (18)$$

where e is the error or surface, r is the exponent defined for the traditional STA regulator, and K_i and K_p are positive values.

In this work, a new look is given to the STA algorithm in order to further simplify the algorithm and increase its dynamic response. Equation (19) represents the proposed method for the traditional STA algorithm which has been called the simplified STA algorithm (SSTA). The Lyapunov theory is used in order to check the stability of this proposed SSTA controller. This proposed SSTA algorithm is simpler compared to the traditional STA technique.

$$w = K \times |e|^r \times Sign\ (e) \quad (19)$$

$$e \times \dot{e} < 0 \quad (20)$$

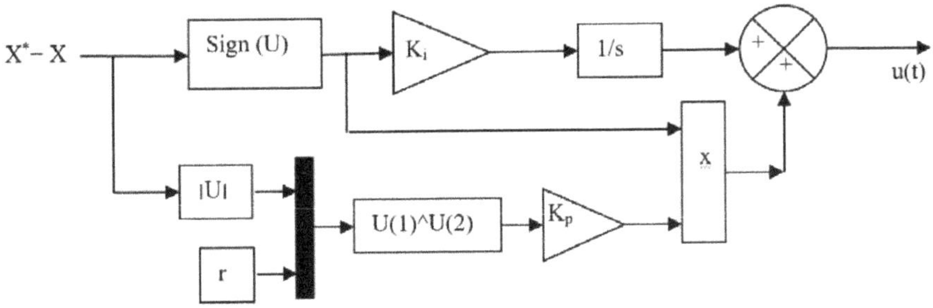

Figure 2. Structure of the traditional STA algorithm.

Figure 3 illustrates the working principle of the proposed SSTA controller. Through this figure, the proposed SSTA controller can be implemented easily and inexpensively and can be applied to any system. Moreover, this controller does not require the mathematical form of the studied system.

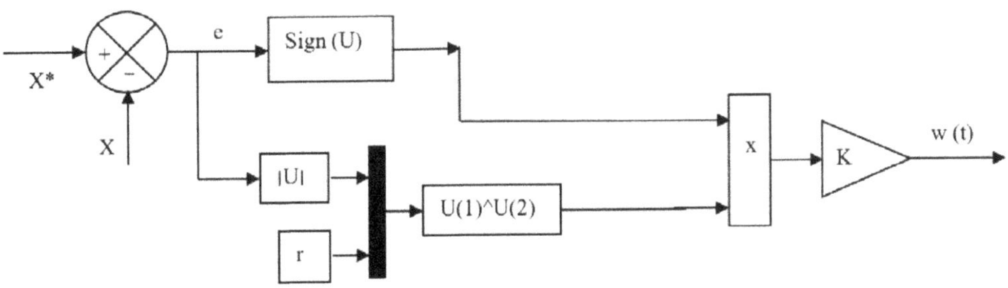

Figure 3. Structure of the proposed SSTA technique.

4. Proposed Five-Level Fuzzy SVM Strategy

Traditionally, the SVM technique is among the most used methods for controlling reflectors and this is because of the results it offers compared to other methods such as the PWM technique [24]. In this method, the calculation of reference voltage and zones is relied upon, which makes this method more complicated, especially in the case of a multi-level inverter [25]. To overcome this problem, a new idea for this method was presented in [26], where the calculation of the maximum and minimum values of the three-phase feeding voltages was used in this proposed method. In this proposed method, the reference tension is not used or calculated as well as the regions where the reference tension is present. This proposed method in [26] is simpler and more intuitive compared to the classical SVM method. In this part, the proposed fuzzy SVM technique is used to control a five-level inverter for an asynchronous generator placed in a multi-turbine wind system. This proposed method of controlling a 5-level inverter is illustrated in Figure 4. The proposed method was used in [27] in order to give the five-level SVM technique.

In this proposed fuzzy SVM technique, twelve hysteresis comparators and four trigonometric signals are used. The use of hysteresis comparators in this proposed SVM method creates a signal at the inverter output of a non-fixed frequency. In order to overcome this drawback, fuzzy logic algorithm (FLA) is used instead of using hysteresis comparators. The use of the FLA leads to a signal at the output of the inverter with a fixed frequency, thus reducing the ripples of both the current and the active power. The SVM method based on the FLA technique proposed in this work is illustrated in Figure 5. With this figure, the proposed method of SVM for controlling the five-level inverter is simpler compared to using the classical five-level SVM method using the calculation of reference voltage and zones. In this method, twelve FLA techniques are used in order to obtain control signals for the inverter transistors (IGBTs).

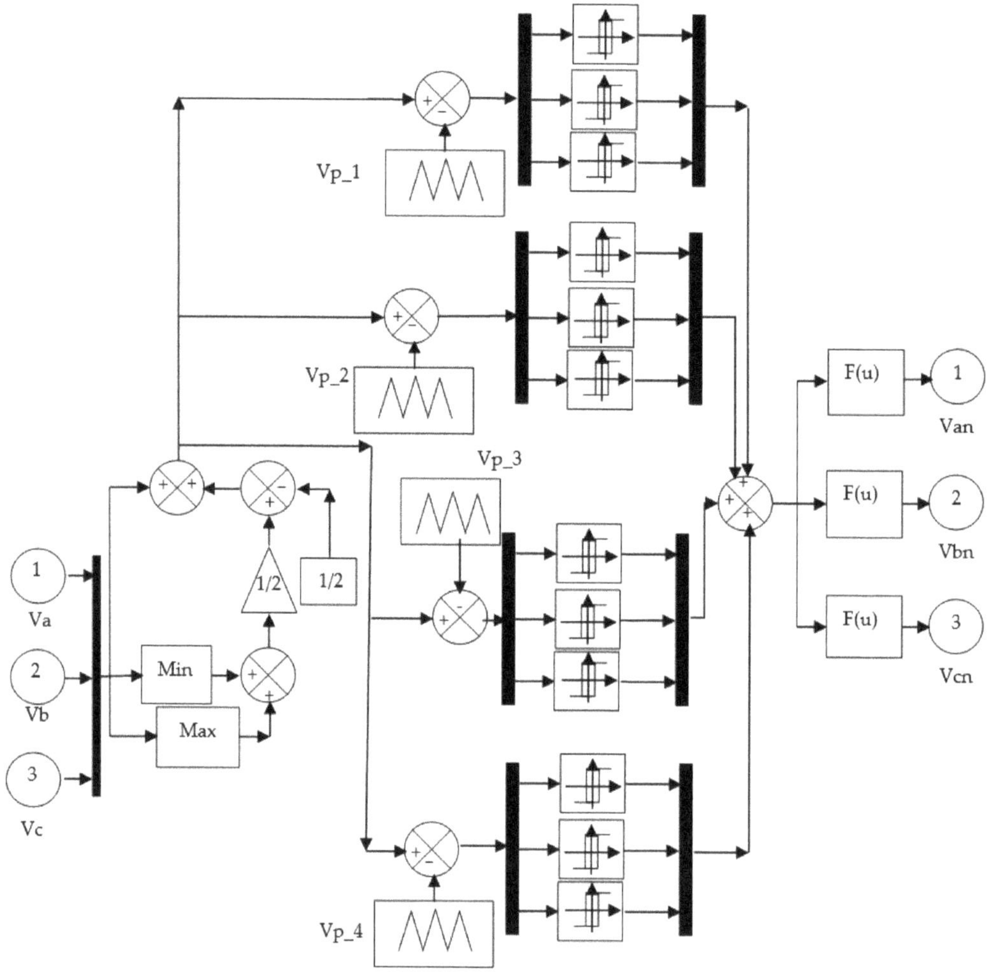

Figure 4. Structure of the five-level SVM technique.

The internal form of the FLA method is shown in Figure 6. From this figure, it is noted that the structure of the FLA technique is simple. This fuzzy controller has two inputs, the error and the change in error, and only one output. Three constant gains (K_1, K_2, and K_3) are used to improve the response and adjust the response of fuzzy logic. The characteristics of the FLA method used to improve the performance and effectiveness of the proposed five-level SVM technique are shown in the bottom of the Figure 6. The type of fuzzy controller used in this work is the Mamdani controller.

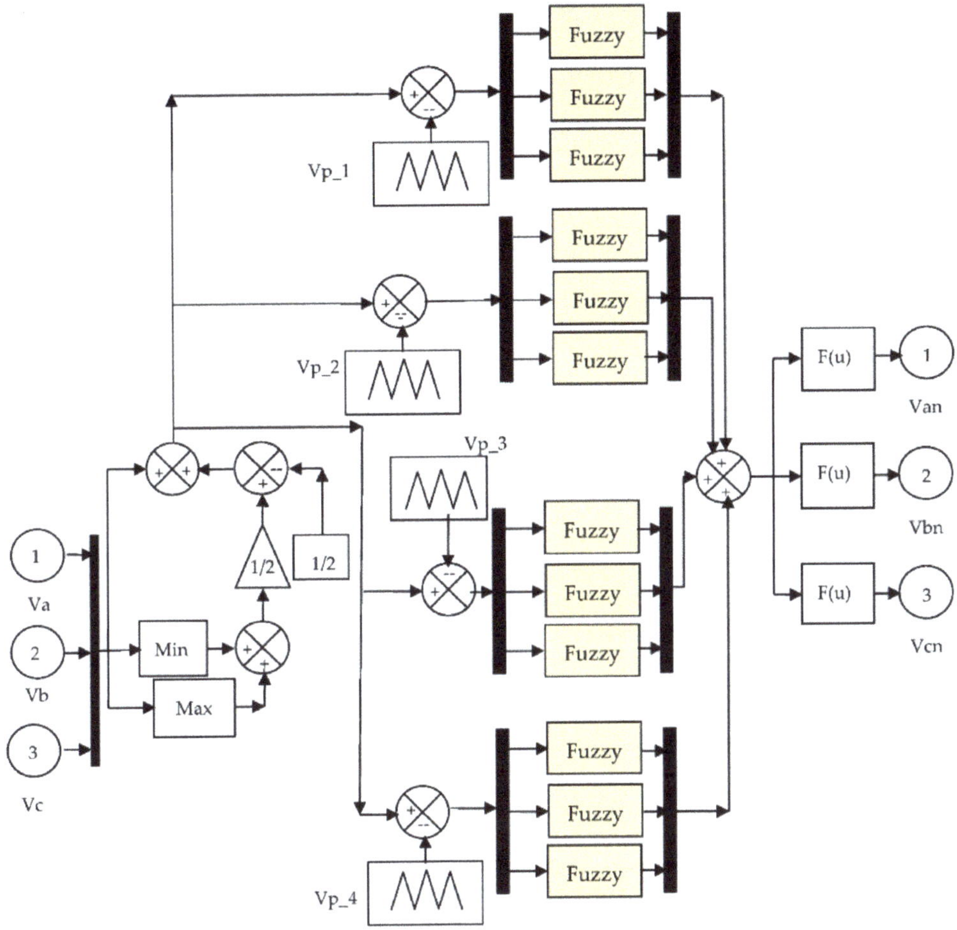

Figure 5. Structure of the proposed five-level fuzzy SVM technique.

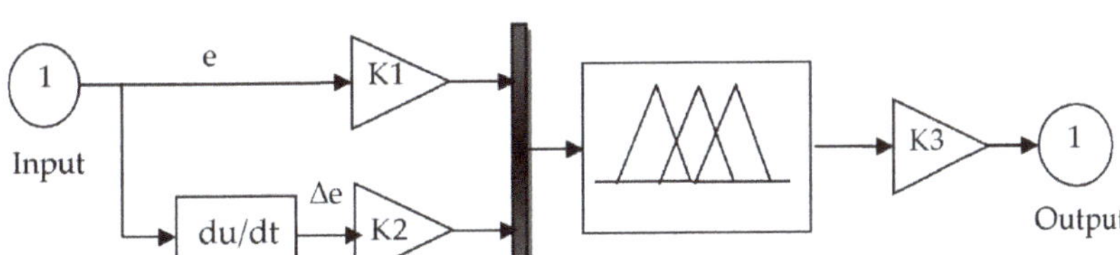

Parameters of the FLA	
Or method	Max
And method	Min
Fis type	Mamdani
Defuzzification	Centroid
Implication	Min
Aggregation	Max

Figure 6. Structure and parameters of the fuzzy logic technique.

Seven membership functions (MFs) are used in the first entry (error) and seven MFs are used in the second input (change in error). These functions used to accomplish fuzzy logic are shown in Figure 7a,b. In order to get a good response and results for the fuzzy logic, the 7 × 7 rule is used, where these rules are represented in Figure 7c.

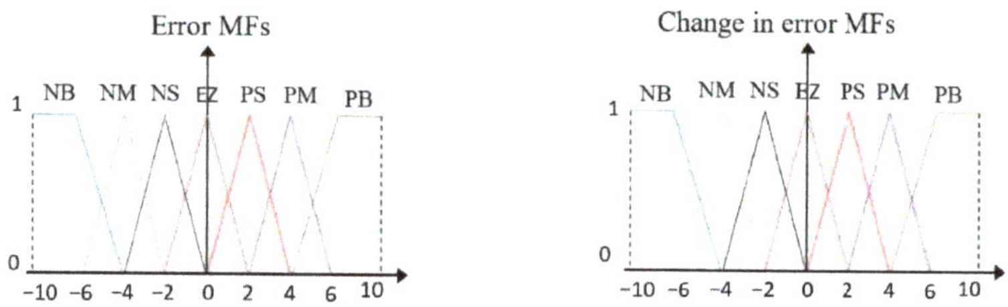

a) Error

b) Change in error

e / Δe	NL	NM	NP	EZ	PS	PM	PL
NL	NL	NL	NL	NL	NM	NP	EZ
NM	NL	NL	NL	NM	NP	EZ	PS
NP	NL	NL	NM	NP	EZ	PS	PM
EZ	NL	NM	NP	EZ	PS	PM	PL
PS	NM	NP	EZ	PS	PM	PL	PL
PM	NP	EZ	PS	PM	PL	PL	PL
PL	EZ	PS	PM	PL	PL	PL	PL

c)Fuzzy rules

Figure 7. Membership functions and fuzzy rules.

Using the Matlab/Simulink software, the surface for the fuzzy logic controller used in this paper is given in order to compensate for the traditional hysteresis comparators. This surface of the fuzzy logic controller is shown in Figure 8. The use of the fuzzy logic technique leads to improving the performance and effectiveness of the proposed five-level SVM technique and thus obtaining a good quality of the electric current, and this is the main objective of this work.

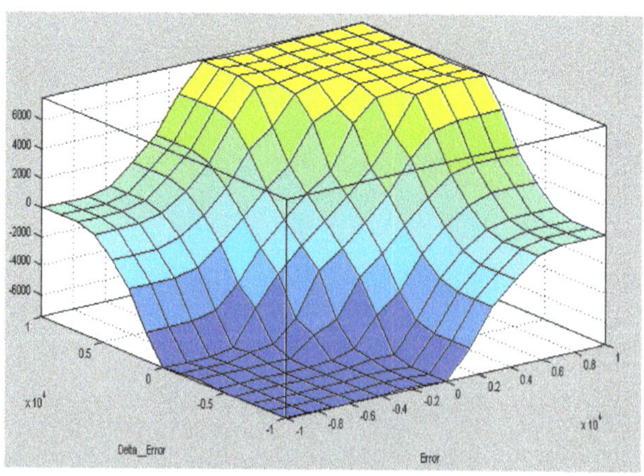

Figure 8. Control surface.

Compared to the PWM technique, the proposed five-level fuzzy SVM (FSVM) method is more complex and contains 12 fuzzy logic controllers which make it not easy and costly compared to the traditional PWM technique. However, in terms of the results obtained, the proposed method is better than the PWM technique. The proposed five-level fuzzy SVM method gives the output of the inverter a high-quality electrical signal (current) with a small value of THD.

5. Traditional IFOC Strategy

The IFOC method is a type of FOC method which offers a fast dynamic response compared to the direct FOC method [28]. The indirect FOC method is a method that differs from the direct FOC method in terms of the principle of work and in terms of internal structure. Several scientific works dealt with this method [29–31], where four PI controllers are used in this method, which makes the dynamic response much faster. Thus, the indirect FOC method is more complex than the direct FOC method [32], but the indirect FOC method provides a better dynamic response than the direct FOC method. Moreover, the indirect FOC method reduces the ripples of torque, reactive power, and flux compared to the direct FOC method [33]. In order to control the generator inverter, the PWM technique is used. This method is based on the principle shown in Equation (21) [34].

$$\Psi_{qs} = 0 \text{ and } \Psi_{ds} = \Psi_s \tag{21}$$

Using Equation (21), the direct and quadrature stator voltages of the generator become as follows:

$$\begin{cases} V_{ds} = V_s = w_s \Psi_s \\ V_{qs} = 0 \end{cases} \tag{22}$$

Using Equations (13) and (22), the direct and quadrature stator currents of the generator become as follows:

$$\begin{cases} I_{ds} = -\frac{M}{L_s} I_{dr} + \frac{\Psi_s}{L_s} \\ I_{qs} = -\frac{M}{L_s} I_{qr} \end{cases} \tag{23}$$

Relying on Equations (11) and (23), the direct and indirect rotor voltages of the generator become as follows [34]:

$$\begin{cases} V_{qr} = R_{dr} I_{qr} + \left(L_r - \frac{M^2}{L_s} \right) w_r I_{qr} + g \frac{M V_s}{L_s} \\ V_{dr} = R_{dr} I_{dr} - w_r \left(L_r - \frac{M^2}{L_s} \right) I_{qr} \end{cases} \tag{24}$$

The direct and quadrature rotor flux of a generator can be expressed by the following equation [33]:

$$\begin{cases} \Psi_{dr} = \left(L_r - \frac{M^2}{L_s} \right) I_{dr} + \frac{M}{w_s . L_s} V_s \\ \Psi_{qr} = \left(L_r - \frac{M^2}{L_s} \right) I_{qr} \end{cases} \tag{25}$$

By Equation (25), direct rotor flux is related to both stator voltage and quadrature rotor current. As for the quadrature rotor flux, it is related to the quadrature rotor current of the generator.

From Equations (21)–(25), the internal structure of the indirect FOC strategy can be given in Figure 9. Through this figure, we note that this technique controls the reactive and active power by controlling the quadrature and direct rotor voltages (V_{qr}^* and V_{dr}^*).

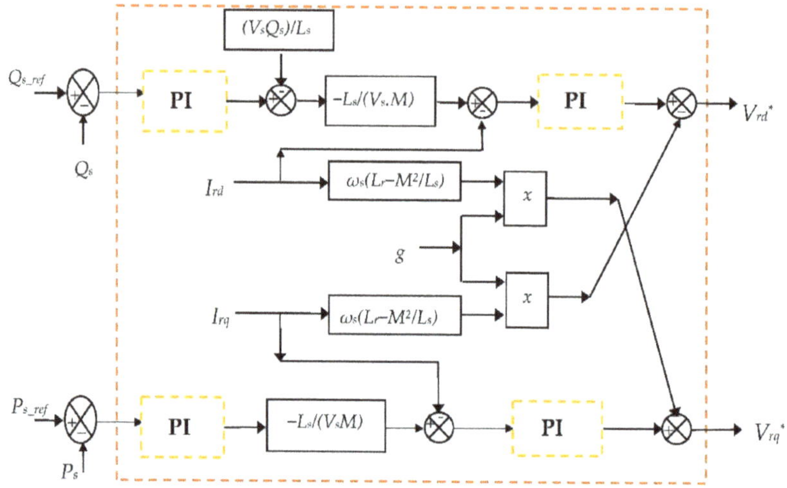

Figure 9. Traditional indirect FOC strategy.

In order to estimate the active and reactive power, Equation (26) is used. In order to estimate the two values, we need to measure both rotor voltage and rotor current.

$$\begin{cases} Q_s = -1.5\left(\frac{\omega_s \Psi_s M}{L_s} I_{dr} - \frac{\omega_s \Psi_s{}^2}{L_s}\right) \\ P_s = -1.5\frac{\omega_s \Psi_s M}{L_s} I_{qr} \end{cases} \tag{26}$$

Figure 10 represents the total system used in this paper, where a multi-rotor wind turbine (MRWT) was used to rotate the generator (DPIG). The latter is controlled by the traditional IFOC method. In this work, the reference value of the reactive power (Qs^*) is set to 0 Var. The reference value of the active power (Ps^*) is obtained using the MPPT technique.

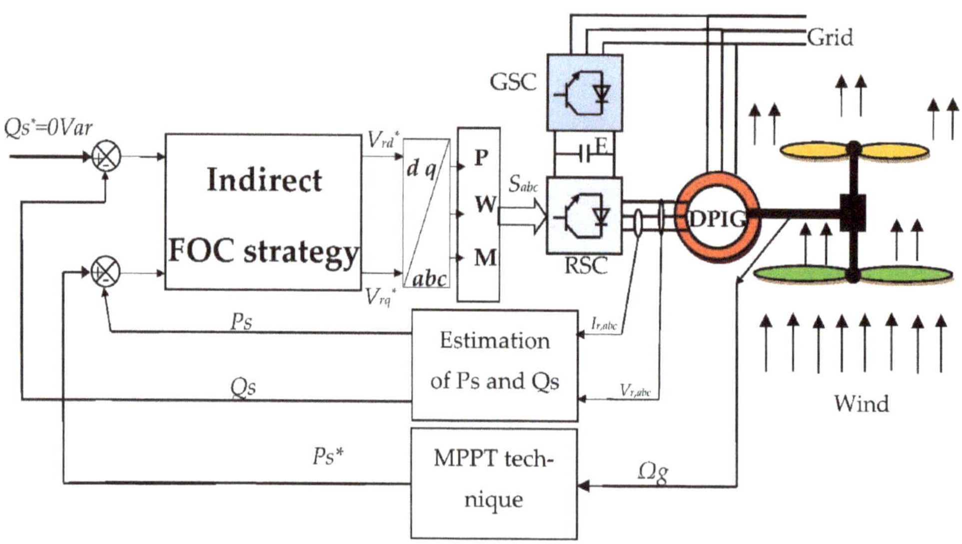

Figure 10. Traditional IFOC technique of the DPIG-based MRWT system.

The indirect FOC strategy gives more ripples in the current, torque, active power, and flux compared to DPC and DTC. Moreover, it has a long dynamic response and gives a poor quality of electric current and active power.

The reason for these shortcomings in this traditional indirect FOC strategy is due to the use of both the traditional PWM technique and conventional PI controller. The use of these classic methods (PI and PWM) makes the indirect FOC strategy not robust and characterized by a long dynamic response compared to some methods such as the DTC strategy.

In order to improve the performance and efficacy of the traditional indirect FOC strategy, a novel scheme for the traditional indirect FOC strategy is proposed in Section 6. This proposed indirect FOC strategy is based on the use of both the proposed SSTA algorithm and the multilevel fuzzy SVM strategy.

6. Proposed Indirect FOC Strategy

In this part, a new idea for indirect FOC strategy is presented based on the proposed simplified STA controller and the five-level fuzzy SVM strategy. The proposed indirect FOC method is a change from the classic indirect FOC method, where the proposed SSTA controller is used in place of the traditional PI controller and the five-level fuzzy SVM technique is used instead of the PWM technique. This proposed indirect FOC method aims to control the active and reactive power of the generator placed in the multi-rotor wind turbine system. On the other hand, this proposed indirect FOC method reduces the ripples of torque, active power, and electric current compared to the classical indirect FOC method. The proposed indirect FOC strategy is shown in Figure 11. From this figure, to control the active/reactive power, two proposed SSTA controllers are used and the same with the reactive power.

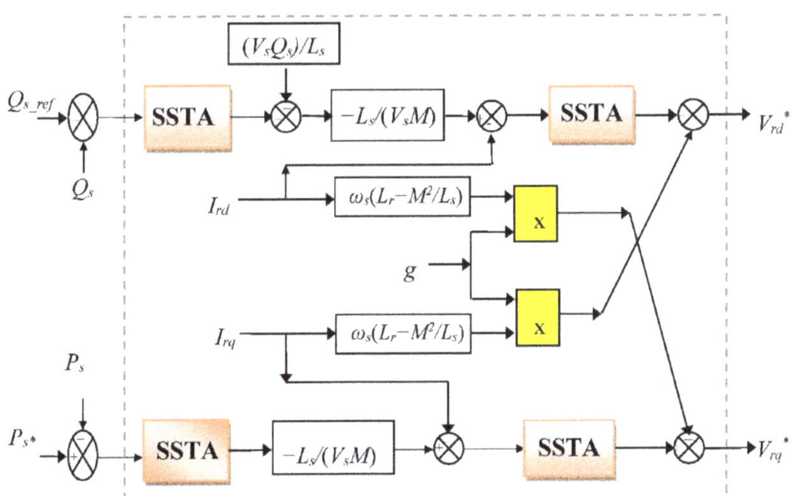

Figure 11. Proposed indirect FOC strategy.

In this proposed indirect FOC method, the same equations used to estimate both the active and reactive power are used in the classical indirect FOC strategy. Therefore, it can be said that this designed indirect FOC strategy is more robust than the rest of the controls such as the classical indirect FOC strategy and DTC. The objective of this designed indirect FOC strategy is to obtain high-quality V_{qr}^* and V_{dr}^* from active and reactive power references for the inverter DPIG control. Controlling the latter very well leads to obtaining a high quality of stator current and active power. On the other hand, the reactive power reference is set to zero. As for the reference value of the active power, it is obtained using the maximum power point tracking (MPPT) technique. The system studied using the proposed indirect FOC method is represented in Figure 12, where almost the same structure as the classical indirect FOC method is preserved.

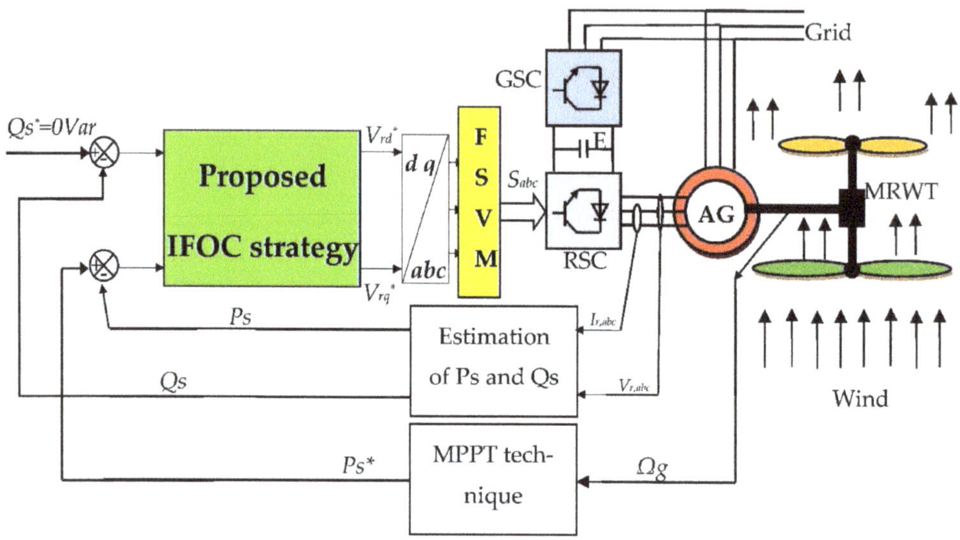

Figure 12. Proposed indirect FOC strategy of DPIG-based MRWT system.

In Table 1, a comparison is given between the classical indirect FOC (IFOC) technique and the proposed indirect FOC method. Through this table, the proposed indirect FOC strategy is more robust and reduces torque and current ripples compared to the classical indirect FOC technique. Moreover, the proposed indirect FOC method improves the rise time, response dynamic, THD value of current, and power quality compared to the classical indirect FOC technique. However, the proposed indirect FOC method is more complicated than the classical indirect FOC method due to the use of the five-level fuzzy SVM (FSVM) technique (instead, the classical indirect FOC method uses the PWM technique).

Table 1. A comparative study between the classical method and the proposed IFOC strategy.

	Traditional Indirect FOC Technique	Proposed Indirect FOC Technique
Type controller used	PI controller	SSTA controller
Rise time	High	Low
Modulation	PWM	Fuzzy SVM
Degree of complexity	Medium	High
Ease	Medium	Complicated
Simplicity of implementation	Medium	Complicated
Response dynamic	Slow	Quick
Robustness	Low	High
THD	High	Low
Power ripple	High	Low

The result of using the five-level fuzzy SVM technique in the proposed IFOC method makes the proposed IFOC method not simple and somewhat complicated compared to the classical IFOC method, where the latter uses PWM, which leads to problems in the case of achieving this proposed IFOC-SSTA-FSVM method.

To implement the proposed method (IFOC-SSTA-FSVM) empirically, there are several problems of implementation related to the large financial cost of completing this project due to the complexity of controlling the MRWT system. On the other hand, there is the complexity of the use of the five-level fuzzy SVM technique to control the inverter of the DPIG-based MRWT system. Moreover, the use of the maximum power point tracking technique increases the complexity and financial cost of the studied MRWT system. Nonetheless, given the results obtained in improving the quality of electric power and the importance of the MRWT system in improving the performance of classical wind turbines and in reducing the size of wind farms, this system is very necessary for the near future.

In the next part, the results are confirmed and the robustness of the proposed indirect FOC method is verified using the Matlab/Simulink software.

7. Results

In order to verify the proposed indirect FOC method, the Matlab/Simulink software is used. The results of the proposed indirect FOC method are compared with the classical indirect FOC method in terms of the ratio of ripples at the level of torque, current, effective power, and reactive power. The two methods are also compared in terms of the THD value of the electric current.

In this work, a generator with the following data is used: 50 Hz, 380/696 V, R_s = 0.012 Ω, L_r = 0.0136 H, L_m = 0.0135 H, p = 2, J = 1000 kg·m^2, P_{sn} = 1.5 MW, R_r = 0.021 Ω, L_s = 0.0137 H, and f_r = 0.0024 N·m/s [35].

In this work, a multi-rotor wind turbine with the following data is used: R_1 = 13.2 m, R_2 = 25.5 m, r_1 = 1 m, r_2 = 0.5 m, r_g = 0.75 m, J_1 = 500 kg·m^2, J_2 = 1000 kg·m^2, $G_1 = r_1/r_g$, and $G_2 = r_2/r_g$.

In this work, the proposed indirect FOC method is tested in the case of two tests, the first test is a tracking test and the second test is to study the behavior of the proposed indirect FOC method in comparison with the classical indirect FOC method in the event of a change in the generator parameters. This is in order to know the robustness of the proposed indirect FOC method with the classical indirect FOC method.

7.1. First Test

In this test, the behavior of the reference tracking is studied, for both the proposed IFOC-SSTA method and the classical IFOC method, where the obtained results are shown in Figure 13. Through this figure, the active and reactive power follow the references perfectly, and this is for the two IFOC methods with a preference for the proposed method in the dynamic response (see Figure 13a,b). The proposed IFOC-SSTA method gave better results in terms of ripples for both active and reactive power compared to the classical IFOC method (see Figure 14a,b). In Figure 13c, the generated torque has the same shape as the active power, where it can be seen that the increase in the active power corresponds to the increase in the torque. In addition, the proposed IFOC-SSTA method reduced torque ripples compared to the classical IFOC method (Figure 14c).

Regarding the current generated by the generator, it is shown in Figure 15d. Through this figure, the electric current takes the form of the active power, where it is noted that the behavior of the current is the same as the behavior of the active power. Moreover, the proposed IFOC-SSTA method gave excellent results in terms of electric current ripples and quality compared to the classical IFOC method (see Figure 14d).

Figure 13e,f represent the THD value of the proposed and classical IFOC method, respectively. Through these two forms, the proposed IFOC-SSTA method reduced the THD value of the electric current excellently compared to the classical IFOC method and the reduction ratio was about 96.72%. These results confirm the robustness of the proposed IFOC method in improving the quality of the effective power and electric current.

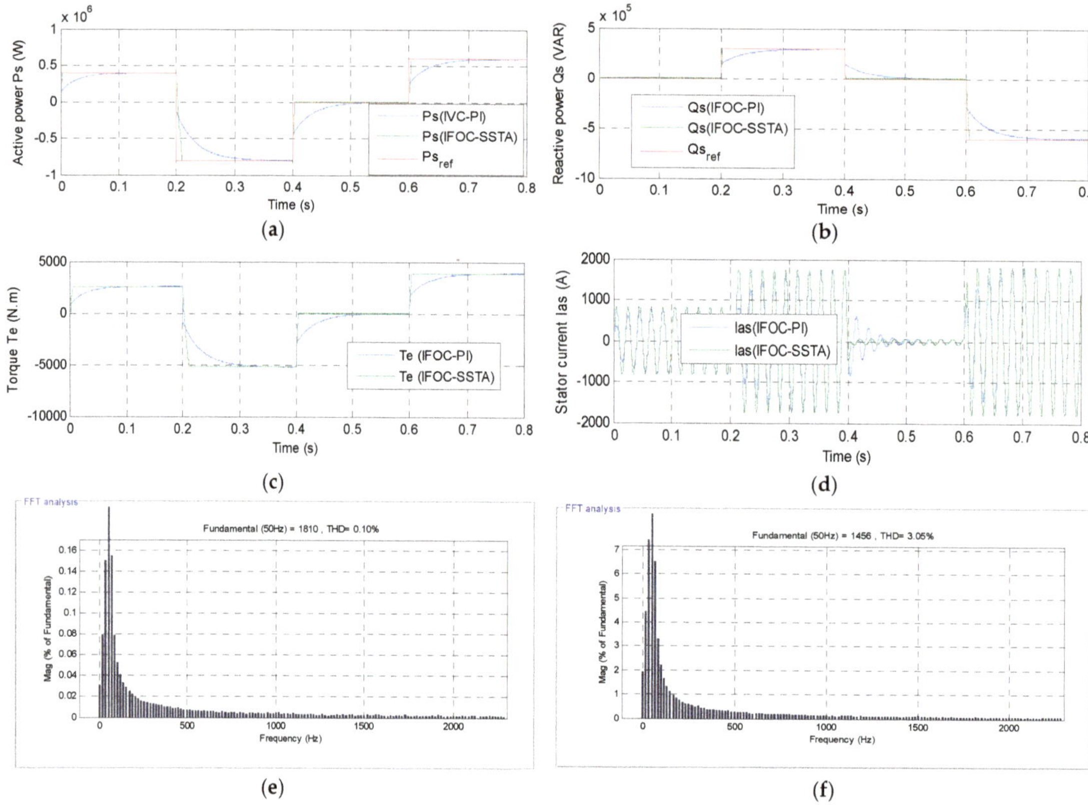

Figure 13. First test results. (**a**) Active power; (**b**) Reactive power; (**c**) Torque; (**d**) Stator current; (**e**) THD value of stator current (IFOC-SSTA strategy); (**f**) THD value of stator current (IFOC strategy).

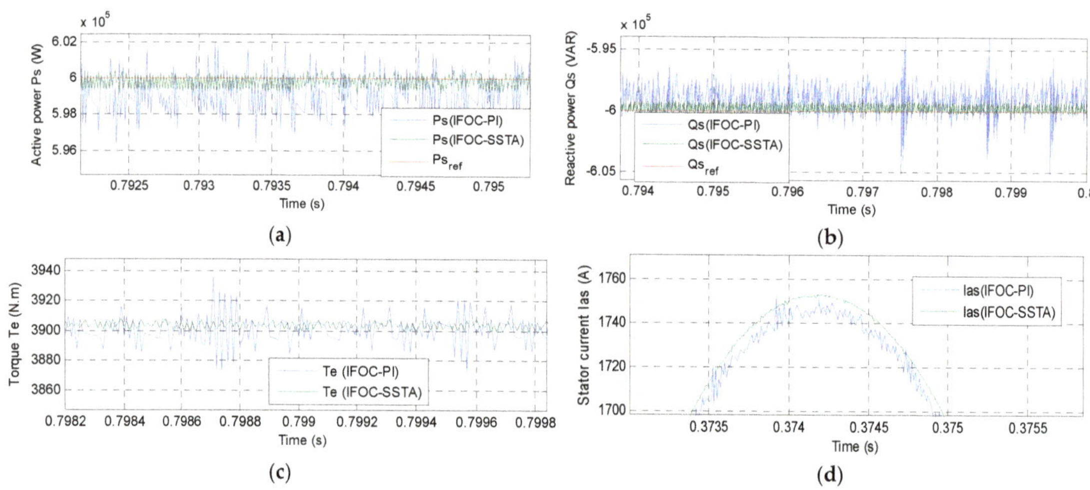

Figure 14. Zoom in the first test results. (**a**) Active power; (**b**) Reactive power; (**c**) Torque; (**d**) Stator current.

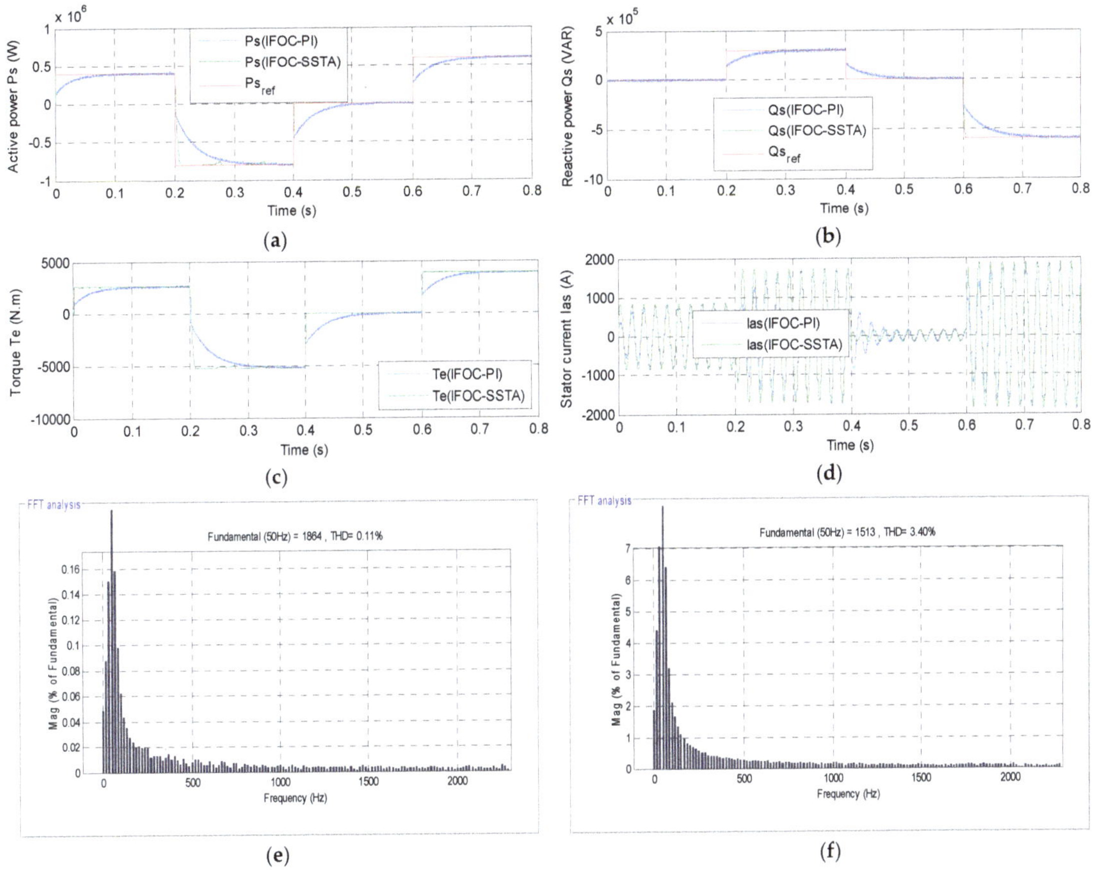

Figure 15. Second test results. (**a**) Active power; (**b**) Reactive power; (**c**) Torque; (**d**) Stator current; (**e**) THD value of current (IFOC-SSTA strategy); (**f**) THD value of current (IFOC strategy).

In the next test, the proposed IFOC method is further confirmed if the parameters of the generator installed in the multi-rotor wind turbine system are changed.

The results obtained from this test are shown in Table 2. Table 2 represents the value of ripples in each of torque, current, active power, and reactive power, and this is for the two IFOC methods. Through this table, the proposed IFOC-SSTA method reduced the ratio of ripples of torque, active power, current, and reactive power by excellent ratios, and the ratios were as follows: 91.66%, 91.20%, 84.21%, and 93.75%, respectively.

Table 2. Comparative ripples obtained from the traditional IFOC with the proposed IFOC strategy.

	Classical IFOC Technique	Proposed IFOC Technique	Ratios
Torque ripple (N·m)	Around 60	Around 5	91.66%
Reactive power ripple (VAR)	Around 8000	Around 500	93.75%
Active power ripple (W)	Around 10,000	Around 880	91.20%
Stator current (A)	Around 19	Around 3	84.21%

In Table 3, the response time is extracted for each of the active power, torque, and reactive power of the two IFOC methods proposed in this work. Through this table, the proposed IFOC-SSTA method gave a small response time compared to the classical IFOC method, and the improvement ratio was about 97.54%, 97.54%, and 98.23% for the active power, torque, and reactive power, respectively. Moreover, the proposed method gave a good response time in comparison with both the DPC strategy and neuro-second order sliding mode control (NSOSMC) completed in [36] (see Table 4).

Table 3. Response time.

	Response Time		
	Torque	**Reactive Power**	**Active Power**
Classical IFOC strategy	0.12 s	0.13 s	0.12 s
Proposed IFOC strategy	2.95 ms	0.0023 s	2.95 ms
Ratios	97.54%	98.23%	97.54%

Table 4. Comparative analysis of response time.

		Response Time		
		Torque	**Reactive Power**	**Active Power**
Proposed technique: IFOC-SSTA		2.95 ms	0.0023 s	2.95 ms
[36]	Direct power control	18 ms	17 ms	18 ms
	Neuro-second order SMC technique	5 ms	9 ms	5 ms

7.2. Second Test

The results of the second test are shown in Figure 15. In this test, the generated parameter values were changed in order to know the change in the behavior of the proposed IFOC-SSTA method compared to the classical IFOC method, as well as its robustness. In this test, Rs, Ls, Rr, Lm, and Lr were changed to the values 0.024 Ω, 0.00685 H, 0.042 Ω, 0.00675 H, and 0.0068 H, respectively. Zoom is given for torque, current, effective power, and reactive power in Figure 16. In Figures 15 and 16, there is a noticeable effect on the level of torque, current, active, and reactive power, where the classical IFOC method was affected by changing the generator parameters more than the IFOC-SSTA method. Moreover, active and reactive power keep following the references well in this test for both the proposed and the classical IFOC method (see Figure 15a,b). Both current and torque take the same form as active power (see Figure 15c,d) with ripples in both torque and current levels. The proposed IFOC-SSTA method reduced these ripples as compared to the classical IFOC method (see Figure 16c,d). Moreover, the proposed IFOC-SSTA method also reduced the ripples in both the active and reactive power compared to the classical IFOC method (see Figure 16a,b). The THD value of the electric current is shown in Figure 15e,f and this is for both the proposed and the classical IFOC method, respectively. Through the two figures, the proposed IFOC-SSTA method reduced the THD value of the electric current compared with the classical IFOC method.

The results obtained from this test are shown in Table 5. Through this table, the proposed IFOC-SSTA method reduced the ripples of torque, reactive power, current, and active power by 93.35%, 98%, 87.50%, and 83.33%, respectively.

The proposed IFOC-SSTA-FSVM method in this work provided excellent results compared to the classical IFOC method in terms of reducing the ripples of torque, current, and active power, as well as improving the quality of electric power. All of these factors help reduce malfunctions and thus reduce maintenance costs and help extend the life of the system as a whole.

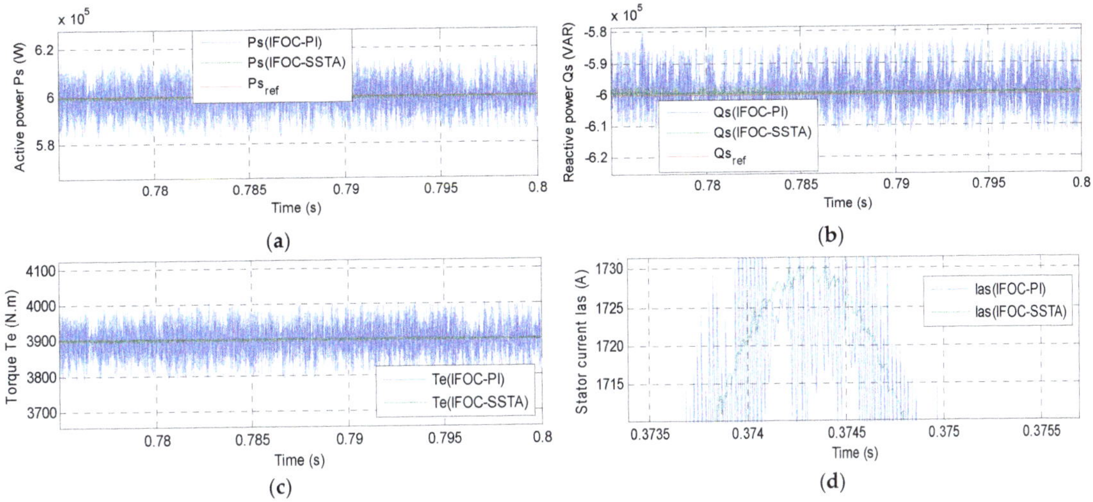

Figure 16. Zoom in the second test results. (**a**) Active power; (**b**) Reactive power; (**c**) Torque; (**d**) Stator current.

Table 5. Comparative ripples obtained from both techniques.

	Traditional IFOC Strategy	Designed IFOC Technique	Ratios
Torque ripple (N·m)	Around 172	Around 11.60	93.35%
Reactive power ripple (VAR)	Around 25,000	Around 500	98%
Active power ripple (W)	Around 30,000	Around 5000	83.33%
Stator current (A)	Around 40	Around 5	87.50%

The following is a comparison between the proposed IFOC-SSTA-FSVM method and some published works in terms of the THD value of the electric current. The results of the comparison are recorded in Table 6. Through this table, the proposed IFOC-SSTA-FSVM method significantly reduced the THD value compared to several published methods, which indicates the robustness of the proposed IFOC-SSTA-FSVM method and the effectiveness of the proposed IFOC-SSTA-FSVM method in improving the quality of electrical energy.

Table 6. Comparative results with other techniques.

References	Techniques	THD (%)
[37]	Field-oriented control with PI controllers	0.77
	Super twisting algorithm (STA)	0.28
[38]	Fuzzy DTC strategy	2.40
[39]	Fractional-order sliding mode control	1.31
[40]	Integral SMC technique	9.71
	Multi-resonant-based sliding mode controller (MRSMC)	3.14
[41]	Backstepping control	2.19
[6]	Field-oriented control	3.7
[12]	DPC control with PI controllers	0.46
	DPC control with terminal synergetic controllers	0.25
[42]	Direct torque control with second-order continuous SMC technique	0.78
[28]	DPC control with integral-proportional controllers	0.43
[9]	DFOC control with PI controllers	1.45
	DFOC control with synergetic- SMC technique	0.50

Table 6. *Cont.*

References	Techniques	THD (%)
[13]	Field-oriented control with neuro-fuzzy controller	0.78
	Field-oriented control with type-2 fuzzy logic controllers	1.14
[43]	Multilevel DTC strategy	1.57
[32]	Vector control	2.20
	Adaptive backstepping sliding mode control	1.15
[44]	Traditional direct vector control	1.65
[45]	DPC with sliding mode controller	1.66
	DPC with super twisting sliding mode controller	0.11
Proposed IFOC strategy	First test	0.10
	Second test	0.11

8. Conclusions

In this work, a new idea was given for the IFOC method based on both a simplified STA controller and a five-level fuzzy SVM technique. This proposed method was verified using the Matlab/Simulink software, comparing the results obtained with the traditional method. The application of the proposed method in the wind system led to the improvement of the dynamic response of the generator and the improvement of the quality of the electric current with the electric energy.

The points drawn from this work are illustrated in the following points:

- A new fuzzy SVM technology was introduced to control the five-level inverter to give a constant frequency at the inverter output, with this method confirmed by numerical simulation.
- A new simplified STA controller was proposed in this paper.
- A new indirect FOC strategy based on the simplified STA controller and the proposed five-level fuzzy SVM technique was presented and confirmed with numerical simulation.
- The robustness of the proposed indirect FOC strategy was presented.
- The characteristics of the designed indirect FOC strategy was analyzed, showing that the undulations of the reactive power, stator current, torque, and active power were minimized.

In a future paper, to ameliorate the quality of the active power and current, the DPIG will be controlled using another robust control scheme, such as fractional order synergetic control and feedback PI controller [46,47].

Author Contributions: Validation, N.B., P.T. and N.T.; conceptualization, H.B.; software, H.B.; methodology, H.B.; investigation, H.B. and I.C.; resources, N.B., P.T. and N.T.; project administration, N.B.; data curation, N.B., P.T. and N.T.; writing—original draft preparation, H.B.; supervision, N.B. and I.C.; visualization: I.C. and N.B.; formal analysis: I.C. and N.B.; funding acquisition: N.B., P.T. and N.T.; writing—review and editing: I.C., N.B., P.T., N.T. and H.B. All authors have read and agreed to the published version of the manuscript.

Funding: This work was supported in part by the Framework Agreement between University of Pitesti (Romania) and King Mongkut's University of Technology North Bangkok (Thailand), in part by an International Research Partnership "Electrical Engineering–Thai French Research Center (EE-TFRC)" under the project framework of the Lorraine Université d'Excellence (LUE) in cooperation between Université de Lorraine and King Mongkut's University of Technology North Bangkok, in part by the National Research Council of Thailand (NRCT) under Senior Research Scholar Program under Grant No. N42A640328, and in part by National Science, Research, and Innovation Fund (NSRF) under King Mongkut's University of Technology North Bangkok under Grant no. KMUTNB-FF-65-20.

Institutional Review Board Statement: Not applicable.

Informed Consent Statement: Not applicable.

Data Availability Statement: Not applicable.

Conflicts of Interest: The authors declare no conflict of interest.

References

1. Mohamed Ahmed, S.A.; Montaser Abd El Sattar, M.A. Dynamic Performance and Effectiveness of Voltage Disturbances on the Improvement of Power Quality for Grid-Connected DFIG System Based Wind Farm. *J. Electr. Eng. Electron. Control. Comput. Sci.* **2020**, *5*, 25–30. Available online: https://jeeeccs.net/index.php/journal/article/view/126 (accessed on 9 April 2022).
2. Talla, J.; Leu, V.Q.; Šmídl, V.; Peroutka, Z. Adaptive Speed Control of Induction Motor Drive With Inaccurate Model. *IEEE Trans. Ind. Electron.* **2018**, *65*, 8532–8542. [CrossRef]
3. Wolkiewicz, M.; Tarchała, G.; Orłowska-Kowalska, T.; Kowalski, C.T. Online Stator Interturn Short Circuits Monitoring in the DFOC Induction-Motor Drive. *IEEE Trans. Ind. Electron.* **2016**, *63*, 2517–2528. [CrossRef]
4. Candelo-Zuluaga, C.; Riba, J.-R.; Garcia, A. PMSM Parameter Estimation for Sensorless FOC Based on Differential Power Factor. *IEEE Trans. Instrum. Meas.* **2021**, *70*, 1–12. [CrossRef]
5. Alexandrou, A.D.; Adamopoulos, N.K.; Kladas, A.G. Development of a Constant Switching Frequency Deadbeat Predictive Control Technique for Field-Oriented Synchronous Permanent-Magnet Motor Drive. *IEEE Trans. Ind. Electron.* **2016**, *63*, 5167–5175. [CrossRef]
6. Amrane, F.; Chaiba, A.; Babes, B.E.; Mekhilef, S. Design and implementation of high performance field oriented control for grid-connected doubly fed induction generator via hysteresis rotor current controller. *Rev. Roum. Sci. Tech.-Electrotech. Energ* **2016**, *61*, 319–324.
7. Eltamaly, A.M.; Al-Saud, M.; Sayed, K.; Abo-Khalil, A.G. Sensorless active and reactive control for DFIG wind turbines using opposition-based learning technique. *Sustainability* **2020**, *12*, 3583. [CrossRef]
8. Habib, B.; Lemdani, S. Combining synergetic control and super twisting algorithm to reduce the active power undulations of doubly fed induction generator for dual-rotor wind turbine system. *Electr. Eng. Electromech.* **2021**, *2021*, 8–17.
9. Benbouhenni, H.; Bizon, N. A Synergetic Sliding Mode Controller Applied to Direct Field-Oriented Control of Induction Generator-Based Variable Speed Dual-Rotor Wind Turbines. *Energies* **2021**, *14*, 4437. [CrossRef]
10. Habib, B. Application of STA methods and modified SVM strategy in direct vector control system of ASG integrated to dual-rotor wind power: Simulation studies. *Int. J. Smart Grid* **2021**, *5*, 62–72.
11. Habib, B. Amelioration effectiveness of torque and rotor flux control applied to the asynchronous generator (AG) for dual-rotor wind turbine using neural third-order sliding mode approaches. *Int. J. Eng. Trans. C: Asp.* **2022**, *35*, 517–530.
12. Benbouhenni, H.; Bizon, N. Terminal Synergetic Control for Direct Active and Reactive Powers in Asynchronous Generator-Based Dual-Rotor Wind Power Systems. *Electronics* **2021**, *10*, 1880. [CrossRef]
13. Amrane, F.; Chaiba, A. A novel direct power control for grid-connected doubly fed induction generator based on hybrid artificial intelligent control with space vector modulation. *Rev. Sci. Tech.-Electrotech. Et Energ.* **2016**, *61*, 263–268.
14. Brando, G.; Dannier, A.; Spina, I. Performance Analysis of a Full Order Sensorless Control Adaptive Observer for Doubly-Fed Induction Generator in Grid Connected Operation. *Energies* **2021**, *14*, 1254. [CrossRef]
15. Yahdou, A.; Hemici, B.; Boudjema, Z. Second order sliding mode control of a dual-rotor wind turbine system by employing a matrix converter. *J. Electr. Eng.* **2016**, *16*, 1–11.
16. Erturk, E.; Sivrioglu, S.; Bolat, F.C. Analysis Model of a Small Scale Counter-Rotating Dual Rotor Wind Turbine with Double Rotational Generator Armature. *Int. J. Renew. Energy Res.* **2018**, *8*, 1849–1858.
17. Fukami, T.; Momiyama, M.; Shima, K.; Hanaoka, R.; Takata, S. Steady-State Analysis of a Dual-Winding Reluctance Generator With a Multiple-Barrier Rotor. *IEEE Trans. Energy Convers.* **2008**, *23*, 492–498. [CrossRef]
18. Worku, M.Y.; Hassan, M.A.; Abido, M.A. Real Time-Based under Frequency Control and Energy Management of Microgrids. *Electronics* **2020**, *9*, 1487. [CrossRef]
19. Wang, L.; Kerrouche, K.D.E.; Mezouar, A.; Van Den Bossche, A.; Draou, A.; Boumediene, L. Feasibility Study of Wind Farm Grid-Connected Project in Algeria under Grid Fault Conditions Using D-Facts Devices. *Appl. Sci.* **2018**, *8*, 2250. [CrossRef]
20. Han, Y.; Ma, R. Adaptive-Gain Second-Order Sliding Mode Direct Power Control for Wind-Turbine-Driven DFIG under Balanced and Unbalanced Grid Voltage. *Energies* **2019**, *12*, 3886. [CrossRef]
21. Alhato, M.M.; Ibrahim, M.N.; Rezk, H.; Bouallègue, S. An Enhanced DC-Link Voltage Response for Wind-Driven Doubly Fed Induction Generator Using Adaptive Fuzzy Extended State Observer and Sliding Mode Control. *Mathematics* **2021**, *9*, 963. [CrossRef]
22. Alhato, M.M.; Bouallègue, S.; Rezk, H. Modeling and Performance Improvement of Direct Power Control of Doubly-Fed Induction Generator Based Wind Turbine through Second-Order Sliding Mode Control Approach. *Mathematics* **2020**, *8*, 2012. [CrossRef]
23. Djilali, L.; Badillo-Olvera, A.; Rios, Y.Y.; López-Beltrán, H.; Saihi, L. Neural High Order Sliding Mode Control for Doubly Fed Induction Generator based Wind Turbines. *IEEE Lat. Am. Trans.* **2022**, *20*, 223–232. [CrossRef]
24. Halabi, L.M.; Alsofyani, I.M.; Lee, K.-B. Multi Open-/Short-Circuit Fault-Tolerance Using Modified SVM Technique for Three-Level HANPC Converters. *IEEE Trans. Power Electron.* **2021**, *36*, 13621–13633. [CrossRef]
25. Habib, B. Utilization of an ANFIS-STSM algorithm to minimize total harmonic distortion. *Int. J. Smart Grid* **2020**, *4*, 56–67.

26. Habib, B.; Boudjema, Z.; Belaidi, A. Direct power control with NSTSM algorithm for DFIG using SVPWM technique. *Iranian J. Electr. Electron. Eng.* **2021**, *17*, 1–11.
27. Habib, B.; Boudjema, Z.; Belaidi, A. Neuro-second order sliding mode control of a DFIG supplied by a two-level NSVM inverter for wind turbine system. *Iran. J. Electr. Electron. Eng.* **2018**, *14*, 362–373.
28. Fayssal, A.; Bruno, F.; Azeddine, C. Experimental investigation of efficient and simple wind-turbine based on DFIG-direct power control using LCL-filter for stand-alone mode. *ISA Trans.* **2021**, 1–34, *in press*. [CrossRef]
29. Evangelista, C.; Valenciaga, F.; Puleston, P. Active and Reactive Power Control for Wind Turbine Based on a MIMO 2-Sliding Mode Algorithm With Variable Gains. *IEEE Trans. Energy Convers.* **2013**, *28*, 682–689. [CrossRef]
30. Habib, B.; Boudjema, Z.; Belaidi, A. Indirect vector control of a DFIG supplied by a two-level FSVM inverter for wind turbine system. *Majlesi J. Electr. Eng.* **2019**, *13*, 45–54.
31. Tang, Z.; Akin, B. A New LMS Algorithm Based Deadtime Compensation Method for PMSM FOC Drives. *IEEE Trans. Ind. Appl.* **2018**, *54*, 6472–6484. [CrossRef]
32. Yahdou, A.; Djilali, A.B.; Boudjema, Z.; Mehedi, F. Improved Vector Control of a Counter-Rotating Wind Turbine System Using Adaptive Backstepping Sliding Mode. *J. Eur. Des Systèmes Autom.* **2020**, *53*, 645–651. [CrossRef]
33. Boulaam, K.; Mekhilef, A. Output power control of a variable wind energy conversion system. *Rev. Sci. Techn.-Electrotech. Energ.* **2017**, *62*, 197–202.
34. Pan, L.; Zhu, Z.; Xiong, Y.; Shao, J. Integral Sliding Mode Control for Maximum Power Point Tracking in DFIG Based Floating Offshore Wind Turbine and Power to Gas. *Processes* **2021**, *9*, 1016. [CrossRef]
35. Yahdou, A.; Hemici, B.; Boudjema, Z. Sliding mode control of dual rotor wind turbine system. *Mediterr. J. Meas. Control* **2015**, *11*, 412–419.
36. Ibrahim, Y.; Semmah, A.; Patrice, W. Neuro-Second Order Sliding Mode Control of a DFIG based Wind Turbine System. *J. Electr. Electron. Eng.* **2020**, *13*, 63–68.
37. Hamid, C.; Aziz, D.; Eddine, C.S.; Othmane, Z.; Mohammed, T.; Hasnae, E. Integral sliding mode control for DFIG based WECS with MPPT based on artificial neural network under a real wind profile. *Energy Rep.* **2021**, *7*, 4809–4824. [CrossRef]
38. Ayrira, W.; Ourahoua, M.; El Hassounia, B.; Haddib, A. Direct torque control improvement of a variable speed DFIG based on a fuzzy inference system. *Math. Comput. Simul.* **2020**, *167*, 308–324. [CrossRef]
39. Beniss, M.A.; El Moussaoui, H.; Lamhamdi, T.; El Markhi, H. Improvement of power quality injected into the grid by using a FOSMC-DPC for doubly fed induction generator. *Int. J. Intell. Eng. Syst.* **2021**, *14*, 556–567. [CrossRef]
40. Quan, Y.; Hang, L.; He, Y.; Zhang, Y. Multi-Resonant-Based Sliding Mode Control of DFIG-Based Wind System under Unbalanced and Harmonic Network Conditions. *Appl. Sci.* **2019**, *9*, 1124. [CrossRef]
41. Zeghdi, Z.; Barazane, L.; Bekakra, Y.; Larbi, A. Improved backstepping control of a DFIG based wind energy conversion system using ant lion optimizer algorithm. *Period. Polytech. Electr. Eng. Comput. Sci.* **2022**, *66*, 1–17. [CrossRef]
42. Boudjema, Z.; Taleb, R.; Djerriri, Y.; Yahdou, A. A novel direct torque control using second order continuous sliding mode of a doubly fed induction generator for a wind energy conversion system. *Turk. J. Elec-Trical Eng. Comput. Sci.* **2017**, *25*, 965–975. [CrossRef]
43. El Ouanjli, N.; Aziz, D.; El Ghzizal, A.; Mohammed, T.; Youness, E.; Khalid, M.; Badre, B. Direct torque control of doubly fed induction motor using three-level NPC inverter. *Prot. Control Mod. Power Syst.* **2019**, *17*, 1–9. [CrossRef]
44. Benbouhenni, H.; Bizon, N. Advanced direct vector control method for optimizing the operation of a double-powered induction generator-based dual-rotor wind turbine system. *Mathematics* **2021**, *9*, 2297. [CrossRef]
45. Yaichi, I.; Semmah, A.; Wira, P.; Djeriri, Y. Super-twisting sliding mode control of a doubly-fed induction generator based on the SVM strategy. *Period. Polytech. Electr. Eng. Comput. Sci.* **2019**, *63*, 178–190. [CrossRef]
46. Benbouhenni, H. Synergetic control theory scheme for asynchronous generator based dual-rotor wind power. *J. Electr. Eng. Electron. Control. Comput. Sci.* **2021**, *7*, 19–28. Available online: https://jeeeccs.net/index.php/journal/article/view/215 (accessed on 9 April 2022).
47. Nnem, L.N.; Lonla, B.L.; Sonfack, G.B.; Mbih, J. Review of a Multipurpose Duty-Cycle Modulation Technology in Electrical and Electronics Engineering. *J. Electr. Eng. Electron. Control Comput. Sci.* **2018**, *4*, 9–18. Available online: https://jeeeccs.net/index.php/journal/article/view/101 (accessed on 9 April 2022).

sustainability

MDPI

Article

Technical Performance Prediction and Employment Potential of Solar PV Systems in Cold Countries

Ephraim Bonah Agyekum [1,*], Usman Mehmood [2,3], Salah Kamel [4], Mokhtar Shouran [5], Elmazeg Elgamli [5,*] and Tomiwa Sunday Adebayo [6,7]

1 Department of Nuclear and Renewable Energy, Ural Federal University Named after the First President of Russia Boris Yeltsin, 19 Mira Street, 620002 Ekaterinburg, Russia
2 Remote Sensing, GIS and Climatic Research Lab, National Center of GIS and Space Applications, Centre for Remote Sensing, University of the Punjab, Lahore 54590, Pakistan; usmanmehmood.umt@gmail.com
3 Department of Political Science, University of Management and Technology, Lahore 54770, Pakistan
4 Electrical Engineering Department, Faculty of Engineering, Aswan University, Aswan 81542, Egypt; skamel@aswu.edu.eg
5 Wolfson Centre for Magnetics, School of Engineering, Cardiff University, Cardiff CF24 3AA, UK; shouranma@cardiff.ac.uk
6 Department of Business Administration, Faculty of Economics and Administrative Science, Cyprus International University, Nicosia 99040, Turkey; twaikline@gmail.com
7 Department of Finance & Accounting, AKFA University, Tashkent 100012, Uzbekistan
* Correspondence: agyekumephraim@yahoo.com (E.B.A.); elgamlies@cardiff.ac.uk (E.E.)

Citation: Agyekum, E.B.; Mehmood, U.; Kamel, S.; Shouran, M.; Elgamli, E.; Adebayo, T.S. Technical Performance Prediction and Employment Potential of Solar PV Systems in Cold Countries. *Sustainability* **2022**, *14*, 3546. https://doi.org/10.3390/su14063546

Academic Editors: Nicu Bizon, Bhargav Appasani and Mamadou Baïlo Camara

Received: 11 February 2022
Accepted: 12 March 2022
Published: 17 March 2022

Abstract: Power distribution to decentralized and remote communities secluded from centralized grid connections has always been a problem for utilities and governments worldwide. This situation is even more critical for the isolated communities in Russia due to the vast nature of the country. Therefore, the Russian government is formulating and implementing several strategies to develop its renewable energy sector. However, very little information is available on the possible performance of solar photovoltaic (PV) modules under Russian weather conditions for all year round. Thus, this study has been designed to fill that research gap by assessing the performance ratio (*PR*), degradation, energy loss prediction, and employment potential of PV modules in the Sverdlovsk region of Russia using the PVsyst simulation model. A side-by-side comparison of the fixed tilted plane and tracking horizontal axis East–West were analyzed. According to the results, the annual production probability (P) for the fixed PV module for a P50, P75, and P90 is 39.68 MWh, 37.72 MWh, and 35.94 MWh, respectively, with a variability of 2.91 MWh. In the case of the tracking PV module, the annual production probability for the P50, P75, and P90 is 43.18 MWh, 41.05 MWh, and 39.12 MWh, respectively, with a variability of 3.17 MWh. A *PR* of 82.3% and 82.6% is obtained for the fixed and tracking systems, respectively, while the PV array losses for the fixed and tracking orientations are 15.1% and 14.9%, respectively. The months of May to August recorded the highest array losses due to the high temperatures that are usually recorded within that period.

Keywords: renewable energy; solar photovoltaic energy; degradation rate; PVsyst software; energy loss prediction

1. Introduction

One challenge confronting the world today is how to generate energy in a more sustainable way to meet its energy needs while maintaining environmental security. The world's demand for electricity is growing due to increasing global population, improved lifestyle, and industrialization [1]. It has been estimated that fossil fuel forms about 80% of the world's primary energy, and energy consumption globally is expected to increase by 2.3% each year from 2015 to 2040 [2]. The concentration of atmospheric carbon dioxide (CO_2) equivalent is said to have almost doubled since the inception of the Industrial

Revolution [3]. This has increased the average global temperature, which is negatively impacting global climate [4,5].

Solar photovoltaic (PV) technology can generate power by directly converting incident solar radiation to electrical power [6,7]. PV technology is one of the renewable energy (RE) options that can help to decarbonize the world to decrease greenhouse gas (GHG) emissions. The continual drop in the cost of PV systems and the formulation of policies by various governments to promote the development and use of RE technologies has led to the rapid growth of the PV industry [8]. The PV industry has witnessed a composite yearly growth rate of more than 40% during the last 15 years, making it one of the fast-growing industries globally. This has necessitated the need to improve project designs and continuous monitoring and prediction of the performance of the PV systems that have either been installed or yet to be installed to ensure their reliability and performance [8].

Therefore, several researchers have performed studies that either assess the performance of already installed or yet to be installed PV power plants in several countries. Malvoni et al. [8] examined a 960 kWp monocrystalline silicon PV system's performance in southern Italy. They obtained a capacity factor (*CF*) and performance ratio (*PR*) of 15.6% and 84.4%, respectively. Kumar et al. [9] studied the performance, degradation, and energy loss of a 200 kW roof-integrated crystalline PV system in northern India. According to their results, the PV system is projected to operate with an annual *PR*, *CF*, and energy loss of 77.27%, 16.72%, and −26.5%, respectively. Ramanan et al. [10] evaluated the performance of two colocated grid-connected PV power plants, which consist of copper indium selenium (CIS) and polycrystalline silicon (p-Si) arrays. A yearly *PR* of 86.73% for the CIS and 78.48% for p-Si were obtained, while the capacity utilization factor ranges from 17.99% for the p-Si to 19.57% for the CIS systems. Similarly, Ameur et al. [11] analyzed and compared several indices that affect the performance of different grid-connected PV technologies, i.e., polycrystalline silicon (pc-Si), amorphous silicon (a-Si), and monocrystalline silicon (mc-Si) with capacities of 2 kWp each. The study's outcome suggests that the polycrystalline, monocrystalline, and amorphous technologies generated a four-year yearly alternating current (*AC*) energy average of 3239 kWh, 3246 kWh, and 2797 kWh with yearly *PR* of 77%, 77%, and 73%, respectively. Dahmoun et al. [12] explored and assessed the performance of a 23.92 MWp polycrystalline PV power plant located in El Bayadh in Algeria. Their study shows that the degradation of the *PR* during the study period is estimated to be 0.76% per year.

Furthermore, Kittner et al. [13] evaluated the economic investment, embodied energy, and CO_2 payback for amorphous silicon thin film and single crystalline systems. They reported that the amorphous silicon thin-film panels have higher net environmental and economic benefits. Padmavathi and Daniel [14] worked on a 3 MW grid-connected polycrystalline PV power plant in India. They evaluated normalized technical performance parameters for the system for the year 2011. The generated yearly average energy by the plant was 1372 kWh per kWp. Radue and van Dyk [15] reported losses of up to 30% for thin-film PV power plants sited in South Africa after a period of 14 months. Kymakis et al. [16] also estimated the performance of a 171 kWp polycrystalline silicon PV power plant installed on the island of Crete. After one year, the average annual *PR* and *CF* were 67.36% and 15.26%, respectively. Belmahdi and El Bouardi [17] evaluated the performance of a 1 MW solar PV power plant under Moroccan weather conditions using the PVsyst software. The optimal angle for the installation of the PV module for the study area in summer was identified to be 32° for fixed tilt and 48° for winter on seasonal adjustment tilts. A 60 kWp PV module was modelled for the Uttar-Pradesh area in India using the PVsyst software by [18]. Their system generated a total of 89.5 MWh per year with a performance ratio of 73.73%. In the Philippines, Dellosa et al. [19] assessed the technical and economic performance of 5 MWp PV system for that country using the PVsyst software. The results from their study revealed that the temperature of the PV panel accounted for the highest energy loss; it accounted for 8% of the loss. A payback period of 4.23 years was recorded for the system.

Finally, Chabachi et al. [20] assessed the performance of a poly-Si/6 MWp in Southwest Algeria. An average monthly efficiency for their PV array was 12.68% with a *PR* of 84%. Yadav et al. [21] employed the PVsyst software to examine the performance of a 295 Wp Si-poly PV module SYP295S under Nepal's weather conditions at Tribhuvan University. Their simulation results revealed that a total of 110 kWp of PV power plant's output would be enough for the whole campus. The designed system produced a surplus energy of 115.1 MWh/year, which can be exported to grid. A 2.4 kWp monocrystalline PV module was studied by [22] at the Mulhouse campus, France. The studied power plant generated a total of 5597.65 kWh of energy. Chandel and Chandel [23] conducted a performance assessment on a 19-MWp (17-MWac) PV plant installed with seasonal adjustable tilt (AT), fixed tilt (FT), and horizontal single-axis solar tracking (HSAT) configurations in India. Key findings from their study suggest that the studied FT system recorded an annual *PR* of 79% with a *CF* of 19%. The AT also recorded a *CF* of 20%, while the HSAT recorded 22%. Kumar et al. [24] also assessed a 10 kWp poly-Si PV power plant for the remote islands of Andaman and Nicobar in India. A yearly average *CF* and *PR* range of (13.73–14.61%) and (64.70–64.93%), respectively were obtained. Other studies, such as [25–27] investigated power plants with capacities ranging from 1.72–5 kWp and determined their performance parameters.

RE development and application in the Russian Federation are relatively lower than in other European countries. The country has mainly relied on fossil fuels and nuclear energy for its electricity and heating demands. Hydropower constitutes a large portion of its installed RE capacity in terms of renewables. The country's total installed RE capacity increased to 53.5 GW; hydro alone constitutes 51.5 GW, with bioenergy taking only 1.35 GW. The solar PV has only 460 MW, while onshore wind also has 111 MW as of 2015 [28]. The country's share of installed solar and wind energy capacities as of 2021 are 0.7 percent and 0.42 percent, respectively [29].

The government of the Russian Federation has therefore committed to the development of the entire RE sector. The government's target is to achieve a renewable energy share of 4.5% by 2030. The country's Energy Strategy stipulates that the percentage of RE in the energy mix must be at least 4.5% in the period from 2020–2030, producing 80–100 billion kWh/year [30,31]. However, there is a lack of detailed information on the performance of solar PV modules under Russian weather conditions for all year round. This does not help encourage individuals and private investors to use or invest in the sector. The objective of this study is to bridge this gap by assessing solar PV's technical performance and social aspect in the Sverdlovsk region of Russia. The current study predicts solar PV system's energy performance, degradation, and energy loss under different orientations, i.e., fixed tilted plane and tracking horizontal axis E–W using the PVsyst simulation software. This is the first time such a study has been conducted in the study area in the Russian Federation, to the best of our knowledge. Unlike the previously reviewed literature, this study assesses the employment potential of such systems in Russia. The results obtained in this study are expected to shape the technical and social aspects of solar PV energy in the country.

This work is organized into four sections; Section 2 covers the materials and method used for the study. The outcomes are obtainable in Section 3, the conclusion and future research recommendations are obtainable in Section 4.

2. Materials and Methods

The methodology adopted for the study, specifications of the PV modules studied, geographical information of the study area, and mathematical relations used for the evaluation of the various performance indicators are presented in this section.

2.1. PVsyst Simulation Model

The PVsyst simulation tool is a commonly used software in designing solar power plants optimally and assessing the energy yields of the plants. It uses meteorological irradiation resources, extensive knowledge of PV technology, and PV system components for the simulation. As a result, the PVsyst tool can assist researchers and engineers to comprehend the PV system's workings to improve the system's design. The proposed grid-connected PV system was simulated using the following steps [1]:

- Identification of the study area (location)
- Downloading of the weather data characteristics for the study area (i.e., solar irradiance, wind speed, and ambient temperature)
- Selection of the orientation of the PV module (i.e., tilt and azimuth angles)
- Selection of the system components of the PV and inverter systems in relation to the requirements of the system
- Discretionary user needs relative to grid-tied system requirements
- The discretionary choice to alter the values for the loss types.

2.2. Meteorological Data of the Study Area

The study area is Yekaterinburg, which is in the Sverdlovsk region in Russia. It is located on latitude 56°50′34.4″ N and longitude 60°39′19.0″ E. The weather data used for the analysis were obtained from Meteonorm 8.0. The study area has an average global horizontal irradiation (GHI) average of 2.82 kWh/m^2/day and horizontal diffuse irradiation of 1.37 kWh/m^2/day. The average annual temperature for the area is 2.6 °C, with an average wind speed of 3.7 m/s. The relative humidity of the study area is also 74.1%, with a global horizontal irradiation year-to-year variability of 3.7%. The weather characteristics of the study area are presented in Figure 1. The lowest temperature and solar irradiation usually occur during the winter; similarly, May–August record relatively high temperatures and insolation. The month of July generally receives the highest solar irradiation and temperatures.

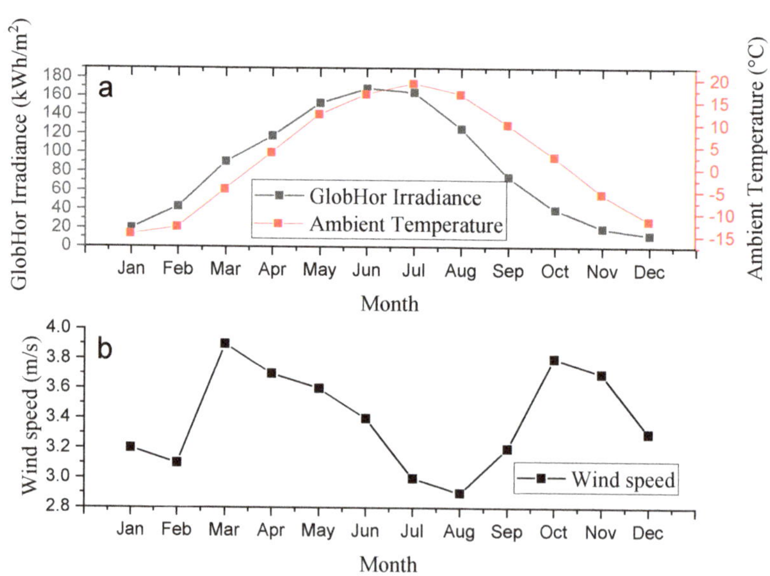

Figure 1. (**a**) GHI and temperature (**b**) wind speed characteristics of the study area (Meteonorm 8.0).

2.3. Solar PV Performance Assessment

The system yields are classified into an array, a reference, and final yields. The yields show the actual array operations in relation to its rated capacity. The array yield Y_A is said to be the output energy DC (direct current) produced from the PV array within a certain time frame normalized by the PV system's rated power [32]. It can be represented mathematically as indicated in Equation (1) [33].

$$Y_A = \frac{E_{DC}}{P_{PV,\ rated}} \left(\text{kWh}/\text{kW}_\text{p} \right) \quad (1)$$

where E_{DC} is the PV array's DC energy output in (kWh), and $P_{PV,\ rated}$ is the PV system rated power in (kW_p).

The alternating current (AC) energy (total) produced by the PV module for a specific time is divided by the installed PV's system rated output power [32]. The final yield Y_F can be defined as the inverter side output, which is AC energy produced daily $Y_{F,\ d}$ or monthly $Y_{F,\ m}$ by the system, which is normalized by its nominal or rated power of the installed PV array. It can be calculated using Equation (2) [34].

$$Y_F = \frac{E_{AC}}{P_{PV,\ rated}} \quad (2)$$

where the AC energy output (kWh) is denoted by E_{AC}.

The reference yield Y_R describes the solar irradiance for the PV system; it is the total in-plane solar irradiance divided by the reference irradiance at standard test conditions (STCs). It is a function of weather conditions, array orientation, and location, as indicated in Equation (3) [9,33].

$$Y_R = \frac{G_I}{G_{STC}} \ (\text{kWh}/\text{kW}) \quad (3)$$

where the total in-plane solar irradiance is denoted with G_I (kWh/m^2), and the reference irradiance at STCs is represented by G_{STC} ($1\ \text{kW}/\text{m}^2$).

The PR of a solar PV system measures the total outcome of losses on the PV system's rated output. The PV system's PR shows how close its performance approaches the ideal performance under real-life operations; it helps compare a PV system's independence of a location, orientation, tilt angle, and nominal rated power capacity. It can be calculated using Equation (4) [32,35].

$$PR = \frac{100 \times Y_F}{Y_R} (\%) \quad (4)$$

CF can be defined as the ratio of the AC energy that is generated by the PV system within a specified period of time (mostly one year) to the system's output energy, which would have been produced if the power plant were to operate at full capacity for the entire period. The yearly CF can calculated using Equation (5) [6].

$$CF = \frac{E_{AC}}{P_{PV,\ rated} \times 8760} \quad (5)$$

The array capture losses L_A shows the losses occasioned by the array's operation, which highlights the array's failure to use the available irradiance [26] completely. The difference between the array yield and the reference yield is the array capture losses; this can be calculated using Equations (6) and (7) [32].

$$L_A = Y_R - Y_A \ (\text{kWh}/\text{kW}_\text{p}) \quad (6)$$

where the losses L_S are caused by losses in changing the inverter's DC output power from the PV system to AC power. This is mathematically represented as follows:

$$L_S = Y_A - Y_F \ (\text{kWh}/\text{kW}_\text{p}) \quad (7)$$

The thermal losses result from the effect of temperature on the performances of the *PV* modules. This can be expressed mathematically as indicated in Equation (8) [36].

$$E_{therm} = E_{PV}\left(1 - \frac{1}{1 - \gamma(T_C - T_0)}\right)\ (\text{MWh}) \quad (8)$$

where the temperature coefficient of the maximal power is denoted by γ, the module temperature under STCs (25 °C) is represented by T_0, and the module temperature is denoted by T_C.

2.4. Energy Production

The daily, monthly, and annually total energy (*AC* or *DC*) produced by the *PV* system can be attained through simple summations as shown in Equation (9) [37,38].

$$\begin{aligned} E_d &= \sum_{1}^{24} E_h (\text{kWh}) \\ E_m &= \sum_{1}^{n} E_d\ (\text{kWh}) \\ E_y &= \sum_{1}^{12} E_m\ (\text{kWh}) \end{aligned} \quad (9)$$

where the number of days in a month is represented by n, and the hourly energy is denoted by E_h. In addition, the daily, monthly, and yearly cumulative energy values are represented with E_d, E_m, and E_y, respectively.

2.5. Characteristics of the Various Components

The *PV* module technology used for the analysis was Si-mono. The AE 300DGM6-60 (1500) solar module manufactured by AE solar was used for the analysis. A total of 120 units were used to achieve the designed power output. The technical description of the *PV* module is presented in Table 1. Increasing the temperature of a *PV* cell decreases its output performance as a result of the increase in the rate of the internal recombination in the *PV* cell, which is caused by increased carrier concentrations. Both the electrical efficiency and the power output of the *PV* power plant relate linearly with its operating temperature. As a result, when the operating temperature of the *PV* module rises beyond 25 °C, it leads to a reduction in the semiconductor material's band-gap, which results in the reduction in the open circuit voltage [39,40]. The current-voltage (I–V) graph for the selected *PV* module for different cell temperatures is presented in Figure 2; in this graph, the negative effect of high temperatures on the performance of *PV* cells is clearly shown. The power–voltage graph, which also demonstrates the impact of solar irradiance on the output of the *PV* cell, is also represented in Figure 3.

Table 1. *PV* module system description.

Model	AE 300DGM6-60 (1500)
Module type	Si-mono
Unit nominal power	300 Wp
Nominal (*STC*)	36.0 kWp
P_{mpp}	32.3 kWp
Cell area	177 m^2
Module area	197 m^2
Efficiency	18.38%

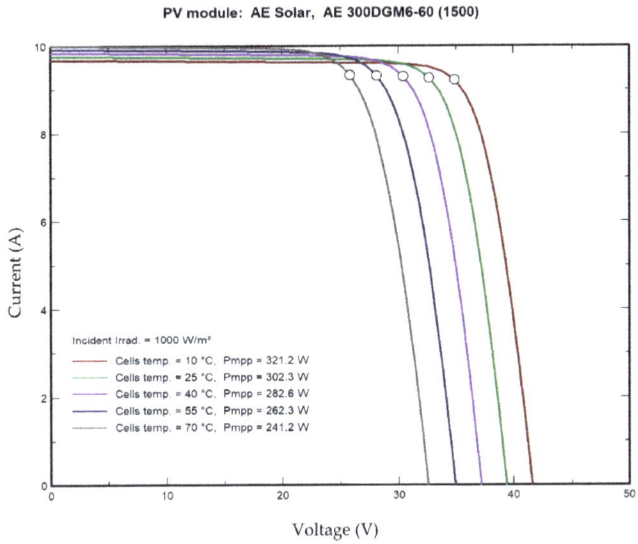

Figure 2. I-V graph for the used *PV* module under different temperature conditions (Obtained from *PV*syst software).

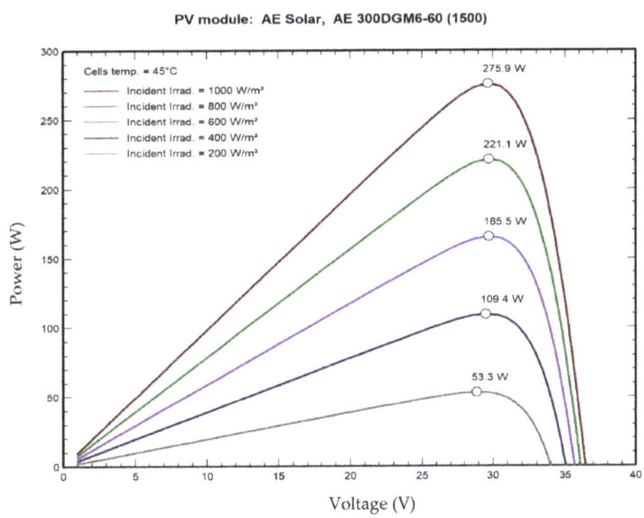

Figure 3. Power–voltage characteristics under different incident irradiance (Obtained from *PV*syst software).

In the inverter case, the Voltwerk model was used for the analysis; the model is 11 kW 400–800 V TL 50 Hz VS11 with a minimum MPP voltage of 400 V and a maximum MPP voltage of 800 V. It has a maximum efficiency of 98%. A total of 3 inverters were used to design the power plant.

Figure 4 depicts the schematic for the proposed *PV* system; it is made up of the *PV* array section where the *PV* modules are connected, the inverter section where the conversion of *DC* to *AC* is done, the load section, and the grid section. In the scheme, the E_{Array} denotes the *PV* array's energy output, while the inverter's energy output is represented by $E_{out\ inv}$. E_{used} denotes the energy used by the user load.

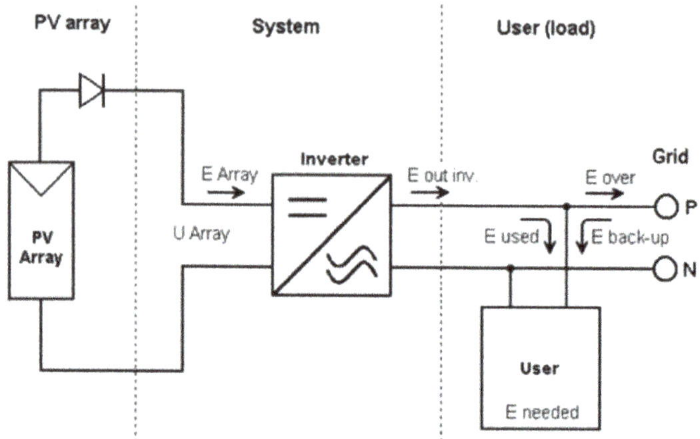

Figure 4. Schematic diagram of the proposed *PV* system.

2.6. Field Mechanisms Used in the Analysis

To take maximum advantage of the irradiation from the sun, *PV* panels are mostly tilted in the direction of the equator at an optimal tilt angle. The optimum tracking angle at a location is determined by both the latitude and the climatic conditions at the said location. There is the option to use the fixed tilted plane, which is commonly used; however, there is an option to use trackers. The single solar tracking and double-axis tracking option can be employed; this is subject to the degree of the freedom movement. The single solar tracking uses a single pivot point for rotation to track the sun's path from one point to another. In the case of the double-axis tracking, it tracks the sun's path in two different axes using two pivot points for rotation; it has both the horizontal and vertical axes [41,42]. In this study, two different field types were compared to assess their effect on the performance of the *PV* power plant at the study site. These are the fixed tilted plane and the tracking horizontal axis E–W. The optimum tilt angle for the fixed *PV* system is 45° for the study area. The two different mechanisms are represented in Figure 5.

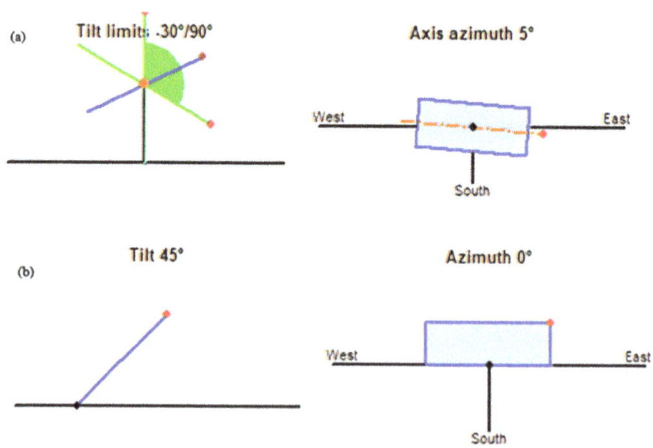

Figure 5. The orientation used, (**a**) tracking horizontal axis E–W and (**b**) fixed tilted plane.

3. Results and Discussion

According to a summary of the results from the simulations, the fixed tilted *PV* system will produce an energy of 39.5 MWh/year with an estimated specific production of 1097 kWh/kWp/year and a *PR* of 82.3%, as shown in Figure 6. Similarly, the tracking horizontal axis E–W system will generate a total of 43.0 MWh/year, which is about 3.5 MWh/year more than that of the fixed tilted plane; it is also expected to record a specific production of 1194 kWh/kWp/year with a *PR* of 82.6%. As widely published in the literature, the performance of *PV* systems depends on several factors; some of these include the ambient temperature, clearness index, and the level of solar irradiation, among others. As reviewed in the introduction section of this work, the obtained *PR* in this study falls within the range of values obtained by studies, such as Malvoni et al. [8] under Mediterranean weather conditions, which obtained 84.4%. Eke and Demircan [43] obtained a *PR* of 72% under Turkey climatic conditions; similarly, Okello et al. [44] obtained 84.3% as the *PR* for a 3.2 kWp grid-connected *PV* system in South Africa. As presented earlier in this section, the slight differences in the *PR* values can be attributed to the factors that affect the *PV* system's performance. The high temperatures during the summer period affected the *PR* during those periods. The output performance of the *PV* decreases with an increase in *PV* temperature [39,45]. Hence the output performance of the system reduces to some level even if there is enough solar radiation. A malfunction in the *PV* system can also be detected based on the *PR* values. Months with lower *PR* can be ascribed to a malfunction in the inverter and the incorrect functioning of the system. The IEC norm describes the normalized production and represents the standardized parameter for the *PV* system's performance assessment. It can therefore be assessed to compare the characteristics of *PV* architectures that are constructed under similar climatic conditions [46]. The useful produced energy per installed kWp/day, system losses, and the collection losses for both orientations are estimated and presented in Figure 7. The arrays' temperature characteristics against effective irradiance are presented in Figure 8. It can be seen from the figure that the array's temperature ranged between −30 °C during the winter period to as high as about 65 °C in the summer period.

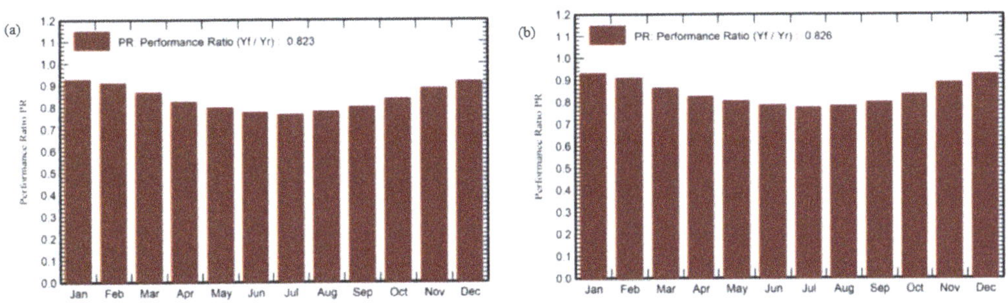

Figure 6. Monthly performance ratio for the (**a**) fixed plane and (**b**) tracking *PV* system.

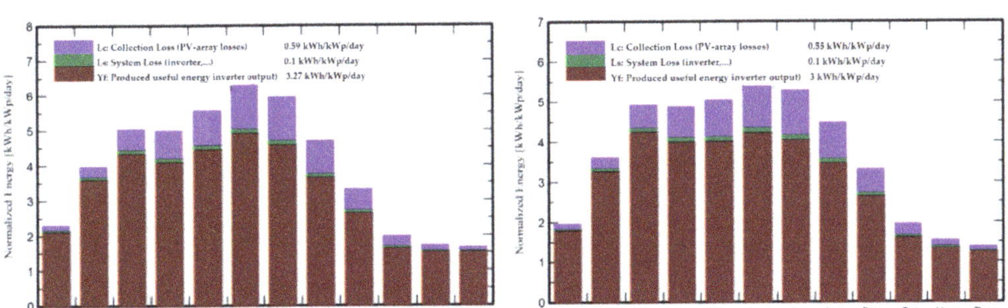

(**a**) (**b**)

Figure 7. Normalized energy production per installed kWp for (**a**) fixed plane (**b**) tracking *PV* system.

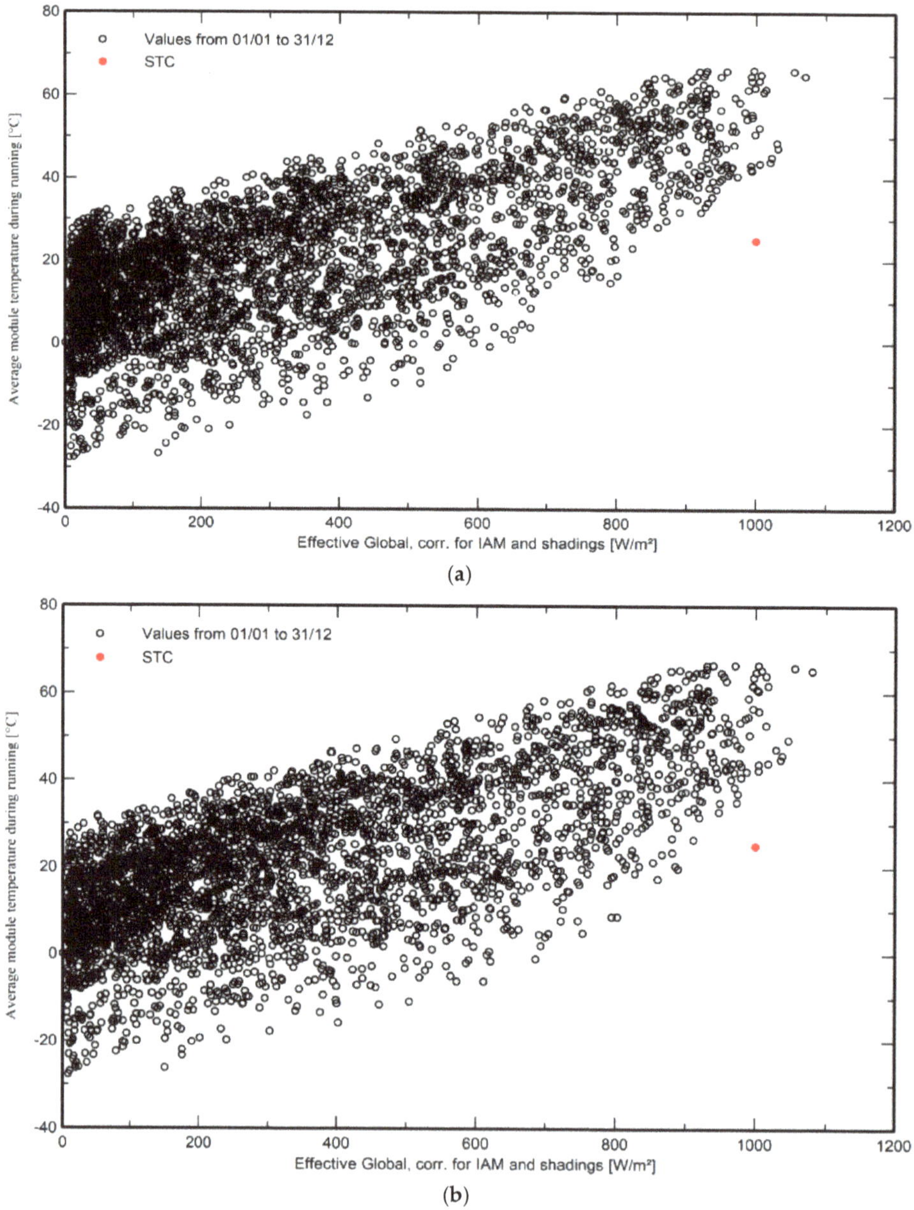

Figure 8. Array temperature vs. effective irradiance (**a**) fixed plane and (**b**) tracking *PV* system.

The expected monthly energy injected into the grid from the *PV* system for both orientations is shown in Figure 9. It is evident that the most energy injected into the grid occurs during the summer period, obviously because of the longer period and high intensity of solar radiation during that time. The balances and main results are presented in Table 2; it includes the global horizontal, horizontal diffuse irradiance, effective global corresponding for shading (GlobEff), ambient temperature (T_amb), global incident in collector plane (GlobInc), the energy output of the *PV* array (EArray), performance ratio

(*PR*), and the energy injected into the grid (E_Grid for each month of the year as well as the overall energy out for the year. Although the highest solar irradiation at the study area occurs during the summer periods, it is clear from the Table that the *PR* at those periods is relatively lower compared to the other periods. This may be due to the relatively high temperatures during those areas, which negatively affect the efficiency of the solar cells. It is clear from the energy values presented in Table 1 that both designs lost energy in the course of the conversion as there is a difference between the *DC* and *AC* sides of the energy produced. This difference is the energy lost due to the losses in the system. The fixed tilted plane system lost 1.249 MWh against 1.277 MW for the tracking system.

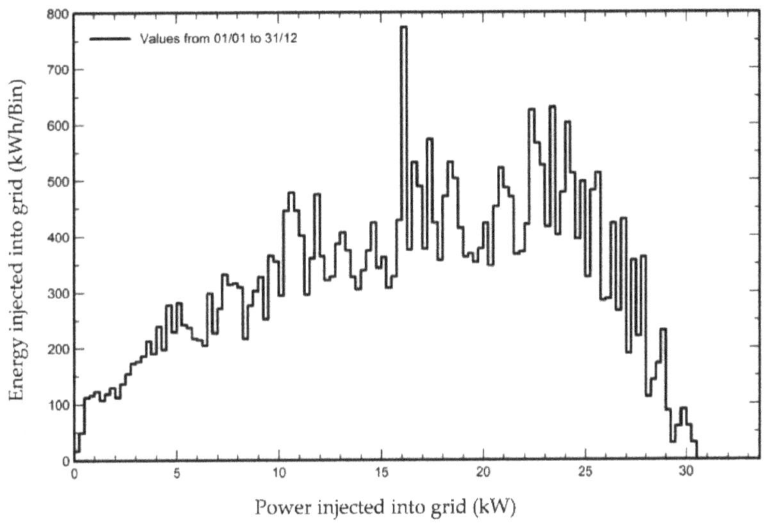

(**a**)

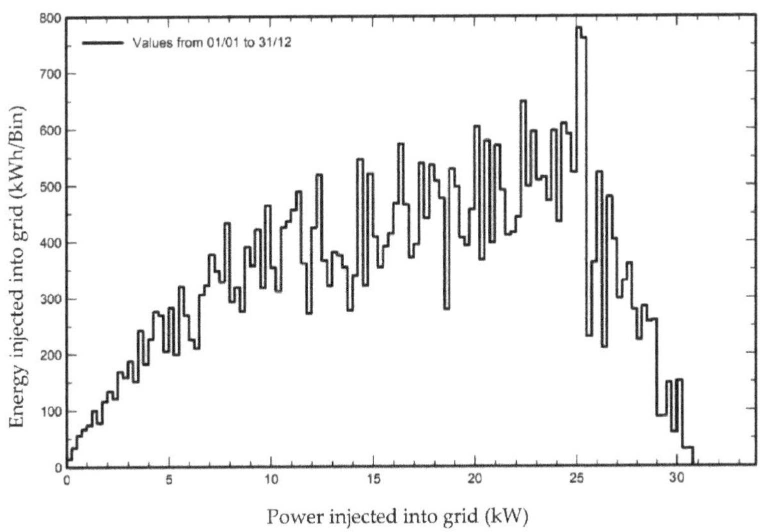

(**b**)

Figure 9. System output power distribution (**a**) fixed plane and (**b**) tracking system.

Table 2. Balances and main results.

	Global Horizontal, kWh/m^2	Diffuse Horizontal, kWh/m^2	T_amb, °C	Global Eff, kWh/m^2		GlobInc, kWh/m^2		Energy Array, MWh		E_Grid, MWh		*PR*	
				Fixed	Tracking	Fixed	Tracking	Fixed	Tracking	Fixed	Tracking	Tracking	Fixed
Jan	19.9	10.17	−14.57	58.0	68.5	60.9	71.3	2.087	2.446	2.032	2.384	0.929	0.926
Feb	42.8	16.23	−13.04	96.2	106.6	101.0	111.3	3.394	3.734	3.310	3.642	0.909	0.911
Mar	90.8	33.36	−4.61	144.6	148.5	152.4	156.4	4.880	4.994	4.754	4.865	0.864	0.867
Apr	117.8	52.17	3.92	137.9	141.2	146.1	149.9	4.467	4.579	4.333	4.446	0.824	0.824
May	152.6	77.64	12.05	147.5	162.6	156.4	172.6	4.634	5.136	4.489	4.991	0.803	0.797
Jun	168.1	76.44	16.66	154.5	178.0	163.9	188.6	4.725	5.481	4.576	5.329	0.785	0.776
Jul	163.9	78.29	19.14	154.9	174.1	164.3	184.7	4.680	5.291	4.531	5.140	0.773	0.766
Aug	125.8	69.21	16.66	130.7	137.1	138.6	145.6	4.014	4.222	3.885	4.093	0.781	0.779
Sep	74.5	39.26	9.94	93.8	94.3	99.3	99.8	2.965	2.974	2.862	2.870	0.799	0.801
Oct	39.8	26.93	2.66	57.3	58.4	60.5	61.4	1.896	1.920	1.821	1.844	0.834	0.837
Nov	19.7	11.85	−5.47	44.0	49.7	46.3	51.8	1.527	1.711	1.473	1.655	0.887	0.884
Dec	12.8	7.59	−11.64	40.9	49.4	43.0	51.3	1.463	1.758	1.417	1.710	0.925	0.916
Year	1028.4	499.14	2.73	1260.3	1368.5	1332.6	1444.9	40.732	44.247	39.483	42.970	0.826	0.823

The *PV*syst also runs a probability distribution analysis for the total yearly energy produced from the system, which could be transferred into the grid system. The probability distribution variance for the plant's production forecast depends on several factors; some of these include inverter efficiency uncertainty, *PV* module modeling/parameters, meteo data, degradation uncertainty, and soiling and mismatch uncertainties [12]. The probability law supposes that during the many years of operation of the *PV* system, the annual yield distribution will follow a statistical law, and this law is assumed to be the normal or Gaussian distribution. The P50-P90 indicates the different levels of yield, for which the probability that a particular year's production is over this value of 50% and 90%, respectively [47]. The probability distribution function for the *PV* plant's energy generation forecast is as shown in Figure 10. According to the results from the simulations, the expected annual production probability for the fixed *PV* module for the P50, P75, and P90 is 39.68 MWh, 37.72 MWh, and 35.94 MWh, respectively, with a variability of 2.91 MWh. In the case of the tracking *PV* module, the annual production probability for the P50, P75, and P90 is 43.18 MWh, 41.05 MWh, and 39.12 MWh, respectively, with a variability of 3.17 MWh.

3.1. System Losses

These are losses in the system that may occur during the conversion of the incident solar energy to electric energy. These losses enable users to know the energy converted into electricity by the system, which can be done by subtracting the total loss from the incident energy on the panel. The decrease in the performance of the *PV* module can be associated with these losses. The normalized production and loss factors for both orientations are presented in Figure 11. The *PV* array losses for both orientations, i.e., fixed and tracking, are 15.1% and 14.9%, respectively. The months of May to August recorded the highest array losses due to the high temperatures and relatively low wind speed during those periods. However, the tracking system positively impacted the *PV* module, as its system losses were relatively less than the fixed *PV* module. The produced useful energy (inverter output) for the fixed and tracked systems are 82.3% and 82.6%, respectively.

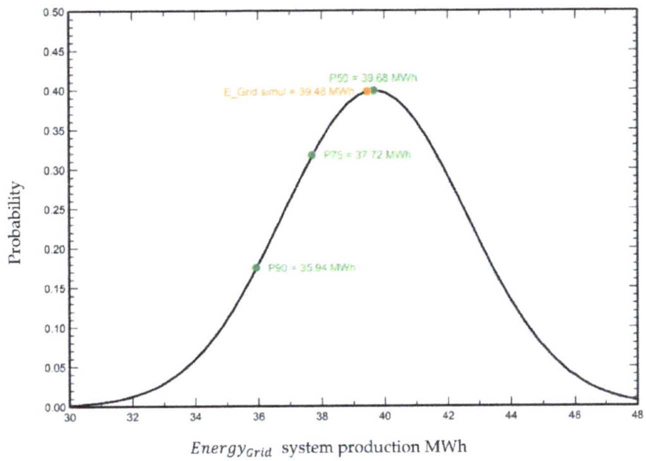

(**a**)

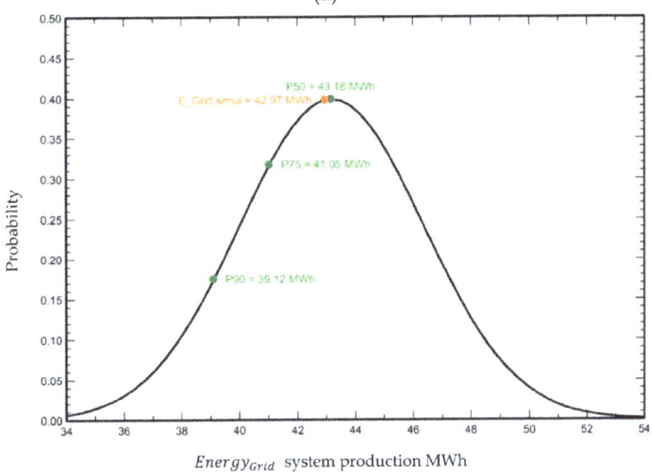

(**b**)

Figure 10. Probability distribution for (**a**) fixed and (**b**) tracking *PV* modules.

The loss diagram for the two orientations is presented in Figure 12. The losses range from the incidence angle modifier (IAM), which is the reflection loss (optical effect) that corresponds to the weakening of the irradiation that is actually reaching the surface of the *PV* cell with respect to the irradiation under normal incidence [48]. Others include manufacture losses, ambient temperature, and ohmic wiring, etc. Soiling loss involves dirt or dust accumulation on the surface of the module, which produces a dimming effect on the incident solar irradiation. The results suggest that both orientations are expected to record a soiling loss of 3%. The soiling loss can be mitigated by periodically washing the surface of the *PV* module; it can, however, add to the operations and maintenance costs, which can affect the plant's economic viability. This is especially for sites with water scarcity. The IAM loss is 2.5% for both designs. The nominal array energy (at *STC* efficiency) is 45.72 MWh for the fixed module; it, however, increased significantly by some 3.92 MWh for the tracking system.

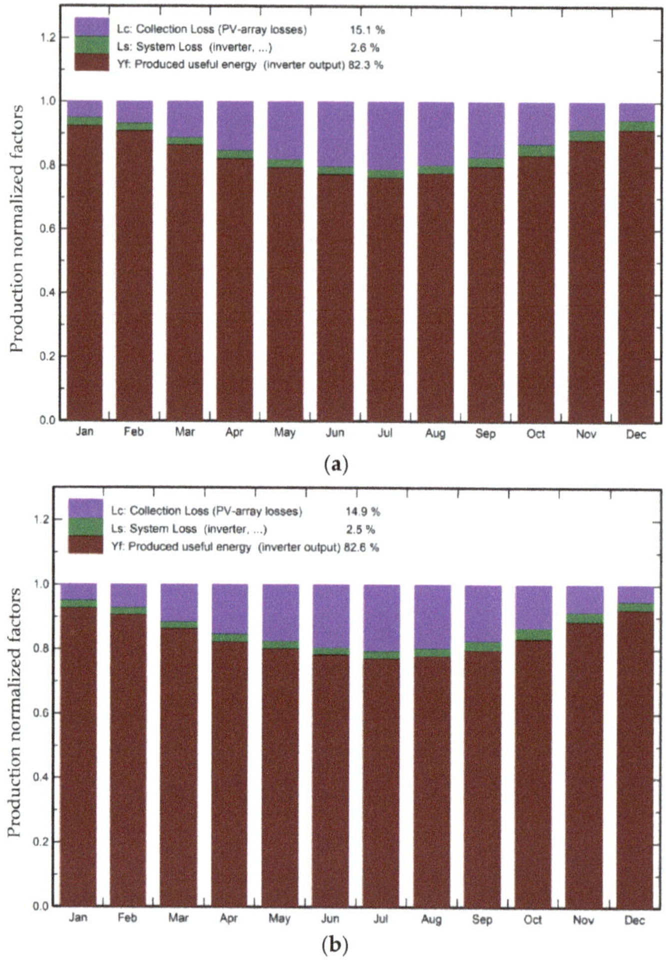

Figure 11. Normalized production and loss factors (**a**) fixed and (**b**) tracking systems.

The module quality loss, the deviation between the nominal capacity indicated on the manufacturer's datasheet, and the real module capacity, signifies the loss of module quality. A module quality loss of 0.4% was recorded. The *PV* array mismatch loss for both designs is the same; they both recorded 3.66%. This power loss is known as electrical mismatch loss. According to a study by Koirala et al. [49], a mismatch loss of up to 12% in the series string may arise but can be reduced to between 0.4–2.4% using suitable series-parallel connections. Presorting according to the max power current is identified as the most effective method for optimizing *PV* array performance [50,51]. The module degradation loss for the two designs is the same for all; they all recorded 3.82%. A total of 1028 kWh/m^2 global horizontal solar irradiation was received during the analysis period.

The two *PV* system's predicted performances were compared with other literature to assess their performance. Results from the other works on other countries presented in Table 3 can be said to be relatively similar to what is presented in this study. The *PR* falls within the range obtained by most studies.

Table 3. Comparison with other studies.

Location	Plant Capacity	Solar Irradiation (kWh/m^2)	Energy to Grid (MWh)	*PR*, %	Degradation Loss, %	Ref
India	10 MW	1940	15 798.192	86.12	-	[52]
India	186 kW	1821	318.52	85.6	-	[53]
Afghanistan	700 kW	1998	1266.1	79.7	1.50	[54]
Chile	8.2 kW	-	6.7780	85.5	-	[46]
Malaysia	380 kW	1631	495.39	80.3	3.81	[55]
India	200 kW	1820.7	292.954	77.27	2.5	[9]
Indonesia	41.1 kW	1732	74.3	82.69	-	[56]
India	230 W	1911	1.068	72.8	1.5	[57]
Vietnam	2 kW	1616	2.5699	76.9	2.0	[58]
Dubai	200 kW	2000	352.62	81.7	-	[59]
Poland	1 kW	-	0.83	60–80	-	[60]
Norway	220–240 W	-	11.92	83.03	-	[61]
Ireland	1.72 kW	1043.1	-	81.5	-	[62]
Serbia	2 kW	-	-	93.6	-	[63]
Norway	2.1 kW	-	1.93	-	-	[64]
Fixed tilted plane	36 kW	1028	39.48	82.3	3.82	Current study
Tracking system	36 kW	1028	42.97	82.6	3.82	Current study

3.2. Social Aspect

Deployment of energy systems at any location is usually accompanied by job creation, as people will be needed from the construction stages to the operation and decommissioning phase [65]. As a result, the study considered the social aspect of the energy systems to assess their employment creation potentials for the study area. The estimated employment potentials for solar *PV* is estimated to be about $0.27549 \times (10^{-7}/\mathrm{kWh/year})$ [65,66]. According to the simulated results, the total energy exported to the grid per year for the fixed axis solar *PV* module is 39,480 kWh against 42,970 kWh for the tracking axis module. According to the mathematical computations, we obtained 0.0011 persons/year for the fixed *PV* module, while 0.0012 persons/year were obtained for the tracking system. Therefore, assuming both modules operated for a lifetime of 25 years, then the fixed module will have an employment potential of 0.028 persons against 0.030 persons for the tracking *PV* module. The employment potential between the two designs does not vary much. It is important to state this analysis is only meant to assess the employment potential of the 36 kWp capacity *PV* modeled. Hence, it indicates the employment potential of a large-scale solar *PV* power plant under Russian weather conditions.

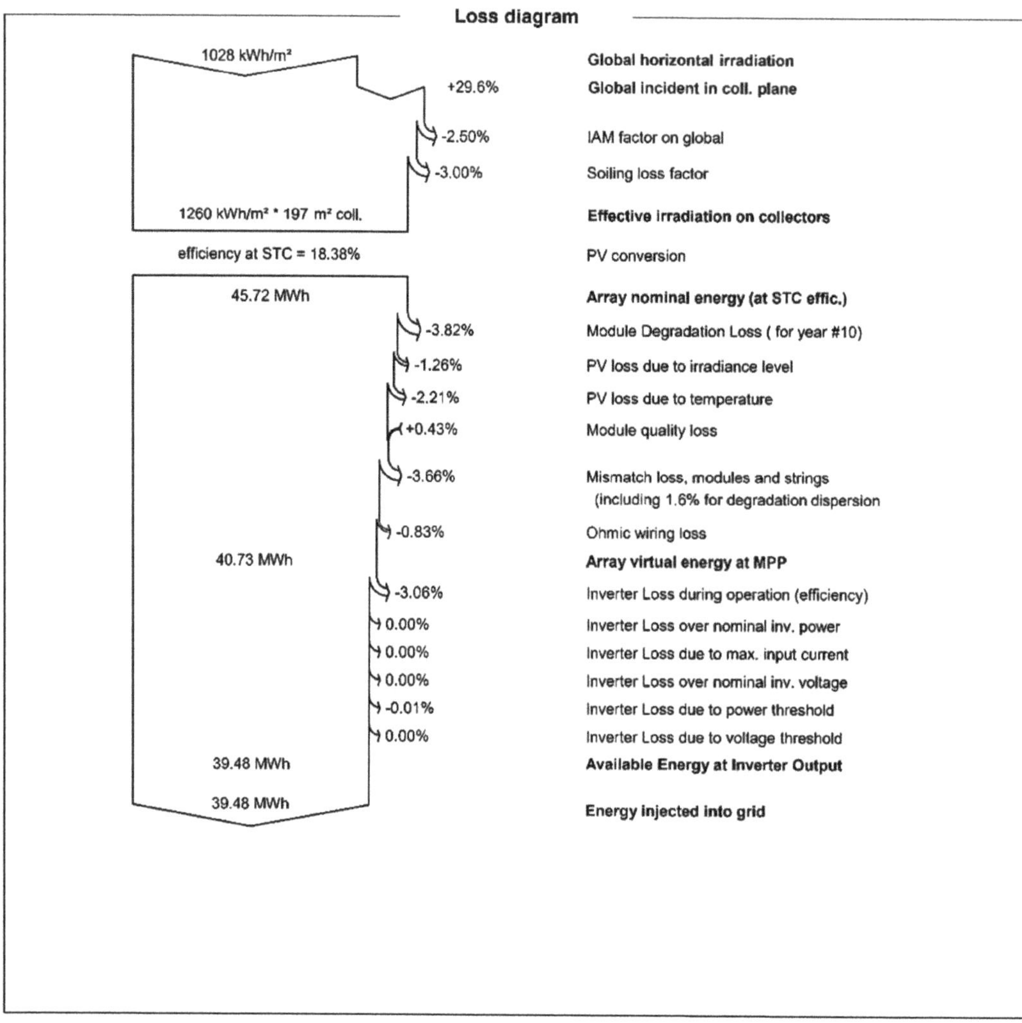

(a)

Figure 12. *Cont.*

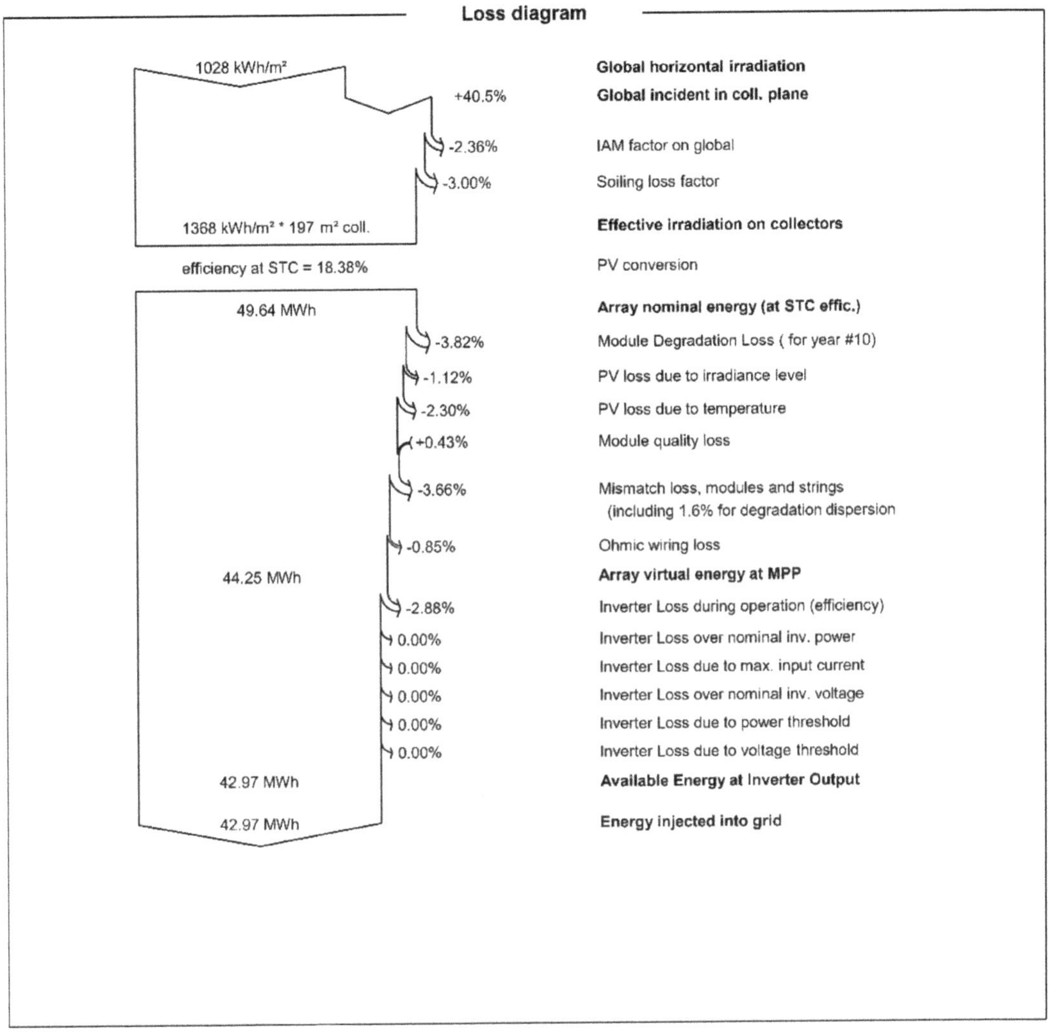

(b)

Figure 12. Loss diagram for (**a**) fixed and (**b**) tracking systems.

4. Conclusions

Russian weather conditions are considered harsh for large-scale solar power plants, especially due to the high negative temperatures during its long winter. Therefore, this study simulated two different designs (i.e., fixed tilted plane and tracking, horizontal axis E–W) to assess their performance, energy loss, and employment potential for potential large-scale solar *PV* development in the Sverdlovsk region of Russia. The following conclusions are made from the study:

- A total of 1028 kWh/m^2 global horizontal solar irradiation was received during the analysis period.
- The fixed tilted plane *PV* panel is expected to export 39.48 MWh of electricity to the grid, while the tracking system will export 42.97 MWh for a year. The months of March

to August are the periods within which much of the electricity will be generated due to the high solar irradiations recorded during those periods.

- A *PR* of 82.3% and 82.6% were obtained for the fixed and tracking systems, respectively. It can clearly be seen that the *PR* is inversely proportional to the temperature of the *PV* module; *PR* values in the winter season are higher than those in summer.
- According to the results from the simulations, the expected annual production probability for the fixed *PV* module for the P50, P75, and P90 is39.68 MWh, 37.72 MWh, and 35.94 MWh, respectively, with a variability of 2.91 MWh. In the case of the tracking *PV* module, the projected annual production probability for the P50, P75, and P90 is 43.18 MWh, 41.05 MWh, and 39.12 MWh, respectively, with a variability of 3.17 MWh.
- The *PV* array losses for both orientations, i.e., fixed and tracking, are expected to be 15.1% and 14.9%, respectively. The months of May to August recorded the highest array losses due to the high temperatures during those periods.
- According to the mathematical computations, we obtained 0.0011 persons/year for the fixed *PV* module, while 0.0012 persons/year were obtained for the tracking system. Therefore, assuming both modules operated for a lifetime of 25 years, then the fixed module will have an employment potential of 0.028 persons against 0.030 persons for the tracking *PV* module.

Russia is currently implementing measures that seek to promote, develop, and use its various RE resources to help cut down its GHG emissions. Therefore, this study is expected to serve as a reference material for government, interested parties, individuals, and policymakers in relation to small and large-scale solar *PV* development, using the performance of the two designs. The data provided in this study give useful information on the possible net energy output of such systems. Future studies can assess the economic viability of such projects on large-scale levels and possibly integrate them with other renewable energy resources, such as wind, to assess their viability for rural and far-to-reach areas in Russia. An environmental impact assessment can also be assessed for large-scale *PV* projects in the country to know the potential reduction in GHG emissions that such projects come with. Similarly, future studies can conduct experimental research to compare actual results and the simulated results to give a real understanding of the performance of *PV* modules under Russian weather conditions. This is because it is highly possible that the simulated results may differ from that conducted under real environmental conditions. It is also recommended to evaluate the performance of different *PV* technologies under Russian weather conditions. This would provide critical information on the optimum technology for Russian weather.

Author Contributions: Conceptualization, E.B.A.; methodology, E.B.A.; software, E.B.A.; validation, E.B.A., U.M., S.K., M.S., E.E. and T.S.A.; formal analysis, E.B.A., U.M., S.K. and T.S.A.; investigation, E.B.A.; resources, E.B.A.; data curation, E.B.A., U.M., S.K. and T.S.A.; writing—original draft preparation, E.B.A.; writing—review and editing, E.B.A., U.M., S.K., M.S. and E.E.; visualization, E.B.A., U.M. and S.K.; project administration, E.B.A., U.M., S.K.; funding acquisition, E.B.A., S.K., M.S. and E.E. All authors have read and agreed to the published version of the manuscript.

Funding: This research was funded by Taif University Researchers Supporting Project number (TURSP-2020/61), Taif University, Taif, Saudi Arabia.

Institutional Review Board Statement: Not applicable.

Informed Consent Statement: Not applicable.

Data Availability Statement: Data used, and their sources are provided in the text.

Acknowledgments: The authors would like to thank Cardiff University/School of Engineering for accepting to pay the APC toward publishing this paper. In addition, the authors would like to acknowledge the financial support received from Taif University Researchers Supporting Project Number (TURSP-2020/61), Taif University, Taif, Saudi Arabia.

Conflicts of Interest: The authors declare no conflict of interest.

Abbreviations

AC	Alternating current
CF	Capacity factor
CIS	Copper indium selenium
DC	Direct current
E–W	East to West
GHG	Greenhouse gas
GHI	Global horizontal irradiation
IAM	Incidence angle modifier
MPP	Maximum Power Point
PR	Performance ratio
STCs	Standard test conditions

References

1. Ahmed, N.; Naveed Khan, A.; Ahmed, N.; Aslam, A.; Imran, K.; Sajid, M.B.; Waqas, A. Techno-Economic Potential Assessment of Mega Scale Grid-Connected PV Power Plant in Five Climate Zones of Pakistan. *Energy Convers. Manag.* **2021**, *237*, 114097. [CrossRef]
2. Abdin, Z.; Mérida, W. Hybrid Energy Systems for Off-Grid Power Supply and Hydrogen Production Based on Renewable Energy: A Techno-Economic Analysis. *Energy Convers. Manag.* **2019**, *196*, 1068–1079. [CrossRef]
3. IPCC. Global Warming of 1.5 °C. 2018. Available online: https://www.ipcc.ch/sr15/download/ (accessed on 2 December 2021).
4. Yaqoob, S.J.; Motahhir, S.; Agyekum, E.B. A New Model for a Photovoltaic Panel Using Proteus Software Tool under Arbitrary Environmental Conditions. *J. Clean. Prod.* **2022**, *333*, 130074. [CrossRef]
5. Agyekum, E.B.; Nutakor, C.; Agwa, A.M.; Kamel, S. A Critical Review of Renewable Hydrogen Production Methods: Factors Affecting Their Scale-Up and Its Role in Future Energy Generation. *Membranes* **2022**, *12*, 173. [CrossRef]
6. Agyekum, E.B. Techno-Economic Comparative Analysis of Solar Photovoltaic Power Systems with and without Storage Systems in Three Different Climatic Regions, Ghana. *Sustain. Energy Technol. Assess.* **2021**, *43*, 100906. [CrossRef]
7. Hassan, Q.; Jaszczur, M.; Przenzak, E. Mathematical Model for the Power Generation from Arbitrarily Oriented Photovoltaic Panel. *E3S Web Conf.* **2017**, *14*, 01028. [CrossRef]
8. Malvoni, M.; Leggieri, A.; Maggiotto, G.; Congedo, P.M.; De Giorgi, M.G. Long Term Performance, Losses and Efficiency Analysis of a 960kWp Photovoltaic System in the Mediterranean Climate. *Energy Convers. Manag.* **2017**, *145*, 169–181. [CrossRef]
9. Kumar, N.M.; Gupta, R.P.; Mathew, M.; Jayakumar, A.; Singh, N.K. Performance, Energy Loss, and Degradation Prediction of Roof-Integrated Crystalline Solar PV System Installed in Northern India. *Case Stud. Therm. Eng.* **2019**, *13*, 100409. [CrossRef]
10. Ramanan, P.; Kalidasa, M.K.; Karthick, A. Performance Analysis and Energy Metrics of Grid-Connected Photovoltaic Systems. *Energy Sustain. Dev.* **2019**, *52*, 104–115. [CrossRef]
11. Ameur, A.; Sekkat, A.; Loudiyi, K.; Aggour, M. Performance Evaluation of Different Photovoltaic Technologies in the Region of Ifrane, Morocco. *Energy Sustain. Dev.* **2019**, *52*, 96–103. [CrossRef]
12. Dahmoun, M.E.-H.; Bekkouche, B.; Sudhakar, K.; Guezgouz, M.; Chenafi, A.; Chaouch, A. Performance Evaluation and Analysis of Grid-Tied Large Scale PV Plant in Algeria. *Energy Sustain. Dev.* **2021**, *61*, 181–195. [CrossRef]
13. Kittner, N.; Gheewala, S.H.; Kamens, R.M. An Environmental Life Cycle Comparison of Single-Crystalline and Amorphous-Silicon Thin-Film Photovoltaic Systems in Thailand. *Energy Sustain. Dev.* **2013**, *17*, 605–614. [CrossRef]
14. Padmavathi, K.; Daniel, S.A. Performance Analysis of a 3MWp Grid Connected Solar Photovoltaic Power Plant in India. *Energy Sustain. Dev.* **2013**, *17*, 615–625. [CrossRef]
15. Radue, C.; van Dyk, E.E. A Comparison of Degradation in Three Amorphous Silicon PV Module Technologies. *Sol. Energy Mater. Sol. Cells* **2010**, *94*, 617–622. [CrossRef]
16. Kymakis, E.; Kalykakis, S.; Papazoglou, T.M. Performance Analysis of a Grid Connected Photovoltaic Park on the Island of Crete. *Energy Convers. Manag.* **2009**, *50*, 433–438. [CrossRef]
17. Belmahdi, B.; Bouardi, A.E. Solar Potential Assessment Using PVsyst Software in the Northern Zone of Morocco. *Procedia Manuf.* **2020**, *46*, 738–745. [CrossRef]
18. Kapoor, S.; Sharma, A.K.; Porwal, D. Design and Simulation of 60kWp Solar On-Grid System for Rural Area in Uttar-Pradesh by "PVsyst". *J. Phys. Conf. Ser.* **2021**, *2070*, 012147. [CrossRef]
19. Dellosa, J.T.; Panes, M.J.C.; Espina, R.U. Techno-Economic Analysis of a 5 MWp Solar Photovoltaic System in the Philippines. In Proceedings of the 2021 IEEE International Conference on Environment and Electrical Engineering and 2021 IEEE Industrial and Commercial Power Systems Europe (EEEIC/I CPS Europe), Bari, Italy, 7–10 September 2021; pp. 1–6.
20. Chabachi, S.; Necaibia, A.; Abdelkhalek, O.; Bouraiou, A.; Ziane, A.; Hamouda, M. Performance Analysis of an Experimental and Simulated Grid Connected Photovoltaic System in Southwest Algeria. *Int. J. Energy Environ. Eng.* **2022**. [CrossRef]
21. Yadav, B.K.; Rauniyar, P.K.; Sudhakar, K.; Bajracharya, T.R.; Priya, S.S. Sustainable Green Campus in NEPAL: 3E Analysis. *Int. J. Low-Carbon Technol.* **2021**, *16*, 531–542. [CrossRef]

22. Haffaf, A.; Lakdja, F.; Ould Abdeslam, D.; Meziane, R. Monitoring, Measured and Simulated Performance Analysis of a 2.4 KWp Grid-Connected PV System Installed on the Mulhouse Campus, France. *Energy Sustain. Dev.* **2021**, *62*, 44–55. [CrossRef]
23. Chandel, R.; Chandel, S.S. Performance Analysis Outcome of a 19-MWp Commercial Solar Photovoltaic Plant with Fixed-Tilt, Adjustable-Tilt, and Solar Tracking Configurations. *Prog. Photovolt. Res. Appl.* **2022**, *30*, 27–48. [CrossRef]
24. Kumar, P.; Pal, N.; Sharma, H. Performance Analysis and Evaluation of 10 KWp Solar Photovoltaic Array for Remote Islands of Andaman and Nicobar. *Sustain. Energy Technol. Assess.* **2020**, *42*, 100889. [CrossRef]
25. Korsavi, S.S.; Zomorodian, Z.S.; Tahsildoost, M. Energy and Economic Performance of Rooftop PV Panels in the Hot and Dry Climate of Iran. *J. Clean. Prod.* **2018**, *174*, 1204–1214. [CrossRef]
26. Omar, M.A.; Mahmoud, M.M. Grid Connected PV- Home Systems in Palestine: A Review on Technical Performance, Effects and Economic Feasibility. *Renew. Sustain. Energy Rev.* **2018**, *82*, 2490–2497. [CrossRef]
27. Elibol, E.; Özmen, Ö.T.; Tutkun, N.; Köysal, O. Outdoor Performance Analysis of Different PV Panel Types. *Renew. Sustain. Energy Rev.* **2017**, *67*, 651–661. [CrossRef]
28. Agyekum, E.B.; Kumar, N.M.; Mehmood, U.; Panjwani, M.K.; Haes Alhelou, H.; Adebayo, T.S.; Al-Hinai, A. Decarbonize Russia—A Best–Worst Method Approach for Assessing the Renewable Energy Potentials, Opportunities and Challenges. *Energy Rep.* **2021**, *7*, 4498–4515. [CrossRef]
29. Statista. Russia: Installed Electricity Generation Capacity by Source. 2021. Available online: https://www.statista.com/statistics/1027465/russia-installed-electricity-generating-capacity-by-source/ (accessed on 27 January 2022).
30. IRENA. *REMAP 2030 Renewable Energy Prospects for the Russian Federation*; IRENA: Abu Dhabi, United Arab Emirates, 2017; Available online: www.irena.org/remap (accessed on 15 December 2021).
31. Institute of Energy Strategy. *Energy Strategy of Russia for the Period Up to 2030*; Institute of Energy Strategy: Moscow, Russia, 2010; p. 174.
32. de Lima, L.C.; de Araújo Ferreira, L.; de Lima Morais, F.H.B. Performance Analysis of a Grid Connected Photovoltaic System in Northeastern Brazil. *Energy Sustain. Dev.* **2017**, *37*, 79–85. [CrossRef]
33. Chandrika, V.; Thalib, M.M.; Karthick, A.; Sathyamurthy, R.; Manokar, A.M.; Subramaniam, U.; Stalin, B. Performance Assessment of Free Standing and Building Integrated Grid Connected Photovoltaic System for Southern Part of India. *Build. Serv. Eng. Res. Technol.* **2021**, *42*, 237–248. [CrossRef]
34. Ramanan, P.; Kalidasa Murugavel, K.; Karthick, A.; Sudhakar, K. Performance Evaluation of Building-Integrated Photovoltaic Systems for Residential Buildings in Southern India. *Build. Serv. Eng. Res. Technol.* **2020**, *41*, 492–506. [CrossRef]
35. Oliveira-Pinto, S.; Stokkermans, J. Assessment of the Potential of Different Floating Solar Technologies—Overview and Analysis of Different Case Studies. *Energy Convers. Manag.* **2020**, *211*, 112747. [CrossRef]
36. Elhadj Sidi, C.E.B.; Ndiaye, M.L.; El Bah, M.; Mbodji, A.; Ndiaye, A.; Ndiaye, P.A. Performance Analysis of the First Large-Scale (15MWp) Grid-Connected Photovoltaic Plant in Mauritania. *Energy Convers. Manag.* **2016**, *119*, 411–421. [CrossRef]
37. Li, C. Comparative Performance Analysis of Grid-Connected PV Power Systems with Different PV Technologies in the Hot Summer and Cold Winter Zone. *Int. J. Photoenergy* **2018**, *2018*, 8307563. [CrossRef]
38. AL-Rasheedi, M.; Gueymard, C.A.; Al-Khayat, M.; Ismail, A.; Lee, J.A.; Al-Duaj, H. Performance Evaluation of a Utility-Scale Dual-Technology Photovoltaic Power Plant at the Shagaya Renewable Energy Park in Kuwait. *Renew. Sustain. Energy Rev.* **2020**, *133*, 110139. [CrossRef]
39. Agyekum, E.B.; PraveenKumar, S.; Alwan, N.T.; Velkin, V.I.; Shcheklein, S.E. Effect of Dual Surface Cooling of Solar Photovoltaic Panel on the Efficiency of the Module: Experimental Investigation. *Heliyon* **2021**, *7*, e07920. [CrossRef] [PubMed]
40. Dubey, S.; Sarvaiya, J.N.; Seshadri, B. Temperature Dependent Photovoltaic (PV) Efficiency and Its Effect on PV Production in the World—A Review. *Energy Procedia* **2013**, *33*, 311–321. [CrossRef]
41. Agyekum, E.B.; Afornu, B.K.; Ansah, M.N.S. Effect of Solar Tracking on the Economic Viability of a Large-Scale PV Power Plant. *Environ. Clim. Technol.* **2020**, *24*, 55–65. [CrossRef]
42. Hafez, A.Z.; Yousef, A.M.; Harag, N.M. Solar Tracking Systems: Technologies and Trackers Drive Types–A Review. *Renew. Sustain. Energy Rev.* **2018**, *91*, 754–782. [CrossRef]
43. Eke, R.; Demircan, H. Performance Analysis of a Multi Crystalline Si Photovoltaic Module under Mugla Climatic Conditions in Turkey. *Energy Convers. Manag.* **2013**, *65*, 580–586. [CrossRef]
44. Okello, D.; van Dyk, E.E.; Vorster, F.J. Analysis of Measured and Simulated Performance Data of a 3.2kWp Grid-Connected PV System in Port Elizabeth, South Africa. *Energy Convers. Manag.* **2015**, *100*, 10–15. [CrossRef]
45. Agyekum, E.B.; PraveenKumar, S.; Alwan, N.T.; Velkin, V.I.; Shcheklein, S.E.; Yaqoob, S.J. Experimental Investigation of the Effect of a Combination of Active and Passive Cooling Mechanism on the Thermal Characteristics and Efficiency of Solar PV Module. *Inventions* **2021**, *6*, 63. [CrossRef]
46. Vidal, H.; Rivera, M.; Wheeler, P.; Vicencio, N. The Analysis Performance of a Grid-Connected 8.2 KWp Photovoltaic System in the Patagonia Region. *Sustainability* **2020**, *12*, 9227. [CrossRef]
47. PVsyst Project Design > P50–P90 Evaluations. Available online: https://www.pvsyst.com/help/p50_p90evaluations.htm (accessed on 31 December 2021).
48. PVsyst Project Design > Array and System Losses > Array Losses, General Considerations. Available online: https://www.pvsyst.com/help/array_losses_general.htm (accessed on 1 January 2022).

49. Koirala, B.P.; Sahan, B.; Henze, N. Study on MPP Mismatch Losses in Photovoltaic Applications. In Proceedings of the European Photovoltaic Solar Energy Conference and Exhibition (EU PVSEC), Hamburg, Germany, 21–25 September 2009; pp. 3727–3733.
50. Tapia, M. *Evaluation of Performance Models against Actual Performance of Grid Connected PV Systems*; Carl von Ossietzky Universität Oldenburg, Institute of Physics: Oldenburg, Germany, 2013. Available online: http://oops.uni-oldenburg.de/2433/7/Thesis_TapiaM.pdf (accessed on 2 January 2022).
51. Herrmann, W.; Kämmer, S.; Yusufoglu, U. Circuit Losses in PV Arrays Caused by Electrical Mismatch of PV Modules—Impacts of Temperature Gradients and a Variation of Irradiance. In Proceedings of the 28th European Photovoltaic Solar Energy Conference and Exhibition, Villepinte, France, 30 September–4 October 2013; pp. 4127–4131.
52. Shiva Kumar, B.; Sudhakar, K. Performance Evaluation of 10 MW Grid Connected Solar Photovoltaic Power Plant in India. *Energy Rep.* **2015**, *1*, 184–192. [CrossRef]
53. Arora, R.; Arora, R.; Sridhara, S.N. Performance Assessment of 186 KWp Grid Interactive Solar Photovoltaic Plant in Northern India. *Int. J. Ambient. Energy* **2019**, *43*, 128–141. [CrossRef]
54. Baqir, M.; Channi, H.K. Analysis and Design of Solar PV System Using Pvsyst Software. *Mater. Today Proc.* **2022**, *48*, 1332–1338. [CrossRef]
55. Husain, A.A.F.; Phesal, M.H.A.; Ab Kadir, M.Z.A.; Ungku Amirulddin, U.A. Techno-Economic Analysis of Commercial Size Grid-Connected Rooftop Solar PV Systems in Malaysia under the NEM 3.0 Scheme. *Appl. Sci.* **2021**, *11*, 10118. [CrossRef]
56. Syahindra, K.D.; Ma'arif, S.; Widayat, A.A.; Fauzi, A.F.; Setiawan, E.A. Solar PV System Performance Ratio Evaluation for Electric Vehicles Charging Stations in Transit Oriented Development (TOD) Areas. *E3S Web Conf.* **2021**, *231*, 02002. [CrossRef]
57. Kumar, R.; Rajoria, C.S.; Sharma, A.; Suhag, S. Design and Simulation of Standalone Solar PV System Using PVsyst Software: A Case Study. *Mater. Today Proc.* **2021**, *46*, 5322–5328. [CrossRef]
58. Duong, M.Q.; Tran, N.T.N.; Sava, G.N.; Tanasiev, V. Design, Performance and Economic Efficiency Analysis of the Photovoltaic Rooftop System. *Rev. Roum. Sci. Tech. Électrotech. Énerg.* **2019**, *64*, 229–234.
59. Satish, M.; Santhosh, S.; Yadav, A. Simulation of a Dubai Based 200 KW Power Plant Using PVsyst Software. In Proceedings of the 2020 7th International Conference on Signal Processing and Integrated Networks (SPIN), Noida, India, 27–28 February 2020; pp. 824–827.
60. Pietruszko, S.M.; Gradzki, M. Performance of a Grid Connected Small PV System in Poland. *Appl. Energy* **2003**, *74*, 177–184. [CrossRef]
61. Adaramola, M.S.; Vågnes, E.E.T. Preliminary Assessment of a Small-Scale Rooftop PV-Grid Tied in Norwegian Climatic Conditions. *Energy Convers. Manag.* **2015**, *90*, 458–465. [CrossRef]
62. Ayompe, L.M.; Duffy, A.; McCormack, S.J.; Conlon, M. Measured Performance of a 1.72 kW Rooftop Grid Connected Photovoltaic System in Ireland. *Energy Convers. Manag.* **2011**, *52*, 816–825. [CrossRef]
63. Milosavljević, D.D.; Pavlović, T.M.; Piršl, D.S. Performance Analysis of A Grid-Connected Solar PV Plant in Niš, Republic of Serbia. *Renew. Sustain. Energy Rev.* **2015**, *44*, 423–435. [CrossRef]
64. Adaramola, M.S. Techno-Economic Analysis of a 2.1kW Rooftop Photovoltaic-Grid-Tied System Based on Actual Performance. *Energy Convers. Manag.* **2015**, *101*, 85–93. [CrossRef]
65. Chauhan, A.; Saini, R.P. Techno-Economic Feasibility Study on Integrated Renewable Energy System for an Isolated Community of India. *Renew. Sustain. Energy Rev.* **2016**, *59*, 388–405. [CrossRef]
66. Baruah, A.; Basu, M.; Amuley, D. Modeling of an Autonomous Hybrid Renewable Energy System for Electrification of a Township: A Case Study for Sikkim, India. *Renew. Sustain. Energy Rev.* **2021**, *135*, 110158. [CrossRef]

sustainability

MDPI

Article

Bidirectional Interface Resonant Converter for Wide Voltage Range Storage Applications

Mouncif Arazi , Alireza Payman, Mamadou Baïlo Camara and Brayima Dakyo *

Electrical Engineering, Faculty of Sciences and Technology, University Le Havre Normandie, 76600 Le Havre, France; moncef.arazi.10@gmail.com (M.A.); paymana@univ-lehavre.fr (A.P.); mamadou-bailo.camara@univ-lehavre.fr (M.B.C.)
* Correspondence: brayima.dakyo@univ-lehavre.fr

Abstract: In this paper, a bidirectional zero voltage switching (ZVS) resonant converter with narrow control frequency deviation is proposed. Wide input–output voltage range applications, such as flywheel or supercapacitors storage units are targeted. Due to symmetrical topology of resonant circuit interfaces, the proposed converter has similar behavior in bidirectional operating mode. We call it Dual Active Bridge Converter (DABC). The proposal topology of the converter is subjected to multi resonant circuits which make it necessary to study with multiscale approaches. Thus, first harmonic approximation and use of selective per unit parameters are established in (2) Methods. Then, the forward direction and backward direction of power flux exchange are detailed according to switching sequences. Switching frequency control must be completed within a narrow range. So, the frequency range deterministic parameters are emphasized in the design procedure in (3) Methods. A narrow range of switching frequency and a wide range voltage control must be ensured to suit for energy storage units, power electronic devices capabilities and electromagnetic compatibility. A 3 kW test bench is used to validate operation principles and to proof success of the developed design procedure. The interest of proposed converter is compared to other solutions from the literature in (4) Results.

Keywords: bidirectional resonant converter; zero voltage switching; zero current switching; wide input voltage range; power losses

Citation: Arazi, M.; Payman, A.; Camara, M.B.; Dakyo, B. Bidirectional Interface Resonant Converter for Wide Voltage Range Storage Applications. *Sustainability* **2022**, *14*, 377. https://doi.org/10.3390/su14010377

Academic Editor: Antonio Caggiano

Received: 16 November 2021
Accepted: 27 December 2021
Published: 30 December 2021

1. Introduction

In the last decade, the converters used in electric vehicles, smart grids and renewable energies applications have had significant progress in term of electric performances [1,2]. In numerous cases, the performances are obtained by means of the integration of Energy Storage Systems (ESS) such as batteries, supercapacitors or flywheels in DC-bus using bidirectional DC/DC converters. The converters ensure the ESS charge and discharge operations according to the power balance [3]. Different topologies of bidirectional DC/DC converters are proposed in the literature [4–6]. Non-isolated bidirectional topologies were suggested to interface batteries and supercapacitors as described in [7]. Indeed, many specific systems require galvanic isolation especially for safety reasons. The operation with high switching frequencies allows reducing the size of the passive components [8]. Use of hard switching in turn off operations unfortunately increases the converter losses. To overcome this problem, the soft switching technique must be applied, which increases the efficiency of the converter [9]. Phase-shifted Dual Active Bridge (DAB) converter has been widely investigated in the literature [10,11]. In [12,13], the authors propose a phase shift control strategy combined with duty cycle control to extend the soft switching capability of the DAB converter.

The control strategy is based on the synchronized phase-shift control with duty-cycle one. To improve the efficiency of the DAB converter, a resonant circuit is added to the basic topology to obtain a resonant converter [14,15]. Using this technique, the soft switching zone can be significantly extended. The series resonant circuit can ensure Zero Voltage Switching (ZVS) capability of the converter, but it can be subjected to fail for light load or no-load operating conditions [16,17]. Using parallel resonant circuits allows no-load operation, but this would come at the expense of a high resonant current almost independent of the load. Thus, exchanged internal energy and the conduction losses are unnecessarily large [18]. The combination of series and parallel circuits, such as LCC resonant converter [19,20] is expected to offer a converter with better characteristics. This topology can operate from rated power to no-load with small internal energy circulating. However, it causes high switching losses for wide voltage range applications. Among the different resonant DC/DC converters, LLC resonant converter has attracted the attention of researchers [21–24]. This topology can achieve the soft switching for both sides of the converter, and can operate in buck or boost mode. However, it is still a classic series resonant converter in backward mode because the inductances of transformer do not participate in the resonant operation. This reduces significantly the efficiency in the backward mode and penalizes LLC topology for bidirectional current applications [25], such as interfacing batteries and supercapacitors. CLLC resonant converters are proposed in [26–28]. These converters suffer from high reactive power when the switching frequency deviates from the resonant frequency during low load conditions. Modified LLC resonant converters with hybrid control are proposed in [29,30] to reduce the reactive power. However, they are still not suitable for wide voltage range systems and suffer due to high power losses in power semiconductors switch-off operations. A qualitative comparison between the results achieved by some bidirectional resonant converters in literature is presented in Table 1, and related behaviors are described in [31–33].

Table 1. Comparison of bidirectional resonant converters for wide voltage range applications.

	LCLL [31]	LLC-L [32]	CLLC [33]
Range of the voltage gain	0.62~1.125	3~5.3	0.55~1.12
Switching frequency (kHz)	65~120	60~100	40~145
Obtained frequency range percentage	45%	40%	72%
Soft start at resonant frequency	No	No	No
Soft switching in Forward & Backward	ZVS for full load range; High turn off losses at light load	ZVS for full load range; High turn off losses	ZVS for full load range; High turn off losses

The proposed topology aims to obtain the followings performs compared to referenced solutions in Table 1:

- A bidirectional current operation required for energy storage units;
- A switching frequency control within a narrow range for forward and backward operations;
- A turn-off losses and inner circulating energy limitation using a wide range DC-voltage variation; Soft start at the resonant frequency.

The paper is organized as follow: Section 2 gives the topology, the characteristic and operating principle of the novel converter. Modeling and global sizing based on the first harmonic approach is given. The details of operating sequences and per-unit variables definition are given to help for relevant analysis. The design procedure and soft switching (ZVS) performances of the converter are presented in Section 3. The experimental test bench developed in laboratory and results are presented and discussed in Section 4 to show the feasibility of the proposed solution. Concluding remarks are given in Section 5.

2. Operating Principle and Main Characteristics of the Proposed Novel Converter

The topology of the proposed resonant converter is presented in Figure 1. It is composed of two active full bridges for bidirectional operations, a high frequency transformer and a symmetric resonant circuit. This last one includes two series inductances (L_{r1}, L_{r2}) and a parallel inductance-capacitance (L_p, C_p).

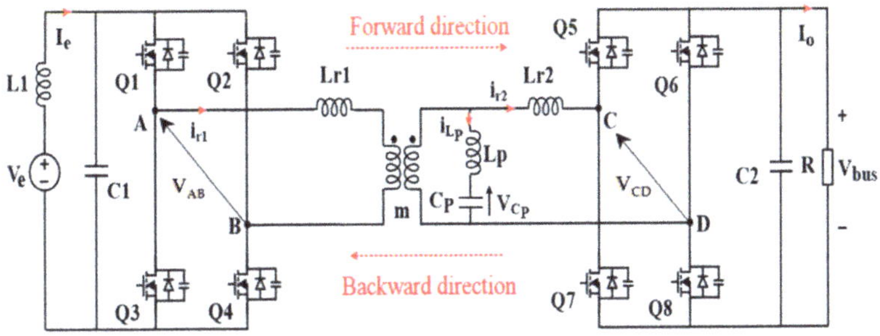

Figure 1. Proposed bidirectional topology of novel resonant converter.

2.1. Forward Mode Analysis

In forward mode, the switches Q1–Q4 are controlled simultaneously using variable frequency PWM signal with a duty cycle of 50%. The switching signals of Q2–Q3 are in complementary logic with Q1–Q4 ones. In this mode, the MOSFET switches of Q5–Q8 and Q6–Q7 at the DC-bus side are turned OFF. So, only the diode rectifier operates in this case. Energy is then forwarded from the DC-source *Ve* to the DC-bus side V_{bus}. The voltage between A and B (V_{AB}) of the Figure 1 is a square waveform. It is analytically expressed by Equation (1), where n is the harmonic frequency order, and f_1 is the fundamental frequency.

$$v_{AB}(t) = \frac{4.V_e}{\pi} \sum_{n=1,3,5\ldots} \frac{1}{n}\sin(n.2\pi f_1.t) \tag{1}$$

First Harmonic Approximation (FHA) is adopted assuming the active energy is mainly attached to the fundamental frequency. This condition is achieved by filtering the current nearby the resonant frequency of the LC–LL circuit. Fundamental of $v_{AB}(t)$ is given in (2). More information about the method can be found in [18].

$$v_{e1}(t) = \frac{4.V_e}{\pi}\sin\left(\omega_1.\mathrm{t}\right) \tag{2}$$

The output voltage $v_{CD}(t)$ of the resonant circuit given in Figure 1 is also assumed as a square wave form, and its fundamental component is given in (3), where φ_v is the phase angle.

$$v_{s1}(t) = \frac{4.V_{bus}}{\pi}\sin(\omega_1.t - \varphi_v) \tag{3}$$

The output impedance through the diodes bridge is reflected by an equivalent resistance R_e expressed in (4). R is a resistance-like load on DC-bus [34], and I_0 is the average current from the rectifier. The FHA model of the converter is presented in Figure 2 and the corresponding analytical model is given in (5).

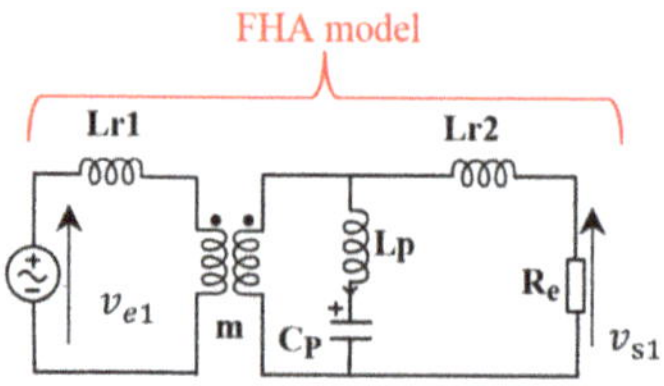

Figure 2. Equivalent model of the proposed converter in forward mode.

$$R_e = \frac{V_{s1}}{I_{s1}} = \frac{8}{\pi^2}\frac{V_{bus}}{I_0} = \frac{8}{\pi^2}.R,\ \omega_1 = \omega \tag{4}$$

$$\begin{bmatrix} V_{e1}(j\omega) \\ I_{e1}(j\omega) \end{bmatrix} = \begin{bmatrix} 1 & j\omega.L_{r1} \\ 0 & 1 \end{bmatrix} \begin{bmatrix} 1 & \frac{j\omega.L_{r2}}{m^2} \\ \frac{m^2}{j\omega.L_p+\frac{1}{j\omega.C_p}} & 1+\frac{j\omega.L_{r2}}{j\omega.L_p+\frac{1}{j\omega.C_p}} \end{bmatrix} \begin{bmatrix} \frac{1}{m} & 0 \\ 0 & m \end{bmatrix} \begin{bmatrix} V_{s1}(j\omega) \\ I_{s1}(j\omega) \end{bmatrix} \tag{5}$$

The voltage gain (G) based on the equivalent model analysis is given in (6).

$$\begin{cases} Z_{in} = j.\omega.L_{r1} + Z_e \\ Z_e = \dfrac{R_e.\left(j.\omega.L_p+\frac{1}{j.\omega.C_p}\right)+\frac{L_{r2}}{C_p}-\omega^2.\ L_{r2}.\ L_p}{m^2\left(\ R_e+j.\omega.L_{r2}+j.\omega.L_p+\frac{1}{j.\omega.C_p}\right)} \end{cases} \tag{6}$$

$$G = \frac{V_{s1}}{m.V_{e1}} = \frac{Z_e}{Z_{in}}\frac{R_e}{R_e + j.\omega.L_{r2}} \tag{7}$$

The FHA response of the proposed circuit can be better analyzed in a per-unit (p.u.) system with the adopted parameters described below:

$$Q = \frac{1}{R_e}\sqrt{\frac{L_p}{C_p}} \tag{8}$$

$$f_n = \frac{f}{f_r} \text{ with } f_r = \frac{1}{2\pi\sqrt{L_p.C_p}} \tag{9}$$

$$\alpha_1 = \frac{m^2.\ L_{r1}}{L_p}\ ;\ \alpha_2 = \frac{L_{r2}}{L_p} \tag{10}$$

where, Q is the quality factor, f_n is the normalized frequency, f_r is the series coupled L_p-C_p resonant frequency. α_1 and α_2 are respectively the normalized values of primary and secondary inductances ratio in per-unit. The resulting output–input DC voltages ratio (Hf) is given in (11).

$$\text{Hf} = \frac{V_{bus}}{V_e} \approx \left| \frac{1}{\left[1+\frac{\alpha_1.f_n{}^2}{f_n{}^2-1}\right] + \left[j.f_n.Q.\left(\alpha_1+\alpha_2+\frac{\alpha_1.\alpha_2.f_n{}^2}{(f_n{}^2-1)}\right)\right]} \right| \tag{11}$$

For $\alpha_1 = \alpha_2 = \alpha$, the voltage ratio Hf versus the normalized frequency f_n calculated for different load conditions (i.e., Q) is shown in Figure 3. The favorite operation zone for ZVS is located in $0.72 < f_n < 1$.

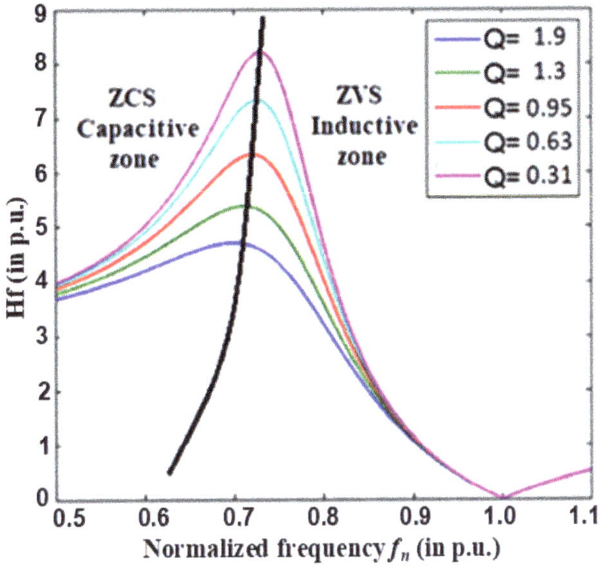

Figure 3. Voltage gain of the converter in forward mode for $\alpha = 1$.

Figure 4 shows the voltage gain versus the normalized frequency f_n for $Q = 1.9$ and different values of inductances ratio α. As displayed in Figure 3, the converter has two potential operation zones (ZCS & ZVS). ZVS can be fully achieved by switching frequency $f_s = f_n$ with $(0.72 * f_r < f_n < f_r)$.

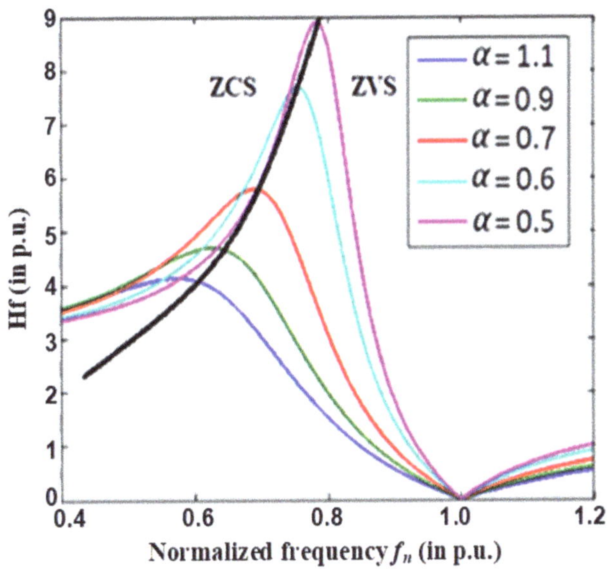

Figure 4. Voltage gain of the converter in forward mode ($Q = 1.9$).

The analysis of Figures 3 and 4 shows the following characteristics: Voltage gain at the resonant frequency is zero; Maximum voltage gain decreases when Q and α increase; and Voltage ratio varies enough in a narrow frequency range.

2.2. Backward Mode Analysis

In backward mode of Figure 1, the switches Q5–Q8 and Q6–Q7 are controlled by variable switching frequency. The MOSFET switches of Q1–Q4 and Q2–Q3 are here turned OFF. In this mode, the energy is transferred from the DC-bus to DC-source which is assumed to be reversible (supercapacitors energy storage unit). The equivalent model of the converter in backward mode is presented in Figure 5 and analytical model is given in (12).

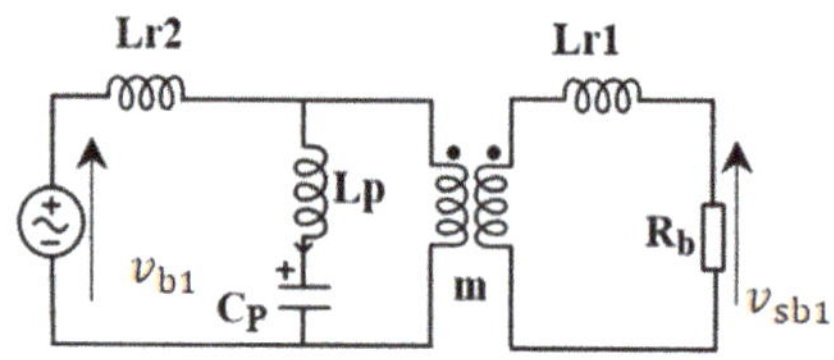

Figure 5. Equivalent model of the converter in backward mode.

$$\begin{bmatrix} V_{b1}(j\omega) \\ I_{b1}(j\omega) \end{bmatrix} = \begin{bmatrix} m & 0 \\ 0 & \frac{1}{m} \end{bmatrix} \begin{bmatrix} 1 & j\omega.L_{r2} \\ \frac{1}{j\omega.L_p + \frac{1}{j\omega.C_p}} & 1 + \frac{j\omega.L_{r2}}{j\omega.L_p + \frac{1}{j\omega.C_p}} \end{bmatrix} \begin{bmatrix} 1 & j\omega.L_{r1} \\ 0 & 1 \end{bmatrix} \begin{bmatrix} V_{sb1}(j\omega) \\ I_{sb1}(j\omega) \end{bmatrix} \tag{12}$$

The input impedance of the resonant circuit is described in (13). The quality factor Q_b is given in (14), and the voltage gain in backward mode (Hb) is presented in (15).

$$\begin{cases} Z_i = j\omega.L_{r2} + Z' \\ Z' = \dfrac{m^2 R_b.\left(j\omega.L_p + \frac{1}{j\omega.C_p}\right) + m^2\left(\frac{L_{r1}}{C_p} - \omega_s{}^2.\ L_{r1}.\ L_p\right)}{\left(R_b + j\omega.L_{r1} + j\omega.L_p + \frac{1}{j\omega.C_p}\right)} \end{cases} \tag{13}$$

$$Q_b = \frac{m^2}{R_b}\sqrt{\frac{L_p}{C_p}} \tag{14}$$

$$Hb = \frac{V_e}{V_{bus}} = \left| \frac{1}{\left[1 + \frac{\alpha.f_n{}^2}{f_n{}^2 - 1}\right] + \left[j.f_n.Q_b.\left(2.\alpha + \frac{\alpha^2.f_n{}^2}{(f_n{}^2 - 1)}\right)\right]} \right| \tag{15}$$

Figure 6 shows the voltage gain of the backward mode versus f_n for different values of the quality factor Q_b. We can see that the converter in the backward mode has exactly the same behavior as the forward mode. This means similar operation capabilities in the same restricted frequency domain. If the same tuned values of $\alpha_1 = \alpha_2 = \alpha$ is assumed, only the quality factor Q_b will determine the operation frequency f_s within $0.72 * f_r < f_1 < f_r$.

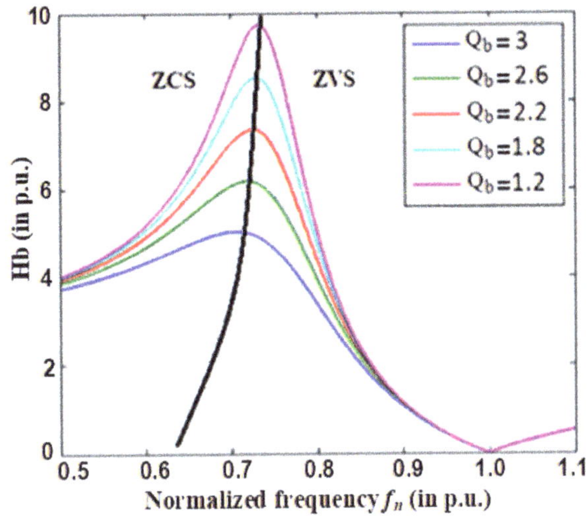

Figure 6. Voltage gain of the proposed converter in backward mode.

2.3. Operation Principles Analysis

The symmetric bidirectional resonant circuit ensures the same behavior of the converter in both forward and backward modes. Thus, only the principle of the forward operating mode will be discussed in this subsection. Operating waveforms in forward mode are illustrated in Figure 7. Switching conditions are also seen.

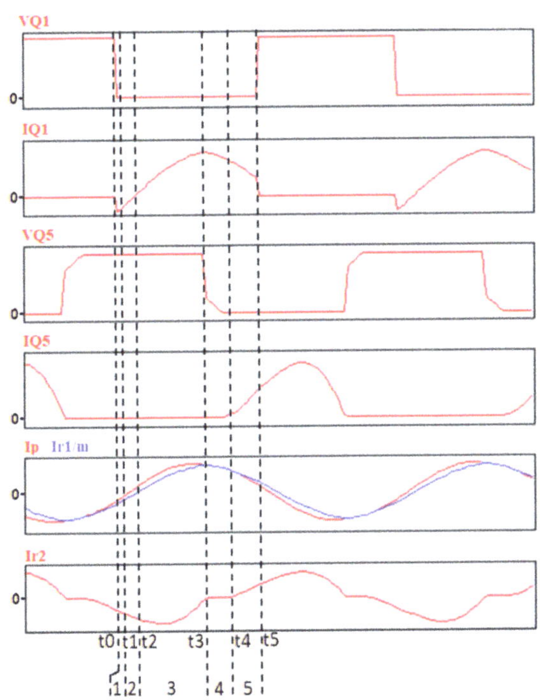

Figure 7. Waveforms of the converter in forward mode.

Referring to this figure, the operating states of the converter in forward mode can be divided into five states for a half switching period. The operating sequences are gathered in Figure 8, which shows the active circuits' paths participating to power exchanges.

State 1 ($t_0 < t < t_1$): This mode begins at t_0 when Q_2 and Q_3 are turned off. The current in the inductance L_{r1} discharges the parallel capacitor of Q_1 and Q_4 and charges the parallel capacitor of Q_2 and Q_3. This state ends when the voltages across Q_1 and Q_4 (V_{Q1}, V_{Q4}) go to zero and those of Q_2 and Q_3 (V_{Q2}, V_{Q3}) reach the input voltage.

State 2 ($t_1 < t < t_2$): After the full discharge of the parallel capacitors of Q_1 and Q_4, the negative resonant current i_{r1} flows through the diodes of Q_1 and Q_4. The energy stored in the resonant circuit is fed back to the input side, to discharge the capacitor C_p. The expressions of the currents and voltages of resonant components in this mode are defined as follows:

$$i_{r1}(t) = \frac{V_i - V_{Lp}(t_1) - V_{Cp}(t_1)}{L_{r1}}(t - t_1) + i_{r1}(t_1) \tag{16}$$

$$i_{Lp}(t) = \left[\frac{-V_s}{m} - V_{Cp}(t_1) + V_{Lr2}(t_1)\right]\sqrt{\frac{C_p}{L_p}}m^2 \sin[\omega_r(t - t_1)] + i_{Lp}(t_1)\cos[\omega_r(t - t_1)] \tag{17}$$

$$V_{Cp}(t) = \frac{-V_s}{m} - V_{Lr2}(t_1) + \frac{1}{m^2}\sqrt{\frac{L_p}{C_p}}\, i_{Lp}(t_1)\sin[\omega_r(t - t_1)] + \left[\frac{V_s}{m} + V_{Cp}(t_1) - V_{Lr2}(t_1)\right]\cos[\omega_r(t - t_1)] \tag{18}$$

$$i_{r2}(t) = \frac{m^2\left[\frac{-V_s}{m} - V_{Lp}(t_1) - V_{Cp}(t_1)\right]}{L_{r2}}(t - t_1) + i_{r2}(t_1) \tag{19}$$

where,

$$\omega_r = \frac{1}{\sqrt{L_p.C_p}}$$

State 3 ($t_2 < t < t_3$): At t_2, the resonant current i_{r1} changes the direction and the current flowing trough Q1 and Q4 becomes positive. Thus, Q1 and Q4 turn on with ZVS. The current i_{r2} is negative. Hence, the diodes of Q6 and Q7 are conducting to transfer the energy from the DC-source to DC-bus side. The expressions of different resonant currents and voltages in this mode are given as follows:

$$i_{r1}(t) = \frac{V_i + V_{Lp}(t_1) - V_{Cp}(t_1)}{L_{r1}}(t - t_2) + i_{r1}(t_2) \tag{20}$$

$$i_{Lp}(t) = \left[\frac{-V_s}{m} - V_{Cp}(t_1) - V_{Lr2}(t_1)\right]\sqrt{\frac{C_p}{L_p}}m^2 \sin[\omega_r(t - t_2)] + i_{Lp}(t_2)\cos[\omega_r(t - t_2)] \tag{21}$$

$$V_{Cp}(t) = \frac{-V_s}{m} - V_{Lr2}(t_2) + \frac{1}{m^2}\sqrt{\frac{L_p}{C_p}}\, i_{Lp}(t_2)\sin[\omega_r(t - t_2)] + \left[\frac{V_s}{m} + V_{Cp}(t_2) + V_{Lr2}(t_2)\right]\cos[\omega_r(t - t_2)] \tag{22}$$

$$i_{r2}(t) = \frac{m^2\left[\frac{-V_s}{m} - V_{Lp}(t_1) - V_{Cp}(t_1)\right]}{L_{r2}}(t - t_2) + i_{r2}(t_2) \tag{23}$$

State 4 ($t_3 < t < t_4$): This state begins when the resonant current reaches the magnetizing current. The secondary current i_{r2} becomes zero and the capacitor C_2 supplies energy to the load. In the primary side, the resonant current i_{r1} charges the resonant circuit. The expressions of the resonant currents and voltages are given in (24) and (25), respectively.

$$i_{r1}(t) = i_{Lp}(t) = [V_i - V_{Cp}(t_3)]\sqrt{\frac{C'_p}{L'_p + L_{r1}}}\sin[\omega'_r(t - t_2)] + i_{r1}(t_3)\cos[\omega'_r(t - t_3)] \tag{24}$$

$$V_{Cp}(t) = V_i + \sqrt{\frac{L'_p + L_{r1}}{C'_p}}\, i_{r1}(t_3) \sin[\,\omega'_r(t - t_3)] + [-V_i - V_{Cp}(t_3)]\ \cos[\,\omega'_r(t - t_3)] \tag{25}$$

where,

$$\omega'_r = \frac{1}{\sqrt{(L_{r1} + L'_p).C'_p}}\,;\ C'_p = C_p.m^2;\ L'_p = \frac{L_p}{m^2}$$

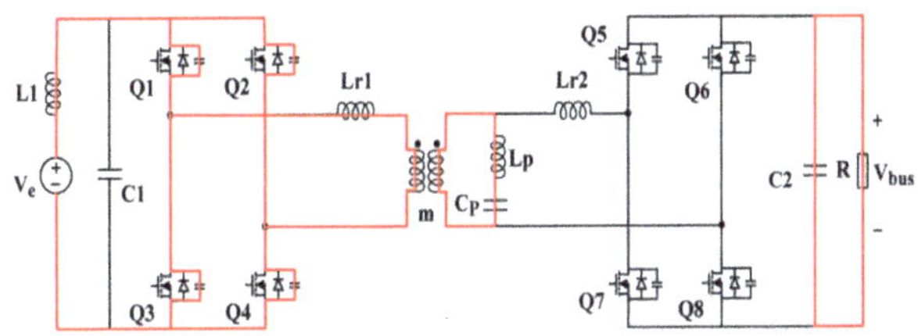

State 1

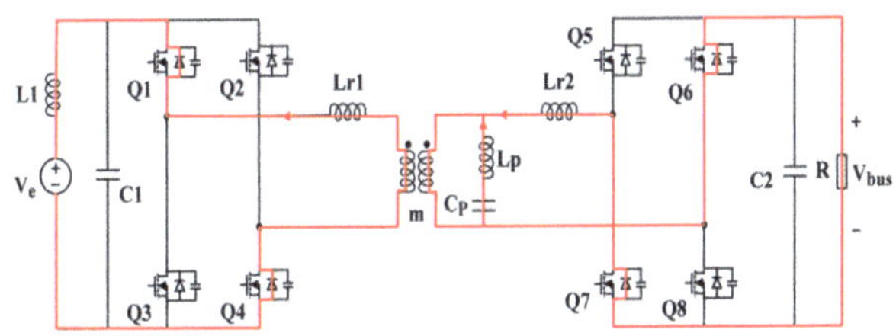

State 2

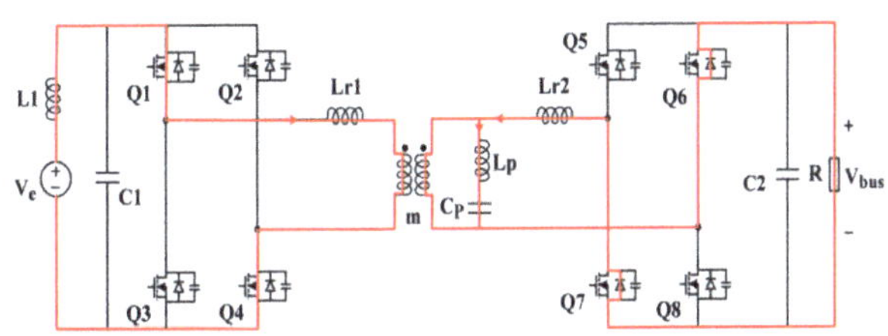

State 3

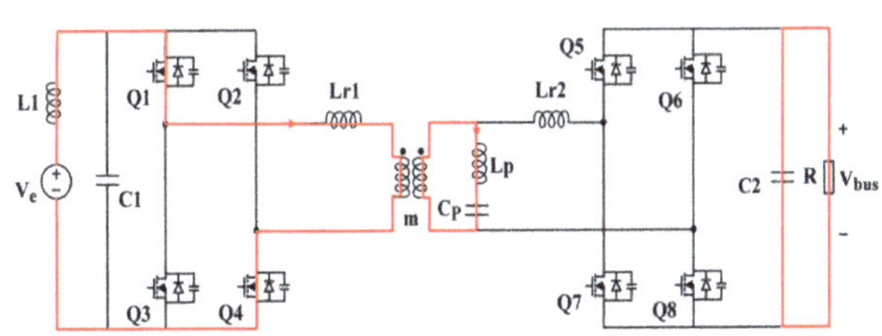

State 4

Figure 8. *Cont.*

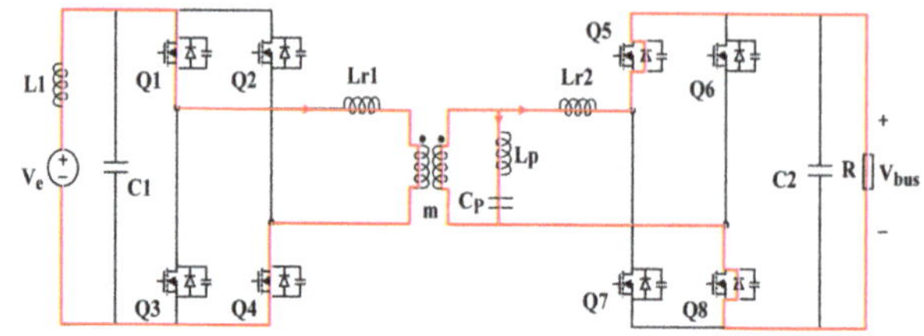

State 5

Figure 8. Operating states of the converter in forward mode.

State 5 ($t_4 < t < t_5$): At t_4, the current i_{r2} changes the sign and it becomes positive. Thus, the diodes of Q_6 and Q_7 turn *off* with ZCS and the body diodes of Q_5 and Q_8 turn *on* to deliver the energy to the load. The resonant currents and voltage in this mode are given by the following equations:

$$i_{r1}(t) = \frac{-V_i - V_{Lp}(t_4) + V_{Cp}(t_4)}{L_{r1}}(t - t_4) + i_{r1}(t_4) \quad (26)$$

$$i_{Lp}(t) = \left[\frac{V_s}{m} - V_{Cp}(t_4) + V_{Lr2}(t_4)\right]\sqrt{\frac{C_p}{L_p}}m^2 \sin[\omega_r(t - t_4)] + i_{Lp}(t_4)\cos[\omega_r(t - t_4)] \quad (27)$$

$$V_{Cp}(t) = \frac{V_s}{m} + V_{Lr2}(t_4) + \frac{1}{m^2}\sqrt{\frac{L_p}{C_p}}\, i_{Lp}(t_4)\sin[\omega_r(t - t_4)] + \left[\frac{-V_s}{m} - V_{Cp}(t_1) + V_{Lr2}(t_1)\right] \cos[\omega_r(t - t_4)] \quad (28)$$

$$i_{r2}(t) = \frac{m^2\left[\frac{-V_s}{m} + V_{Lp}(t_1) + V_{Cp}(t_1)\right]}{L_{r2}}(t - t_4) + i_{r2}(t_4) \quad (29)$$

The operating principles analysis for the next half cycle is the same as detailed above.

3. Converter Design Method

The design of the converter must ensure essential criteria to have a better efficiency of the converter in both forward and backward mode. The criteria to be met are as follows:

- The switching frequency must be high to minimize the size of the magnetic and capacity components.
- ZVS for the input switches (inverter) and ZCS-ZVS for the output rectifier diodes should be ensured to reduce the switching losses. The converter must perform that regardless of the direction of energy transfer and terminal voltages conditions within a predefined range.
- The converter should ensure a sufficient voltage gain for all load and input voltage conditions.
- The switching frequency range should be as narrow as possible regardless of the direction of energy transfer, the load and the input voltage conditions. In these conditions, the circulating energy is to be kept reduced.

To ensure these criteria, a design procedure will be presented for a 3 kW resonant circuit test bench. The proposed converter was designed to serve as energy path between a variable DC-source with a voltage range of 60~240 V and a 270 V DC-bus. The maximum power of the converter corresponds to the resistance R and the maximum quality factor is Q_{max}.

$$R = 24.3\ \Omega;\ Q_{max} = 1.9 \quad (30)$$

Referring to Figure 4, the maximum voltage gain decreases if α increases. The next step is to define the value of α that ensures the maximum voltage gain for Q_{max}. Based on Figure 4, the required maximum voltage gain can be achieved for $\alpha = 0.9$ or less. The switching frequency range and the voltage gain rely on α. A small value of α involves a narrow switching frequency range. However, α should not be very small in order to not induce relatively large size of the inductance L_p. The converter must operate in the ZVS region to ensure soft switching by switches control. The minimum switching frequency is the frequency that gives the maximum voltage gain corresponding to the maximum delivered power. This frequency should be located in the ZVS region, as mentioned previously. The switching frequency increases from the minimum value and the voltage gain decreases until it reaches zero for no-load operation, theoretically corresponding to the resonant frequency f_r. The DC-bus voltage level is controlled to 270 V while the DC-source voltage varies between 60 V and 240 V. So, the converter is with a voltage gain range of 1.125~4.5 in forward mode. In backward mode, the converter is in buck mode depending on the changes of roles through transformer terminals and parameters of consequent quality factor Q_b.

Figure 9 shows the operating regions of the converter in forward and backward modes. The parallel inductance L_p should be chosen to ensure the required voltage gain. A small value of L_p allows to have a high voltage gain with ZVS for all load conditions. However, a very small value of L_p will increase the circulating energy and the conduction losses. Once L_p is defined, L_{r2} is calculated using Equation (10). The critical parameter to respect is the capacitor current to reduce the volume of connected capacitors. For this reason, the parallel components C_p and L_p are placed on the secondary of the transformer with a high transformer turn ratio to decrease the current of the capacitor and to increase the voltage capability. Parallel capacitor is calculated using Equation (31), and the transformer turn ratio is set according to Equation (32). The series inductance L_{r1} is then calculated using Equation (10).

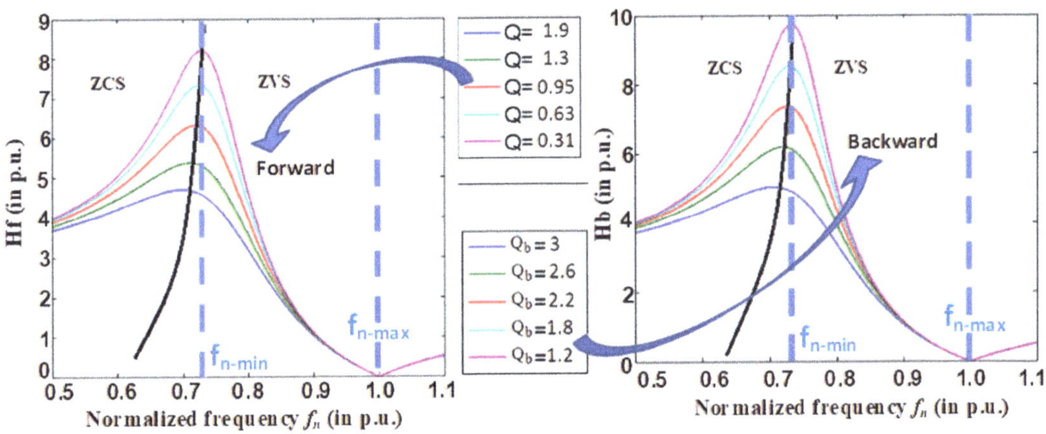

Figure 9. Proposed converter Forward and Backward operating regions.

$$C_p = \frac{1}{L_p.(2\,.\pi\,.f_r)^2} \tag{31}$$

$$m = \frac{V_{bus}}{V_{e\_min}} \tag{32}$$

The design procedure of the proposed converter is described in Figure 10.

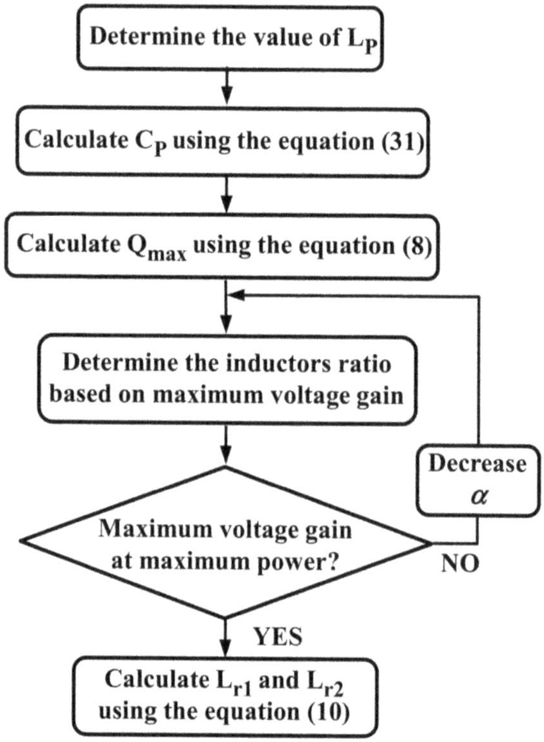

Figure 10. Flowchart of the proposed design procedure.

4. Experimental Test Bench and Results

In order to verify the design method and to test the proposed bidirectional resonant converter topology, a laboratory test bench with a 3 kW resonant circuit was built. It was integrated in a 12 kW Dual Active Bridge converter designed for the other laboratory works. The experimental test bench is shown in Figure 11. The characteristics of this test bench are given in Table 2 and the reference of the 800V/39A MOSFETs is IXFN44N80P.

Table 2. Characteristics of the built test bench.

Parameters	Value
Input voltage range	60~240 V
Output voltage	270 V
Maximum power	3 kW
Maximum input current	50 A
Switching frequency range	50~70 kHz
Resonant frequency	70 kHz
Transformer turn ratio	1:4

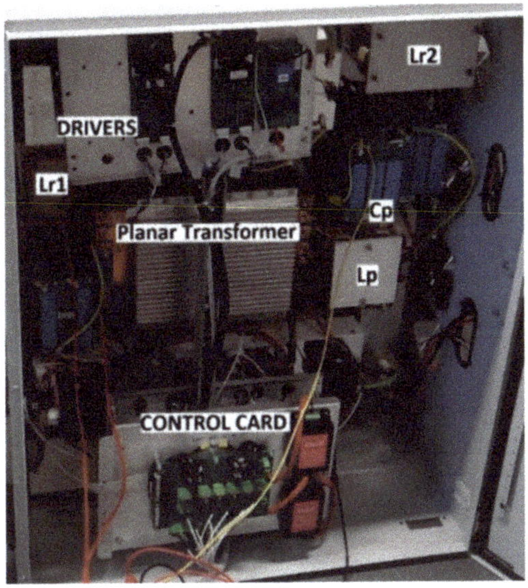

Figure 11. Proposed resonant converter test bench.

Figure 12 presents the open loop control strategy, where the frequency *fs* is an input parameter. This switching frequency is adjusted through DS1103 controller board which generates the control signals to maintain the DC-bus voltage at a desired level. To verify the soft-switching over the full operating range of the converter, some experimental tests are done with various input voltage and for different loads powers in both forward and backward mode.

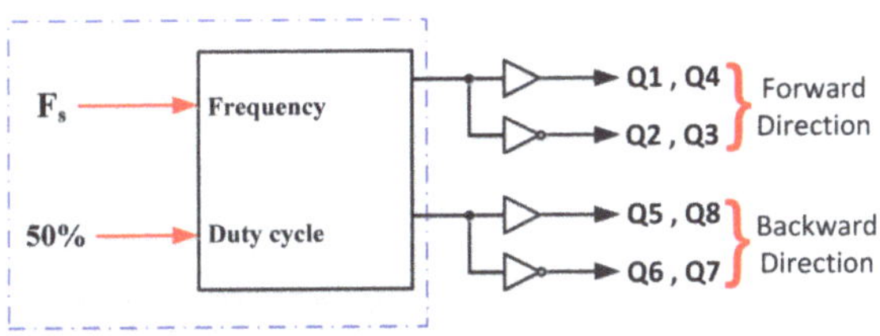

Figure 12. Open loop control strategy of the converter.

Figure 13 shows experimental waveforms in forward mode. It is seen that ZVS is reached for the primary switches of the inverter. Further, the rectifier diodes turn off with ZCS regardless the load variations and the input voltage value. We also notice the narrow variation of the switching frequency which is typically shown in Figures 3 and 4 with the variations of the load and the input voltage such as:

- fs = 51kHz for Ve = 60 V and P = 3 kW (full load);
- fs = 60 kHz for Ve = 140 V and P = 1.5 kW (half load);
- fs = 60 kHz for Ve = 100 V and P = 0.35 kW (low load).
- fs = 63 kHz for Ve = 240 V and P = 0 kW (no-load).

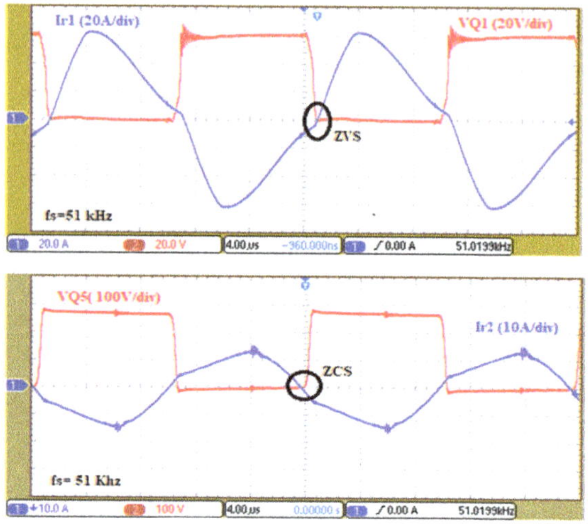

(**a**)

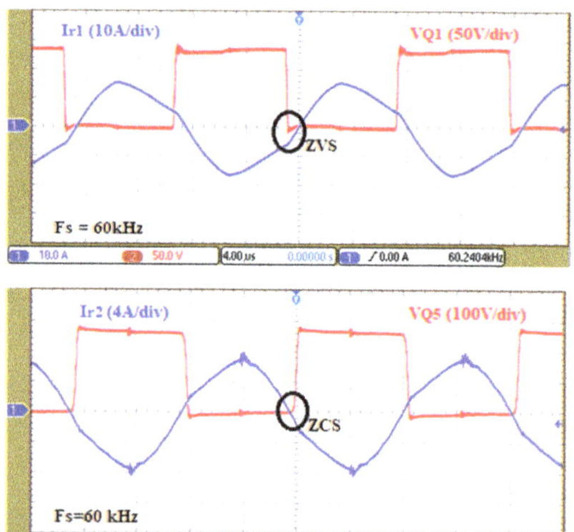

(**b**)

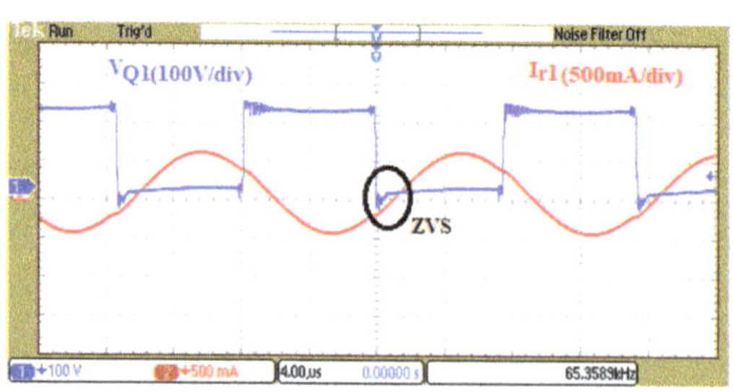

(**c**)

Figure 13. Experimental waveforms of the converter in forward mode: (**a**) Ve = 60 V, P = 3 kW (full load); (**b**) Ve = 140 V, P = 1.5 kW (half load); (**c**): Ve = 240 V at no load.

The narrow frequency range reduces internal energy circulating and the turn off losses. Figure 14 presents the V_{ab} voltage at the primary side in forward mode, and the resonant current i_{r1}. We can see that the current lags the voltage. Then, ZVS is always reached.

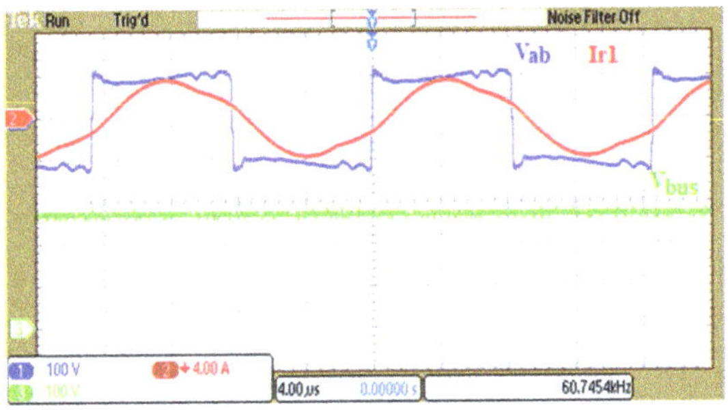

Figure 14. Experimental waveforms at P = 0.35 kW and Ve = 100 V.

Figure 15 shows experimental waveforms in backward mode for different load conditions and DC-source voltage. The active switches Q_5~Q_8 are turned on with ZVS and the diodes of Q_1~Q_4 are turned off with ZCS regardless load and voltage conditions. Note that the switching frequency range in backward mode is also very narrow, which proves also the analyzed behavior in Figure 6.

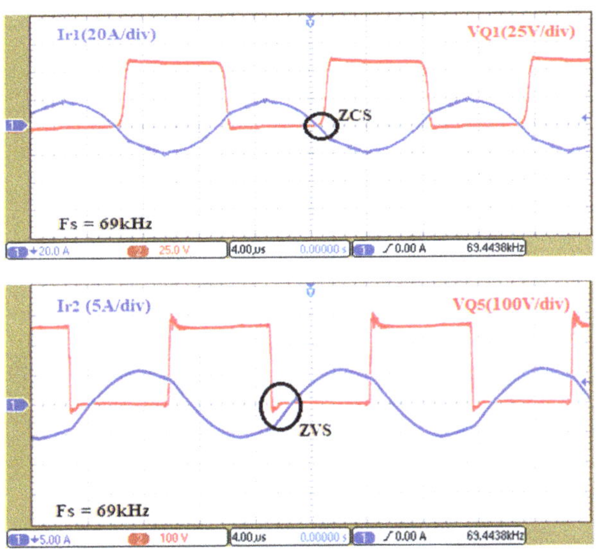

(a)

Figure 15. *Cont.*

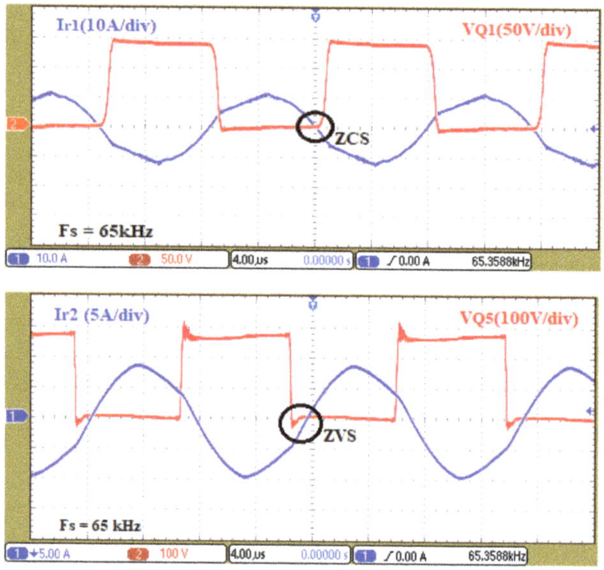

(b)

Figure 15. Experimental waveforms of the converter in backward mode: (**a**) Ve = 60 V, P = 1 kW; (**b**) Ve = 140 V, P = 1.5 kW.

Figure 16 shows the experimental waveforms of V_{bus}, i_{r1} and v_{Cp} during startup. We notice the soft startup of the converter with no voltage or current peaks. This is due to the zero-voltage gain at the starting resonant frequency which demonstrates another advantage of the proposed topology. Figure 17 shows the measured efficiency of the converter in forward mode versus the power for different input voltages. The maximum efficiency is about 96% at 3 kW when the DC-source voltage is more significant (240 V in this case). The converter efficiency reduces when the input voltage decreases because the input current increases.

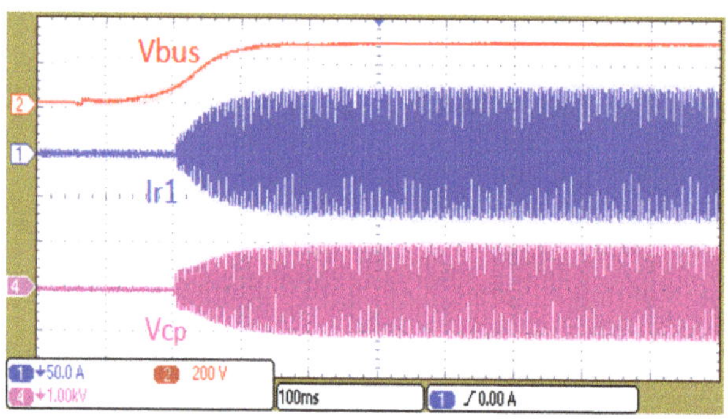

Figure 16. Startup experimental waveforms: where V_{bus} is the DC-bus voltage, i_{r1} is the primary resonant current and v_{Cp} is the capacitor voltage.

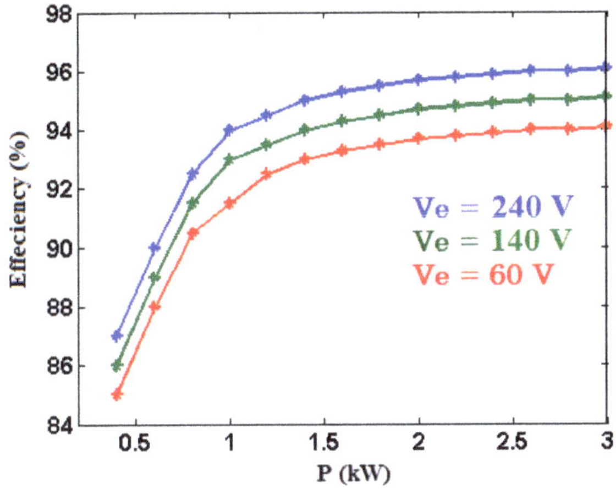

Figure 17. Efficiency of the converter in forward mode versus the power for different input voltages.

Note that the efficiency of the converter can be increased by using low power loss components with the electric wiring optimizing. Figure 18 presents the power losses distribution in the converter for a full load condition, where the losses in the transformer and the resonant circuit reach 49% of the total losses (120 W). On the other hand, the losses in the active inverter are about 27% of the total losses, while the losses in the diode rectifier are almost 15% of the total losses. The conduction losses are higher compared to those of switching, in order of 3.5 times for the inverter and 4 times for the rectifier. The conduction losses in the inverter reach 21% of the total losses and they are 12% of the total losses in the diode rectifier. Due to soft switching, the switching losses in the inverter and in the rectifier diodes are reduced. They are 6% and 3% of the total losses, respectively. Figure 18 gives also the characteristics of the converter compared to the other ones summarized previously in Table 1.

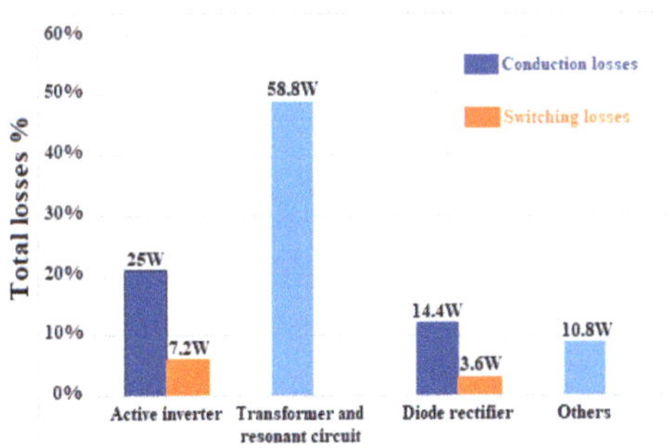

SYMETRICAL RESONANT CONVERTER	LC-LL Proposed
Voltage gain range	0.22 ~ 4.5
Switching frequency [kHz]	51~65
Narrow control frequency range	Yes (22%)
Soft start at resonant frequency	Yes
Soft switching in Forward & Backward	ZVS and Low turn off losses for full load range

Figure 18. Losses distribution and criteria fulfillment of the novel resonant converter.

5. Conclusions

This paper presents a bidirectional resonant converter for wide voltage range applications. Due to the symmetric resonant circuit, the converter has the same behavior in both energies transfer modes. The voltage gain varies from zero to a maximum value in the suitable ZVS region. So, ZVS for active switches of the inverter side and ZCS for the diodes of the rectifier side are achieved regardless the input voltage, the energy transfer direction and the load conditions. Proposed converter topology has the capability to operate under wide voltage range 60~240 V. Control of the power flow is carried out in a predetermined narrow fundamental switching frequency band which increases significantly the efficiency by reducing the circulating energy and the turn off losses. In addition, the proposed converter has no voltage gain at the resonant frequency and so, no specific startup strategy is needed. The experimental results of the 3 kW resonant circuit test bench validate the feasibility of the proposed converter. Moreover, they prove the effectiveness of the proposed design procedure for the developed converter.

Author Contributions: Conceptualization, methodology, validation, formal analysis, investigation, data curation, writing—original draft preparation, writing—review and editing, visualization, M.A., A.P., M.B.C., B.D.; supervision, project administration, funding acquisition, A.P., M.B.C., B.D. All authors have read and agreed to the published version of the manuscript.

Funding: This article has been funded by the council of Normandy Region (France).

Institutional Review Board Statement: Not applicable.

Informed Consent Statement: Not applicable.

Data Availability Statement: The data presented in this study are available on request from the corresponding authors.

Acknowledgments: This work was supported by University of Le Havre Normandy and is funded by Normandy region in France.

Conflicts of Interest: The authors declare no conflict of interest.

References

1. Vinnikov, D.; Chub, A.; Kosenko, R.; Zakis, J.; Liivik, E. Comparison of Performance of Phase-Shift and Asymmetrical Pulsewidth Modulation Techniques for the Novel Galvanically Isolated Buck–Boost DC–DC Converter for Photovoltaic Applications. *IEEE J. Emerg. Sel. Top. Power Electron.* **2016**, *5*, 624–637. [CrossRef]
2. Prabhakaran, P.; Agarwal, V. Novel Four-Port DC–DC Converter for Interfacing Solar PV–Fuel Cell Hybrid Sources with Low-Voltage Bipolar DC Microgrids. *IEEE J. Emerg. Sel. Top. Power Electron.* **2020**, *8*, 1330–1340. [CrossRef]
3. Sha, D.; Chen, D.; Zhang, J. A Bidirectional Three-Level DC–DC Converter with Reduced Circulating Loss and Fully ZVS Achievement for Battery Charging/Discharging. *IEEE J. Emerg. Sel. Top. Power Electron.* **2018**, *6*, 993–1003. [CrossRef]
4. Kardan, F.; Alizadeh, R.; Banaei, M.R. A New Three Input DC/DC Converter for Hybrid PV/FC/Battery Applications. *IEEE J. Emerg. Sel. Top. Power Electron.* **2017**, *5*, 1771–1778. [CrossRef]
5. Bellache, K.; Camara, M.B.; Dakyo, B. Transient Power Control for Diesel-Generator Assistance in Electric Boat Applications Using Supercapacitors and Batteries. *IEEE J. Emerg. Sel. Top. Power Electron.* **2017**, *6*, 416–428. [CrossRef]
6. Faraji, R.; Farzanehfard, H. Soft-Switched Nonisolated High Step-Up Three-Port DC–DC Converter for Hybrid Energy Systems. *IEEE Trans. Power Electron.* **2018**, *33*, 10101–10111. [CrossRef]
7. Pires, V.F.; Foito, D.; Cordeiro, A. A DC–DC Converter with Quadratic Gain and Bidirectional Capability for Batteries/Supercapacitors. *IEEE Trans. Ind. Appl.* **2018**, *54*, 274–285. [CrossRef]
8. Wang, L.; Zhu, Q.; Yu, W.; Huang, A.Q. A Medium-Voltage Medium-Frequency Isolated DC–DC Converter Based on 15-kV SiC MOSFETs. *IEEE J. Emerg. Sel. Top. Power Electron.* **2017**, *5*, 100–109. [CrossRef]
9. Sayed, K. Zero-voltage soft-switching DC–DC converter-based charger for LV battery in hybrid electric vehicles. *IET Power Electron.* **2019**, *12*, 3389–3396. [CrossRef]
10. Akagi, H.; Kinouchi, S.-I.; Miyazaki, Y. Bidirectional Isolated Dual-Active-Bridge (DAB) DC-DC Converters Using 1.2-kV 400-A SiC-MOSFET Dual Modules. *CPSS Trans. Power Electron. Appl.* **2016**, *1*, 33–40. [CrossRef]
11. De Din, E.; Siddique, H.A.B.; Cupelli, M.; Monti, A.; De Doncker, R.W. Voltage Control of Parallel-Connected Dual-Active Bridge Converters for Shipboard Applications. *IEEE J. Emerg. Sel. Top. Power Electron.* **2018**, *6*, 664–673. [CrossRef]
12. Xu, G.; Sha, D.; Xu, Y.; Liao, X. Hybrid-Bridge-Based DAB Converter with Voltage Match Control for Wide Voltage Conversion Gain Application. *IEEE Trans. Power Electron.* **2017**, *33*, 1378–1388. [CrossRef]
13. Dung, N.A.; Chiu, H.; Lin, J.; Hsieh, Y.; Liu, Y. Efficiency optimisation of ZVS isolated bidirectional DAB converters. *IET Power Electron.* **2018**, *11*, 1499–1506. [CrossRef]
14. Mukherjee, S.; Kumar, A.; Chakraborty, S. Comparison of DAB and LLC DC–DC Converters in High-Step-Down Fixed-Conversion-Ratio (DCX) Applications. *IEEE Trans. Power Electron.* **2021**, *36*, 4383–4398. [CrossRef]
15. Du, Y.; Bhat, A.K.S. Analysis and Design of a High-Frequency Isolated Dual-Tank LCL Resonant AC–DC Converter. *IEEE Trans. Ind. Appl.* **2015**, *52*, 1566–1576. [CrossRef]
16. Salem, M.; Jusoh, A.; Idris, N.R.N.; Alhamrouni, I. Performance study of series resonant converter using zero voltage switching. In Proceedings of the 2014 IEEE Conference on Energy Conversion (CENCON), Johor Bahru, Malaysia, 13–14 October 2014; pp. 96–100. [CrossRef]

17. Liu, G.; Jang, Y.; Jovanović, M.M.; Zhang, J.Q. Implementation of a 3.3-kW DC–DC Converter for EV On-Board Charger Employing the Series-Resonant Converter with Reduced-Frequency-Range Control. *IEEE Trans. Power Electron.* **2016**, *32*, 4168–4184. [CrossRef]
18. Outeiro, M.; Buja, G.; Czarkowski, D. Resonant Power Converters: An Overview with Multiple Elements in the Resonant Tank Network. *IEEE Ind. Electron. Mag.* **2016**, *10*, 21–45. [CrossRef]
19. Soeiro, T.; Muhlethaler, J.; Linner, J.; Ranstad, P.; Kolar, J.W. Automated Design of a High-Power High-Frequency LCC Resonant Converter for Electrostatic Precipitators. *IEEE Trans. Ind. Electron.* **2012**, *60*, 4805–4819. [CrossRef]
20. Yang, R.; Ding, H.; Xu, Y.; Yao, L.; Xiang, Y. An Analytical Steady-State Model of LCC type Series–Parallel Resonant Converter with Capacitive Output Filter. *IEEE Trans. Power Electron.* **2013**, *29*, 328–338. [CrossRef]
21. Wei, Y.; Luo, Q.; Mantooth, A. Comprehensive analysis and design of LLC resonant converter with magnetic control. *CPSS Trans. Power Electron. Appl.* **2019**, *4*, 265–275. [CrossRef]
22. Li, M.; Ouyang, Z.; Andersen, M.A.E. High frequency LLC resonant converter with magnetic shunt integrated planar transformer. *IEEE Trans. Power Electron.* **2018**, *34*, 2405–2415. [CrossRef]
23. Kundu, U.; Yenduri, K.; Sensarma, P. Accurate ZVS Analysis for Magnetic Design and Efficiency Improvement of Full-Bridge LLC Resonant Converter. *IEEE Trans. Power Electron.* **2017**, *32*, 1703–1706. [CrossRef]
24. Buccella, C.; Cecati, C.; Latafat, H.; Pepe, P.; Razi, K. Observer-Based Control of LLC DC/DC Resonant Converter Using Extended Describing Functions. *IEEE Trans. Power Electron.* **2015**, *30*, 5881–5891. [CrossRef]
25. Jiang, T.; Chen, X.; Zhang, J.; Wang, Y. Bidirectional LLC resonant converter for energy storage applications. In Proceedings of the 2013 Twenty-Eighth Annual IEEE Applied Power Electronics Conference and Exposition (APEC), Long Beach, CA, USA, 17–21 March 2013. [CrossRef]
26. Sun, J.; Yuan, L.; Gu, Q.; Duan, R.; Lu, Z.; Zhao, Z. Design-Oriented Comprehensive Time-Domain Model for CLLC Class Isolated Bidirectional DC-DC Converter for Various Operation Modes. *IEEE Trans. Power Electron.* **2019**, *35*, 3491–3505. [CrossRef]
27. Liu, Y.; Du, G.; Wang, X.; Lei, Y. Analysis and Design of High-Efficiency Bidirectional GaN-Based CLLC Resonant Converter. *Energies* **2019**, *12*, 3859. [CrossRef]
28. Zou, S.; Lu, J.; Mallik, A.; Khaligh, A. Bi-Directional CLLC Converter with Synchronous Rectification for Plug-In Electric Vehicles. *IEEE Trans. Ind. Appl.* **2017**, *54*, 998–1005. [CrossRef]
29. Jiang, T.; Zhang, J.; Wu, X.; Sheng, K.; Wang, Y. A Bidirectional Three-Level LLC Resonant Converter with PWAM Control. *IEEE Trans. Power Electron.* **2016**, *31*, 2213–2225. [CrossRef]
30. EKim, E.-S.; Oh, J.-S. High-Efficiency Bidirectional LLC Resonant Converter with Primary Auxiliary Windings. *Energies* **2019**, *12*, 4692. [CrossRef]
31. Zhang, Y.; Zhang, D.; Li, J.; Zhu, H. Bidirectional LCLL Resonant Converter with Wide Output Voltage Range. *IEEE Trans. Power Electron.* **2020**, *35*, 11813–11826. [CrossRef]
32. Jiang, T.; Zhang, J.; Wu, X.; Sheng, K.; Wang, Y. A Bidirectional LLC Resonant Converter with Automatic Forward and Backward Mode Transition. *IEEE Trans. Power Electron.* **2015**, *30*, 757–770. [CrossRef]
33. Zahid, Z.; Dalala, Z.M.; Chen, R.; Chen, B.; Lai, J.S. Design of Bidirectional DC–DC Resonant Converter for Vehicle-to-Grid (V2G) Applications. *IEEE Trans. Transporta. Electrif.* **2015**, *1*, 232–244. [CrossRef]
34. Outeiro, M.T.; Buja, G. Comparison of resonant power converters with two, three, and four energy storage elements. In Proceedings of the IECON 2015—41st Annual Conference of the IEEE Industrial Electronics Society, Yokohama, Japan, 9–12 November 2015. [CrossRef]

sustainability

Article

Frequency Stability of AC/DC Interconnected Power Systems with Wind Energy Using Arithmetic Optimization Algorithm-Based Fuzzy-PID Controller

Ahmed H. A. Elkasem [1], Mohamed Khamies [1], Gaber Magdy [2], Ibrahim B. M. Taha [3,*] and Salah Kamel [1,*]

1 Electrical Engineering Department, Faculty of Engineering, Aswan University, Aswan 81542, Egypt; ahmedhamdykasem2016@yahoo.com (A.H.A.E.); mohamedahmedmak@yahoo.com (M.K.)
2 Department of Electrical Engineering, Faculty of Energy Engineering, Aswan University, Aswan 81528, Egypt; gabermagdy@aswu.edu.eg
3 Department of Electrical Engineering, College of Engineering, Taif University, Taif 21944, Saudi Arabia
* Correspondence: i.taha@tu.edu.sa (I.B.M.T.); skamel@aswu.edu.eg (S.K.)

Abstract: This article proposes an intelligent control strategy to enhance the frequency dynamic performance of interconnected multi-source power systems composing of thermal, hydro, and gas power plants and the high penetration level of wind energy. The proposed control strategy is based on a combination of fuzzy logic control with a proportional-integral-derivative (PID) controller to overcome the PID limitations during abnormal conditions. Moreover, a newly adopted optimization technique namely Arithmetic optimization algorithm (AOA) is proposed to fine-tune the proposed fuzzy-PID controller to overcome the disadvantages of conventional and heuristic optimization techniques (i.e., long time in estimating controller parameters-slow convergence curves). Furthermore, the effect of the high voltage direct current link is taken into account in the studied interconnected power system to eliminate the AC transmission disadvantages (i.e., frequent tripping during oscillations in large power systems–high level of fault current). The dynamic performance analysis confirms the superiority of the proposed fuzzy-PID controller based on the AOA compared to the fuzzy-PID controller based on a hybrid local unimodal sampling and teaching learning-based optimization (TLBO) in terms of minimum objective function value and overshoots and undershoots oscillation measurement. Also, the AOA's proficiency has been verified over several other powerful optimization techniques; differential evolution, TLBO using the PID controller. Moreover, the simulation results ensure the effectiveness and robustness of the proposed fuzzy-PID controller using the AOA in achieving better performance under several contingencies; different load variations, the high penetration level of the wind power, and system uncertainties compared to other literature controllers adjusting by various optimization techniques.

Keywords: load frequency control (LFC); multi-source power system; fuzzy logic control (FLC); high wind energy penetration

Citation: Elkasem, A.H.A.; Khamies, M.; Magdy, G.; Taha, I.B.M.; Kamel, S. Frequency Stability of AC/DC Interconnected Power Systems with Wind Energy Using Arithmetic Optimization Algorithm-Based Fuzzy-PID Controller. *Sustainability* **2021**, *13*, 12095. https://doi.org/10.3390/su132112095

Academic Editors: Nicu Bizon, Mamadou Baïlo Camara and Bhargav Appasani

Received: 18 September 2021
Accepted: 28 October 2021
Published: 2 November 2021

1. Introduction

Consistent with the noticeable increase in energy demand, it is necessary to establish new energy sources. However, most efforts concern with establishing renewable energy sources (RESs) instead of conventional energy sources (CESs) due to the negative and harmful effects (e.g., global warming) of CES on our community [1–3]. So, energy planners and researchers make great efforts and strive to establish RESs paralleling with electrical networks for reducing CESs' hazards. In addition, RESs are clean and safe energy which be friendly to the environment [4]. While the establishment of RESs decreases the system inertia and may negatively affect the system stability [5,6]. Based on the aforementioned observations, the modern power grids will face a great challenge in keeping the system's frequency and the tie-line power stable. Therefore, it is important to keep the system stable during these previous conditions, this occurs by applying load frequency control (LFC) to maintain the frequency and tie-line power of the system at their specified values [7]. In this regard, numerous control techniques/strategies have been conducted to make progress in modern power system frequency stability.

One of these strategies is optimal control, which includes linear quadratic regulators that are applied to regulate the frequency of two-area power systems considering AC/DC tie-lines [8]. Also in [9], the linear quadratic regulator is applied to enhance the frequency of the two-area power system in the presence of electrical vehicles. Another strategy known as robust control techniques, such as the second-order sliding mode controller, has been applied to regulate the frequency in multi-area power systems [10]. In [11], one robust control technique known as the μ -synthesis approach is applied to regulate the frequency of the islanded micro-grid frequency containing (diesel engine generator, fuel cell, wind turbine, and PV array). In [12], μ-synthesis approach has been applied to recover the frequency fluctuations under uncertainty weighing selection in power plants. H_∞ Controller has been designed in [13] to regulate the frequency of the power system while accounting for uncertainties. There is another type of these strategies known as model predictive control, which has been applied to enhance the performance of three-area interconnected power system considering the penetration of wind turbines [14]. Also, there are intelligent control strategies such as (e.g., artificial neural network and fuzzy logic control (FLC)) have been applied to counteract the system's frequency deviations during disturbances in the presence of tidal power units, electrical vehicles, energy storage systems, and solar power systems [15,16] and so on. While all these strategies were successful in resolving LFC difficulties, they depended on the designer's knowledge and experience, experimenting, and trial and error procedures in finding controller parameters, and it takes a very long time to approximate their parameters.

On the other hand, the proportional-integral-derivative (PID) controller and its forms are still the most popular controller due to their good characteristics such as simplicity and low cost [17]. However, the PID controllers are sensitive to system parameter variations (i.e., system uncertainties) and nonlinearities [18]. Thus, the FLC strategy has been proposed to support the PID controller performance in enhancing the dynamic frequency stability of modern power systems. The main advantages of the fuzzy controller include its simplicity of execution, high sensitivity to system fluctuations, and the ability to safely handle changes in the operating point or system parameters due to online updating of the controller parameters [19]. The first attempt in using the FLC for solving the LFC problem with the PID controller was conducted in [20]. Additionally, the self-tuning Fuzzy-PID controller was proposed to enhance the frequency stability of the interconnected power system, consisting of two areas [21]. Also, the fuzzy-PID controller has been utilized to stabilize the frequency of a multi-source power system [22]. Where the proposed Fuzzy-PID controller parameters have been selected based on an optimization technique. In addition, the Fuzzy-PID controller has been utilized to improve the stability of power system frequency considering the flexible AC transmission system devices (i.e., the static synchronous series compensator, the unified power flow controller, and the interline power flow controller) effect in [23,24]. The FLC is presented by parameters (i.e., inputs, scaling factors, membership functions, and rule base) that have not specific rules which can be followed to detect their values. Generally, trial and error methods have been used to select the parametric values, but these methods may not give the best performance. In this regard, this study proposes selecting the scaling factors of the fuzzy-PID controller based on a new meta-heuristic technique known as the Arithmetic Optimization Algorithm (AOA).

Therefore, there is another choice to take care of the LFC issue: to utilize various optimization methods and these are successfully utilized to manage the nonlinear functions associated with the LFC design. These several optimization techniques are applied for finding the optimal controller parameters to overcome the LFC problem and achieve more system security. The utilized techniques by researchers in the LFC issue such as; grasshopper optimization algorithm [25], ant colony optimization technique [26], Jaya algorithm [27], particle swarm optimizer [28], firefly algorithm [29], hybrid pattern search shuffled-frog leaping algorithm [30], multi-objective genetic algorithm [31], grey wolf optimizer [32], sine cosine algorithm [33], harris hawks optimizer and salp swarm algorithm [34], lightning-attachment procedure optimization (LAPO) [35] and improved LAPO [36]. However, these techniques achieve exceptional performance by ensuring effectual LFC design. They have a slow convergence rate, restricted search capability, as well as local optimum convergence. On the other hand, the AOA algorithm has been made to overcome previous limitations related to different optimization techniques. The superiority of AOA than other conventional optimization algorithms returns to the gradient-free mechanism and its capability to avoid the local solutions and obtain the global solution with little search agents. Also, many experimental results show that, AOA provides very promising results in solving real-world engineering design. So, this study applied the AOA algorithm for fine-tuning the proposed fuzzy-PID controller parameters due to its promising results in solving several real-world engineering design problems [37]. Also, the AOA recently gave a distinguished performance in medicine field (i.e., evaluate images of COVID-19) [38].

The interconnected multi-source power systems need strong links (tie-lines) utilized in constructural power exchange between different control areas. Also at abnormal conditions, these links provide a support to inter-area. Several studies have been maiden about the topic of LFC with the presence of AC transmission line only without HVDC line connection [39,40]. However, many problems related to the AC interconnection between areas in the power system, especially long-distance power transmission, remain. The problems associated with the AC interconnection can be mentioned as frequent tripping occurs at the instant of large power oscillations, an increase in fault current level, which leads to damage in the power system and transmission oscillations from one area to another, which causes deterioration in system dynamic performance. So this study proposes using of HVDC interconnection besides the AC tie line for transmitting the bulk power over the long distance to eliminate the demerits of the AC transmission lines and according to good features of HVDC transmission. The attractive features of HVDC links are summarized as: there are converters in HVDC lines that give the ability to fast controllability in power between interconnected areas. It can overcome the transient stability problems associated with AC transmission [41]. According to the point of obtaining stabilizing in the electrical power systems such as mentioned previously when adding HVDC lines, there are several studies deal with the issue of predicting processes of wind speed during participation in electrical power systems using different meta-heuristic techniques. These researchers seek to avoid fluctuations when wind speed exceeds the permissible limits [42–45]. On the other hand, there are off-shore wind turbines which characterized by a high average speed compared to on-shore wind turbines. Also, those off-shore wind turbines produce more electricity than on-shore wind turbines. Therefore, many researchers strive to choose the best site of off-shore wind turbines in coastal areas such as Turkey and USA to generate more electricity and link between these turbines and main electrical network to meet the need for citizens [46,47].

In terms of LFC issues, traditional controllers, such as the PID controller, have some challenges in parameter tuning and have not accommodated system stability in the face of uncertainties. Moreover, few studies applied fuzzy-PID controller to diminish the demerits of PID controller, but the parameters of the fuzzy-PID has been selected based in conventional and heuristic optimization algorithms. Furthermore, the renewables penetration effect has been considered in few studies, and not considered in other works [22,24,48]. Additionally, the effect of HVDC has been ignored in most works related to LFC studies [39,40]. So, this study proposes a robust control strategy based on Fuzzy-PID controller to keep the stability of systems involving different types of generating units in addition to renewable sources. In addition, the parameters of the fuzzy-PID controller have been selected based on a novel meta-heuristic algorithm known as AOA algorithm due to its merits. Unlike, previous works which have neglected the parameters variations effect [14,22], the control design consider, system nonlinearities and system uncertainties have been considered during designing the proposed control strategy. Finally, the effect of HVDC has been considered to eliminate the demerits of AC transmission lines. Furthermore, Table 1 introduces a comparison between the motivation of this work and other studies.

Table 1. Comparison between this work and previous mentioned studies.

Properties	[25]	[39]	[40]	[49]	[50]	[50]	This Study
Type of controller	Fuzzy-PID controller	Optimal PI-PD cascaded controller	Optimal PID controller	Optimal PID controller	Optimal PID controller	Fuzzy-PID controller	Fuzzy-PID controller
Adoption of controller design on	Grasshopper optimization algorithm (GOA)	Flower pollination algorithm (FPA)	Grey wolf optimization (GWO)	Differential evolution (DE)	Teaching-learning based optimization (TLBO)	Hybrid local unimodal sampling (LUS) with TLBO	Arithmetic optimization algorithm (AOA)
Penetration of renewable energy sources	Not considered	Not considered	Not considered	Not considered	Not considered	Not considered	Considered with high penetration of wind energy
Effect of system uncertainties	considered	Not considered	considered	Not considered	Not considered	Not considered	considered
Effect of HVDC link	Not considered	Not considered	Not considered	considered	considered	considered	considered

The main contribution of this study can be summarized as follows:

i. Proposing a fuzzy-PID controller for stabilizing the frequency of interconnected multi-source power systems considering high wind power penetration.
ii. The proposed controller parameters have been selected via a new meta-heuristic optimization technique known as AOA algorithm according to its noteworthy features. While, it is the first attempt to apply the AOA algorithm in adjusting and optimizing the frequency controller parameters, thus enhancing the stability of the power system.
iii. Considering, the effect of HVDC links to eliminate the problems related to the AC links.
iv. Considering load disturbance, renewable power penetration (i.e., wind power), and system parameters variations during designing the parameters of the proposed fuzzy-PID controller-based AOA algorithm.
v. Comparing the performance of the AOA algorithm with other optimization algorithms such as differential evolution (DE) and teaching-learning based optimization (TLBO) for selecting the parameters of the PID controller in hybrid two-area power system.
vi. Comparing the performance of the proposed control strategy with other techniques performances such as; PID controller-based differential evolution (PID-DE) [49], PID controller-based teaching-learning based optimization (PID-TLBO) [50], and fuzzy-PID controller-based a hybrid local unimodal sampling (LUS) with TLBO (Fuzzy-PID-LUS-TLBO) [50] in order to ensure the effectiveness and robustness of the proposed controller.

The remainder of this research is summarized as follows: the modeling and configuration of the studied interconnected power system considering wind energy are discussed in Section 2. Section 3 presents the proposed fuzzy-PID controller methodology, the proposed optimization technique AOA, and the construction of the proposed control strategy. Then, Section 4 presents the simulations and investigation results. Finally, the conclusion of this work is mentioned in Section 5.

2. Modeling and Configuration of the Studied System

2.1. A dynamic Model of Two-Area Interconnected Power System

This article discusses the LFC issue of interconnected multi-source power systems. Where the studied system is composite of two-areas which interconnected together by a tie-line. Three power plants (i.e., the reheat thermal unit, the gas unit and the hydro unit) are included in each area of the investigated power system. Moreover, each unit in both areas has its speed governing system, turbine, and generator. The capacity or rating power of the investigated system is 2000 MW [51]. Also, the system dynamic performance has been investigated in the presence and absence of an HVDC link. The fuzzy-PID controller is proposed to be equipped in both areas for each generation unit to minimize the oscillations in both area frequencies and tie-line power between them. The input signals of the proposed fuzzy-PID controller represent the area control error *(ACE)* and its derivative, while the output signal represents the secondary control action on each generated unit. Figure 1 shows the dynamic model of the studied two-area interconnected power system and the schematic diagram is shown in Figure 2. The nominal parameters' values of the studied power system are given in Table 2. The ACEs in both areas can be obtained according to formulas as follows in Equations (1) and (2):

$$ACE_1 = \Delta P_{tie1-2} + B_1 \Delta f_1 \quad (1)$$

$$ACE_2 = \Delta P_{tie2-1} + B_2 \Delta f_2 \quad (2)$$

where, ΔP_{tie1-2} and ΔP_{tie2-1} represent the tie-line power exchange at area 1 and area 2, B_1 and B_2 are the bias frequency factors of area 1 and area 2 respectively, Δf_1 is the deviation in frequency waveform in area 1, and also Δf_2 is the deviation in frequency in area 2.

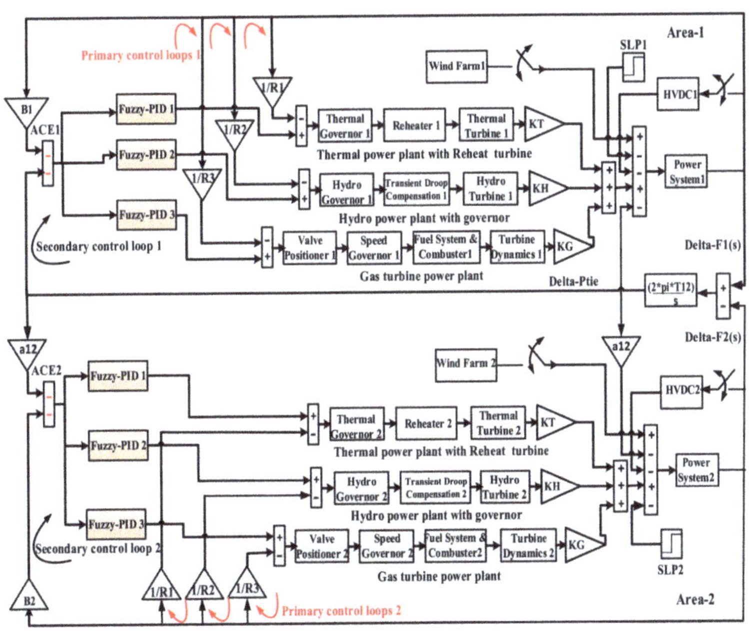

Figure 1. Transfer function model of the two-area interconnected power system. (see Appendix A).

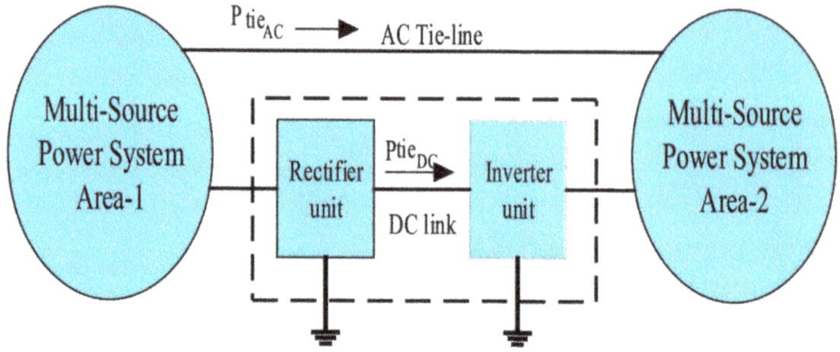

Figure 2. Schematic diagram of the studied AC/DC interconnected power system.

Table 2. The parameters of two identical interconnected areas with standard values [49].

Symbol	Nominal Values
B_i	0.4312 MW/HZ
T_{12}	0.0433 MW
R_1	2.4 HZ/MW
R_2	2.4 HZ/MW
R_3	2.4 HZ/MW
a_{12}	−1
K_T	0.543478
K_H	0.326084
K_G	0.130438
K_{ps}	68.9566
T_{ps}	11.49 s
T_{sg}	0.08 s
T_t	0.3 s
K_r	0.3
T_r	10 s
T_{gh}	0.2 s
T_{rs}	5 s
T_{rh}	28.75 s
T_w	1 s
b_g	0.05
c_g	1
Y_c	1 s
X_c	0.6 s
T_{cr}	0.01 s
T_{fc}	0.23 s
T_{cd}	0.2 s
K_{dc}	1
T_{dc}	0.2 s

2.2. The Wind Farm Configuration

The model of the wind power has been built using the MATLAB/SIMULINK program. The random wind power is integrated with conventional units in both areas. According to the design of the wind power model, a white noise block is used to get a random speed which is multiplied by the wind speed as shown in Figure 3 [52]. The following equations illustrate the captured power from the wind by the rotor of the wind turbine [35].

$$P_{wt} = \frac{1}{2}\rho A_T v_w^3 C_p(\lambda, \beta) \quad (3)$$

$$C_p(\lambda, \beta) = C_1\left(\frac{C_2}{\lambda_i} - C_3\beta - C_4\beta^2 - C_5\right) \times e^{\frac{-C_6}{\lambda_i}} + C_7\lambda_T \quad (4)$$

$$\lambda_T = \lambda_T^{OP} = \frac{\omega_T r_T}{V_W} \tag{5}$$

$$\frac{1}{\lambda_i} = \frac{1}{\lambda_T + 0.08\beta} - \frac{0.035}{\beta^3 + 1} \tag{6}$$

where, P_{wt} represents the captured output power of wind turbine, A_T is the swept area by the blades of turbine in m^2, ρ is the air density (nominally 1.22 Kg/m^3), V_W is the wind speed in m/s.

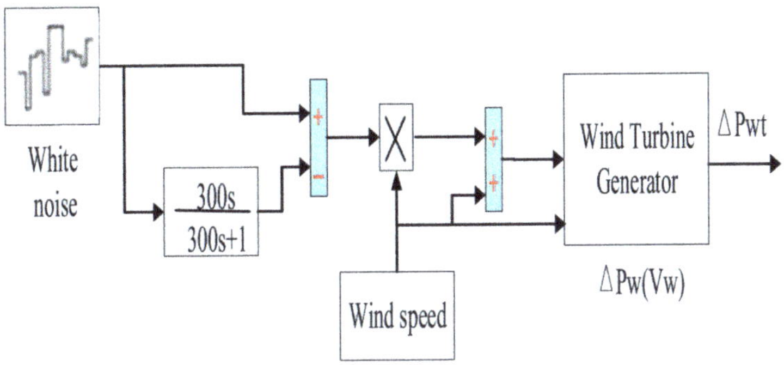

Figure 3. The wind power modeling using MATLAB/Simulink.

The climatic and geographical conditions where the studied wind turbine units located are the same. Thus, all previous parameters are applied to these units. The coefficient of rotor blades C_p based on turbine coefficients C_1–C_7 and it is a function on λ_T which refers to the optimum tip-speed ratio (TSR) and pitch angle β. λ_T is a function on the rotor speed (ω_T) and the blade length of rotor radius (r_T). Moreover, λ_i is referring to the intermittent TSR. Table 3 shows the nominal parameter values of the wind turbine for the wind farm applied with the studied power system. Figure 4 shows the random output power of 130 wind turbine units of 750 KW which have been penetrated at both areas of the studied power system.

Table 3. The wind farm nominal parameters [52].

Parameters	Values
P_{wt}	750 KW
V_W	15 m/s
A_T	1648 m^2
r_T	22.9 m
ω_T	22.5 rpm
C_1	−0.6175
C_2	116
C_3	0.4
C_4	0
C_5	5
C_6	21
C_7	0.1405

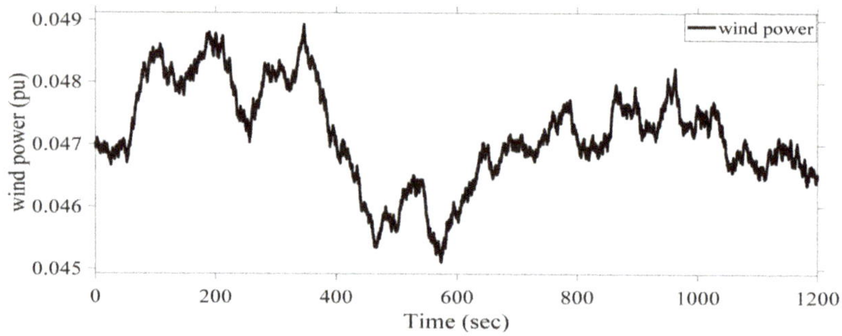

Figure 4. The wind power variation pattern.

3. Control Methodology and Problem Formulation

According to RESs penetration, system nonlinearities, and system uncertainties, it is essential to design good controller to improve the system performance during abnormal conditions. Hence, this study proposes fuzzy-PID controller to overcome any deviations resulted from previous considerations. Moreover, the proposed controller parameters have been selected based AOA algorithm.

3.1. The Proposed Control Strategy

In this article, three fuzzy-PID controllers are proposed as responsible for extracting extra active power from thermal, hydro, and gas turbines respectively, when load disturbance occurs. In this regard, several studies have applied the trial and error runs method to detect the fuzzy-PID controller's input and output scaling factors [53]. Therefore, it is difficult to obtain the optimal parameter values which enhance the system performance through these trial and error methods. Therefore, this paper proposes designing the fuzzy-PID controller with the optimized input and output scaling factors. The AOA has been selected in this work to obtain the optimized values of the proposed controller's input and output scaling factors. Furthermore, Figure 5 shows the structure of the fuzzy-PID controller of the thermal, hydro, and gas units. It has two inputs; ACE and the change in ACE, and one output. The input scaling factors are K_1 and K_2 and the output scaling factors are K_3 and K_4 which are optimized via the proposed AOA.

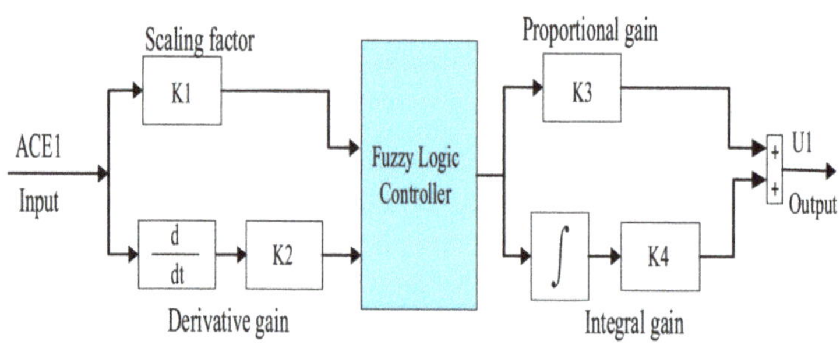

Figure 5. The proposed Fuzzy-PID controller diagram.

The fuzzy member ship may be triangular membership or Gaussian membership. Several studies applied the triangular membership due to its merits, where the triangular membership represents one of the attractive linear memberships of the fuzzy methodology, which is characterized by less computation time and simplicity [54]. The sensitivity increases when moving from linear membership functions to curvilinear membership functions (i.e., Gaussian-sigmoidal). Therefore, from the literature review of fuzzy logic, the triangular membership is shown in more than 90% of practical applications of electrical systems which are used in input and output. In addition, this membership belongs to the first-order mathematical function, which is characterized by reducing the computational load. It is usually applied along with a PID controller and fully symmetric functions in input and output. The selection of fuzzy control parameters depends on the nature of the studied system and the knowledge of the designer of the system. So, the selection of fuzzy membership functions range is based on the prediction of the universe of discourse of input and output in the studied system [55]. Usually, the decision-maker is able to define the risk-free of input and output. Thus, the range intervals have been selected between $[-1, 1]$ intervals, which is expected and doesn't need the input and output to go far away from this period to achieve more system stability. It is advisable to use the symmetric triangular MF with 50% overlap, and then apply a tuning procedure during which we can either change the left and/or right spread and/or overlap. This is to be continued till, getting satisfactory results [56]. In this work, five triangular membership functions of the fuzzy-PID controller are utilized namely negative big (NB), negative small (NS), zero (Z), positive small (PS), and positive big (PB). The triangular member function is shown in Figure 6, which is used for both inputs and output. Accordingly, the five memberships of the input and output variables of the fuzzy-PID controller, the generation of fuzzy output need 25 rules. These rules play an important role in the performance of the fuzzy-PID controller. These rules of the fuzzy logic controller are depicted in Table 4.

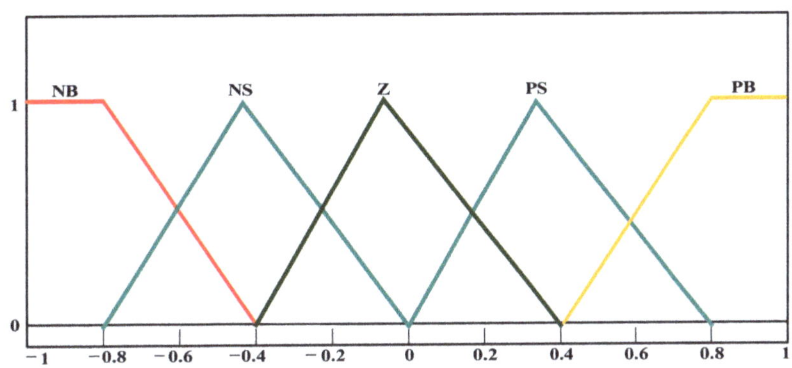

Figure 6. The Fuzzy Logic Controller membership functions of inputs and output.

Table 4. The Fuzzy rule base [50].

ACE	$\dot{ACE}$				
	NB	**NS**	**Z**	**PS**	**PB**
NB	NB	NB	NB	NS	Z
NS	NB	NB	NS	Z	PS
Z	NB	NS	Z	PS	PB
PS	NS	Z	PS	PB	PB
PB	Z	PS	PB	PB	PB

3.2. The Proposed Optimization Technique (AOA)

In 2020, a new meta-heuristic optimization technique namely the Arithmetic optimization algorithm (AOA) was invented by Laith et al. [57]. This technique is characterized by a high exploration search strategy, meaning that they can achieve the global optimum solution with few search agents. The distribution behavior of this technique is based on the main mathematical operations including; (Division (D), Multiplication (M), Addition (A), and Subtraction (S)). According to meta-heuristic techniques, the former of this method can make a wide coverage of searching space by using the number of search agents of the algorithm to avoid the local solution but for obtaining the global one. The AOA follows a detected methodology to obtain the global solution mentioned in the next steps:

Step 1: the proposed controller parameters in this paper have upper and lower boundaries; the population of the AOA method according to the main mathematical operations has been generated between these boundaries to achieve the global goal. The population of the AOA method can be formulated in Equation (7) as follows:

$$s(N,d) = rand(N,d) \times (UB - LB) + LB \tag{7}$$

where; N refers to the number of utilized search agents, d represents the variable dimensions (controller parameters), UB and L represent the upper and lower value of variables.

Step 2: The AOA method begins with candidate solutions (S) generated randomly. The optimal obtained solution in each endeavor represents a solution near the global goal (target). The next matrix illustrates the position of solutions obtained.

$$s = \begin{bmatrix} s_{1,1} & s_{1,2} & \cdots & s_{1,d} \\ s_{2,1} & s_{2,2} & \cdots & s_{2,d} \\ \vdots & \vdots & \ddots & \vdots \\ s_{n,1} & s_{n,2} & \cdots & s_{N,d} \end{bmatrix} \tag{8}$$

Step 3: achieving the fitness solution among the obtained ones as mentioned in step 2. The fitness solution can be formulated in Equation (9) as follows:

$$f_{fitness} = [f1\ f2\ f3 \ldots\ldots\ldots\ldots f_N]^T \tag{9}$$

Step 4: before the AOA role begins, the search phase (exploration and exploitation) must be detected through the next formulation in Equation (10) of Math Optimizer Acceleration (MOA):

$$MOA(C\_Iter) = C\_Iter \times \left(\frac{Max - Min}{M\_Iter}\right) + Min \tag{10}$$

where; $MOA(C\_Iter)$ refers to the value of function at the tth iteration, C_Iter represents the current iteration. The current iteration is between among 1 and the maximum number of iterations (M_Iter) dimensions (controller parameters), Min and Max are the minimum and maximum values of the accelerated function.

Step 5: the mathematical calculation processes as (Division (D) and Multiplication (M)) cannot reach the target easily according to their high dispersion. Accordingly, these strategies (D and M) are utilized in the exploration search process. The first operator D in this phase is conditioned by r2 < 0.5 (r2 is a random number) and the second operator M will be neglected until the first one ends its task in searching for a solution near to the goal. Otherwise, M will lead the process instead of the first operator D. The exploration process is modeled in Equation (11):

$$Xi,j(C\_Iter+1)\begin{cases} best(x_j) \div (MOP+\epsilon) \times ((UBj - LBj) \times \mu + LBj), & r2 < 0.5 \\ best(x_j) \times (MOP) \times ((UBj - LBj) \times \mu + LBj), & otherwise \end{cases} \tag{11}$$

where, $Xi(C\_Iter+1)$ and $Xi,j(C\_Iter+1)$ are the ith solution in the next iteration and the jth position of the ith solution at the current iteration, $best(x_j)$ represents the jth position in the best-obtained solution, ϵ is a small integer number, μ is the control parameter to make adjusting in the process of search which is fixed equal to 0.5.

Step 6: deep search (exploitation) in this strategy using the mathematical operators (Subtraction (S) and Addition (A)) has been applied to be near to the optimal solution and reach it after several iterations. The first operator S in this phase is conditioned by r3 < 0.5 (r3 is a random number) and the second operator A will be neglected until the first one end its task in deep searching for obtaining the best solution. Otherwise, *A* will lead the process instead of the first operator *S*. The exploitation process is modeled in Equation (12):

$$Xi,j(C\_Iter+1)\begin{cases} best(x_j) - (MOP) \times ((UBj - LBj) \times \mu + LBj), \; r3 < 0.5 \\ best(x_j) + (MOP) \times ((UBj - LBj) \times \mu + LBj), \; otherwise \end{cases} \quad (12)$$

Step 7: steps (3) to (6) are repeated until ending all iterations.

Step 8: The last step is to achieve the optimum solution which achieves the objective function.

Furthermore, Figure 7 illustrates the flow chart of the AOA which clarify the previous optimization steps.

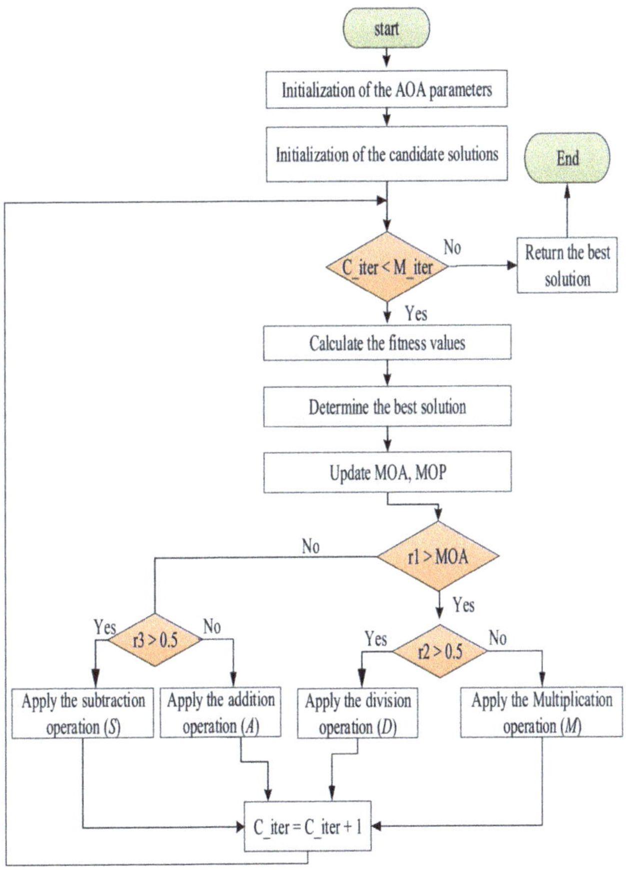

Figure 7. The flowchart of the AOA technique.

3.3. The Proposed Fuzzy-PID Control Strategy Based AOA Algorithm

The parameters of the proposed control strategy have been selected based on AOA algorithm to overcome any deviations related to the considered system. Moreover, the integral time absolute error (ITAE) function is utilized as an objective function (J) of the proposed optimization algorithm. Furthermore, Equation (13) formulates the objective function J to minimize the deviations in system related to the frequency and tie-line power. The ITAE has been selected in this work according to its merits like, it has an additional time multiplies with the error function which makes the system faster than using other objective functions forms (e.g., the integral square error (ISE) and the integral absolute error (IAE)). Also, the ITAE performance index has the advantage of settling the system, which is more quickly compared to other objective functions [49].

$$J = \text{ITAE} = \int_{0}^{Tsim} |\Delta f1| + |\Delta f2| + |\Delta ptie|).t \times d_t \tag{13}$$

where; T_{sim} is the simulation time. The flow chart of the proposed AOA is shown in Figure 8.

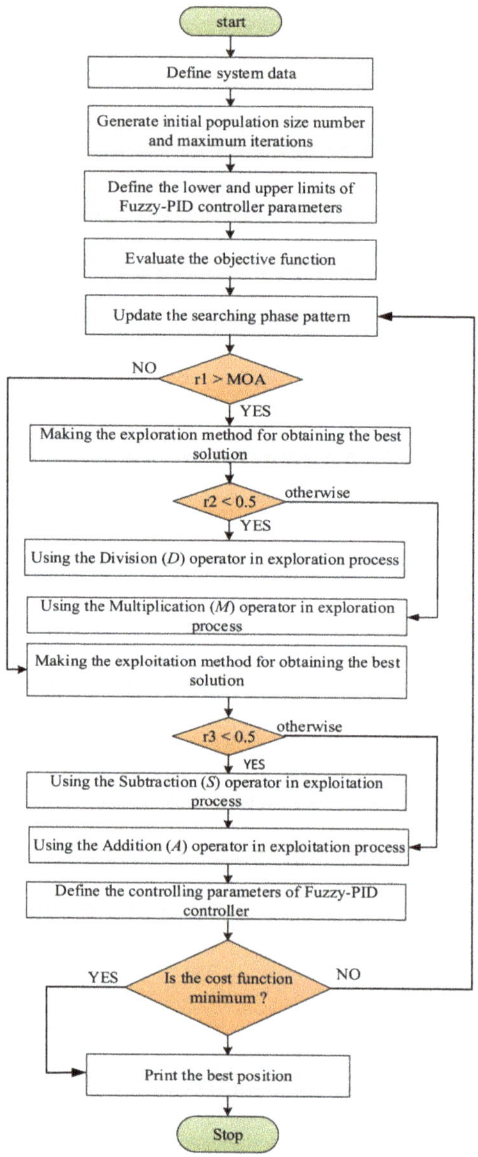

Figure 8. The flowchart of applying AOA technique with the proposed controller.

4. Discussion and Simulation Results

The investigated system was built on a 2.60 GHz Intel (R) PC with 4.00 GB of RAM using the MATLAB/SIMULINK®software (R2019b) environment. In addition, the AOA has been written in the m file in order to tune the proposed controller parameters of the automatic load frequency control process. In this work, the performance of Fuzzy-PID and PID controllers that are applied to enhance the studied system performance using the AOA technique is measured according to the value of the best objective function over iterations. The initial values of the proposed AOA technique utilized in this work are; the number of search agents equals 30 and the number of maximum iterations equals 50. Also, the limitations of the proposed fuzzy-PID controller are in the range of [0, 10]. The convergence curve of the proposed fuzzy-PID controller compared to the PID controller using the AOA is shown in Figure 9. There is a clear difference between the performances of both controllers using the AOA technique. The behavior of the PID convergence curve can be summarized as follows: it begins with a high best function value (near to 0.08) and drops along iterations until it ends its career at the final iteration, reaching the best function value (near to 0.03). As for the Fuzzy-PID convergence curve behaviors, they start with a low best function value until it reaches a value (near to zero) to obtain an optimal objective function with more system stability. In general, the preference is for the Fuzzy-PID controller.

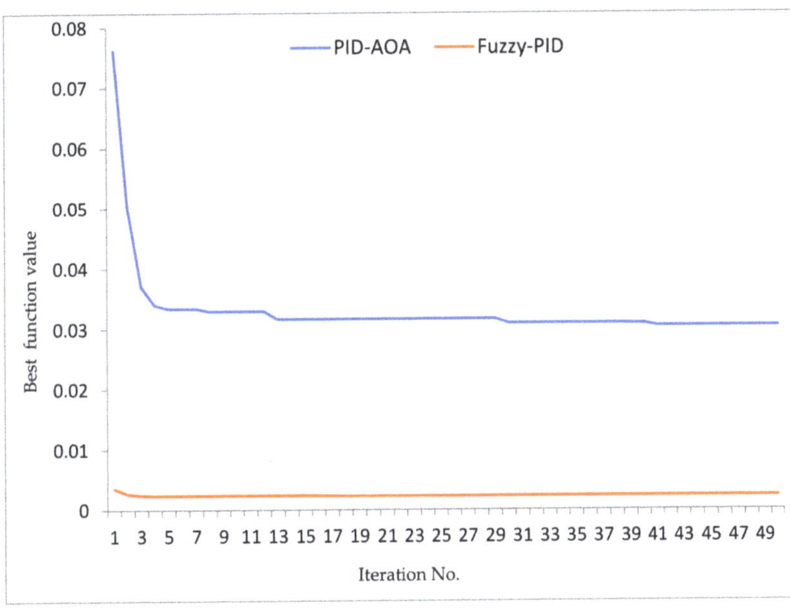

Figure 9. The convergence curve of the designed controllers based on the AOA.

All next simulation results ensure the effectiveness of the AOA in obtaining optimum controller parameters compared to other applied techniques as (PID controller based-DE [49], PID controller based-TLBO [50], and Fuzzy-PID controller based-LUS-TLBO [50]). For programming all mentioned optimization techniques, MATLAB/M-files matched with MATLAB /Simulink. All graphical and numerical numbers of obtained results are discussed in the next scenarios as follows:

4.1. Studied Power System Performance Considering AC-Lines Connection Only

This scenario clarifies the dynamic system performance of the occurring deviation in frequency waveform at both areas and the tie-line power exchange between them with a 1%

step load perturbation (SLP) which applied to the first area only. The AC-line has tied both areas of the studied power system without any HVDC lines. Compared to other controllers applied previously by researchers, the proposed fuzzy-PID controller-based AOA proves its robustness in adjusting and stabilizing the power system frequency. Figures 10 and 11 show the frequency deviation performance at both areas of the studied system. The tie-line power exchange has been cleared in Figure 12. Table 5 indicates the optimal Fuzzy-PID controller parameters compared to other mentioned controllers with different optimization techniques. Additionally, the performance specifications; overshoot (OS), and undershoot (US) of the proposed Fuzzy-PID controller and followed controllers for the studied system are shown in Table 6. The percentage improvements in US and OS with different controllers are denoted in Table 7.

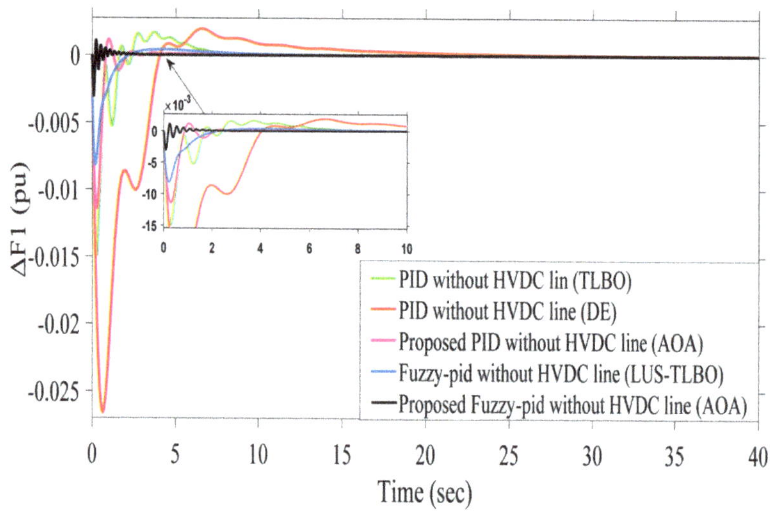

Figure 10. Dynamic response comparison results of Δf_1 for scenario 4.1.

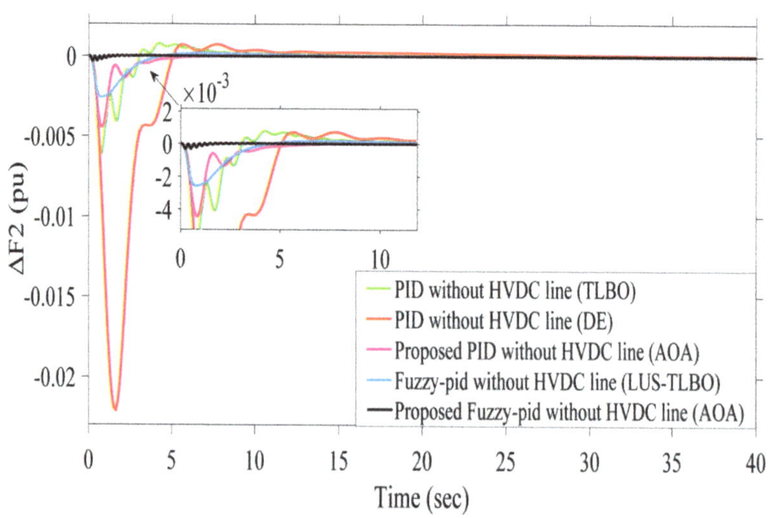

Figure 11. Dynamic response comparison results of Δf_2 for scenario 4.1.

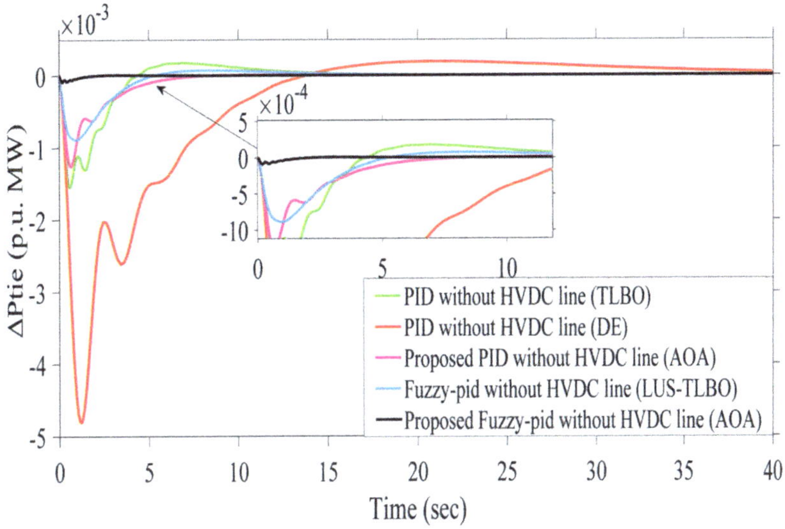

Figure 12. Dynamic response comparison results of Δp_{tie} for scenario 4.1.

Table 5. The optimal controllers' values for scenario 4.1.

ACE	Thermal k1	Thermal k2	Thermal k3	Thermal k4	Hydro k1	Hydro k2	Hydro k3	Hydro k4	Gas k1	Gas k2	Gas k3	Gas k4
Fuzzy-PID (LUS-TLBO) [50]	1.9985	1.9874	1.9679	1.9926	0.1002	1.1278	0.1032	0.7264	1.9782	1.0734	1.979	1.6516
Fuzzy-PID (AOA)	10	4.7015	4.7895	10	10	0.5402	0.01	10	9.4636	10	1.0988	10
	k_p	k_i	k_d		k_p	k_i	k_d		k_p	k_i	k_d	
PID (TLBO) [50]	4.1468	4.0771	2.0157		1.0431	0.6030	2.2866		4.7678	3.7644	4.9498	
PID (DE) [49]	0.779	0.2762	0.6894		0.5805	0.2291	0.7079		0.5023	0.9529	0.6569	
PID (AOA)	10	1.5975	2.7449		1.5975	0.0837	0.0875		10	10	1.2779	

Table 6. Performance evaluation of PID and Fuzzy-PID using all mentioned techniques for scenario 4.1.

Different Dynamic Responses	Fuzzy-PID Based LUS-TLBO OS & US × (10^{-3})	Fuzzy-PID Based AOA OS & US × (10^{-3})	PID Based-TLBO OS & US × (10^{-3})	PID Based DE OS & US × (10^{-3})	PID Based AOA OS & US × (10^{-3})
Dynamic response of (ΔF1)	0.5510 −8.9579	1.09 −3.059	1.7217 −19.7259	2.0347 −26.5777	1.158 −11.42
Dynamic response of (ΔF2)	0.2119 −3.0119	0.03285 −0.321	0.4363 −12.7986	0.7722 −22.1421	0.02096 −4.443
Dynamic response of (ΔPtie)	0.0826 −0.9653	0.008388 −0.08917	0.1712 −3.0782	0.1935 −4.7595	0.01107 −1.249

Table 7. Percentage improvement in US and OS for all previous mentioned controllers with different techniques based on PID controller via DE for scenario 4.1.

Controller	Δf_1		Δf_2		ΔP_{tie}	
	U_{sh}	O_{sh}	U_{sh}	O_{sh}	U_{sh}	O_{sh}
Fuzzy-PID (LUS-TLBO)	66.29	72.92	86.4	72.56	79.72	57.31
Fuzzy-PID (AOA)	88.49	46.43	98.55	95.75	98.13	95.67
PID (TLBO)	25.78	15.38	42.2	43.5	35.33	11.53
PID (AOA)	57.03	43.09	79.93	97.29	73.76	94.28

4.2. Studied Power System Performance Considering AC-DC Lines Connection

The effect of adding an HVDC lines connection in addition to the existing AC lines to transmit the power alternately between both areas in the studied system is introduced in this scenario with a 1% step load perturbation (SLP) is applied to the first area only. The behavior of frequency at both areas and the tie-line power between the both is shown in Figures 13–15 respectively. Table 8 indicates different obtained controller parameters of the studied system in the presence of an HVDC line connection. It is observed that from Table 9, the proposed fuzzy-PID controller-based AOA achieves more system stability (less oscillation) by monitoring the system performance specifications such as OS and US than other mentioned controllers. Table 10 illustrates the percentage improvement in US and OS of area frequencies and tie-line power with different controllers.

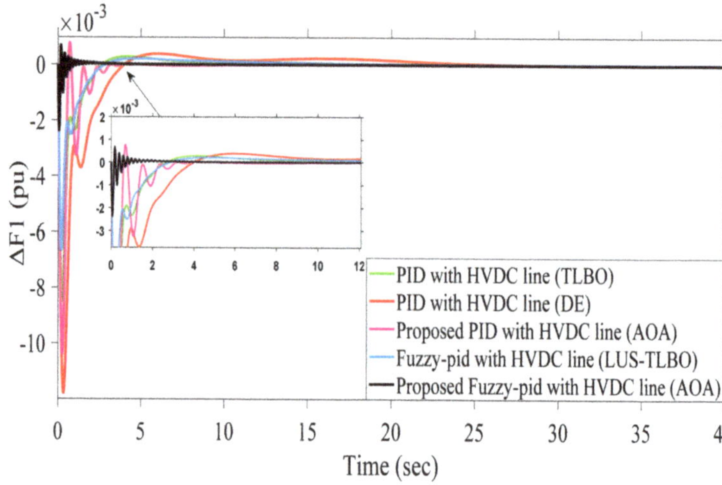

Figure 13. Dynamic response comparison results of Δf_1 for scenario 4.2.

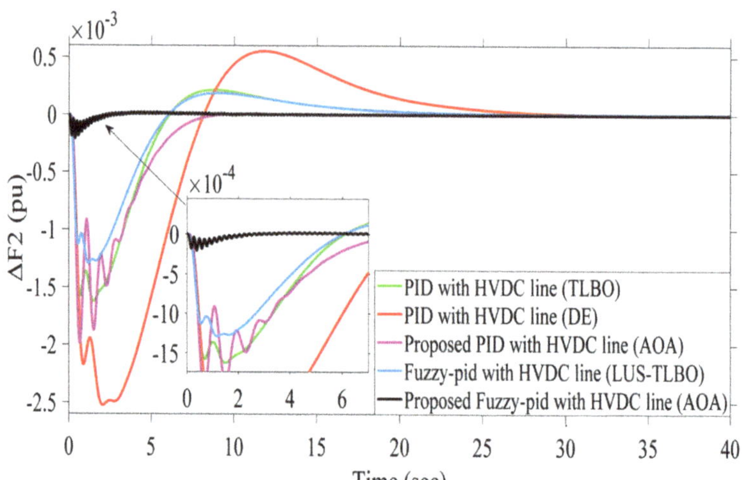

Figure 14. Dynamic response comparison results of Δf_2 for scenario 4.2.

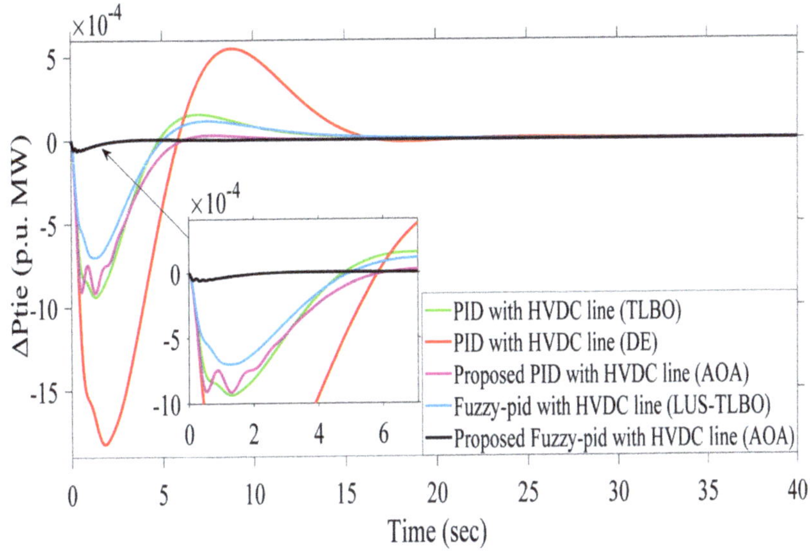

Figure 15. Dynamic response comparison results of Δp_{tie} for scenario 4.2.

Table 8. The optimal controllers' values for scenario 4.2.

ACE	Thermal k1	Thermal k2	Thermal k3	Thermal k4	Hydro k1	Hydro k2	Hydro k3	Hydro k4	Gas k1	Gas k2	Gas k3	Gas k4
Fuzzy-PID (LUS-TLBO) [50]	1.9995	1.9889	1.9975	1.9829	0.9668	1.2913	0.1001	1.9988	1.9969	1.1982	1.9867	1.9882
Fuzzy-PID (AOA)	10	9.9164	4.8295	10	10	0.01	4.9166	10	10	0.01	10	9.8531
		k_p	k_i	k_d		k_p	k_i	k_d		k_p	k_i	k_d
PID (TLBO) [50]	5.0658	3.9658		2.417	0.7032	0.0220		0.0264	8.7211	7.4729		2.4181
PID (DE) [49]	1.6929	1.9923		0.8269	1.77731	0.7091		0.4355	0.9094	1.9425		0.2513
PID (AOA)	9.8739	1.2609		3.5014	10	0.0164		1.9788	1.2609	10		0.490

Table 9. Performance evaluation of PID and Fuzzy-PID using all mentioned techniques for scenario 4.2.

Different Dynamic Responses	Fuzzy-PID Based LUS-TLBO OS & US × (10^{-3})	Fuzzy-PID Based AOA OS & US × (10^{-3})	PID Based-TLBO OS & US × (10^{-3})	PID Based DE OS & US × (10^{-3})	PID Based AOA OS & US × (10^{-3})
Dynamic response of (ΔF1)	0.2809 −6.7244	0.6828 −2.373	0.2798 −8.497	0.3792 −11.6667	0.7707 −10.4
Dynamic response of (ΔF2)	0.2084 −1.4021	0.02112 −0.2083	0.2138 −1.624	0.5491 −2.5199	0.005693 −1.992
Dynamic response of (ΔPtie)	0.1353 −0.7292	0.006201 −0.05983	0.1557 −0.9366	0.5474 −1.8133	0.03018 −0.9134

Table 10. Percentage improvement in US and OS for all previous mentioned controllers with different techniques based on PID controller via DE for scenario 4.2.

Controller	Δf_1 U_{sh}	Δf_1 O_{sh}	Δf_2 U_{sh}	Δf_2 O_{sh}	ΔP_{tie} U_{sh}	ΔP_{tie} O_{sh}
Fuzzy-PID (LUS-TLBO)	42.36	25.92	44.36	62.05	59.79	75.28
Fuzzy-PID (AOA)	79.66	80.06	91.73	96.15	96.7	98.87
PID (TLBO)	27.17	26.21	35.55	61.06	48.35	71.56
PID (AOA)	10.86	103.2	20.95	98.96	49.63	94.48

4.3. Studied Power System Performance Considering AC-DC Lines Connection in Addition to Different Load Disturbances

The perturbation in load at the first area has been increased to be a 5% SLP instead of 1% to ensure the validation of the obtained fuzzy-PID controller parameters that mentioned in Table 4 in stabilizing the studied system frequency. The frequency deviation of both areas and the exchanged power between them is shown in Figures 16–18 respectively. The OSs and USs values of oscillations at frequency waveform at area 1 and area 2 are mentioned in Table 11; in addition to values of tie-line power. Table 12 shows the percentage improvement in OS and US with different controllers in the case of increasing SLP to 0.05 p.u. It can be said that the studied system became more stable using the proposed fuzzy-PID controller-based AOA compared to the fuzzy-PID controller-based LUS-TLBO and PID controller based on TLBO, DE, and AOA.

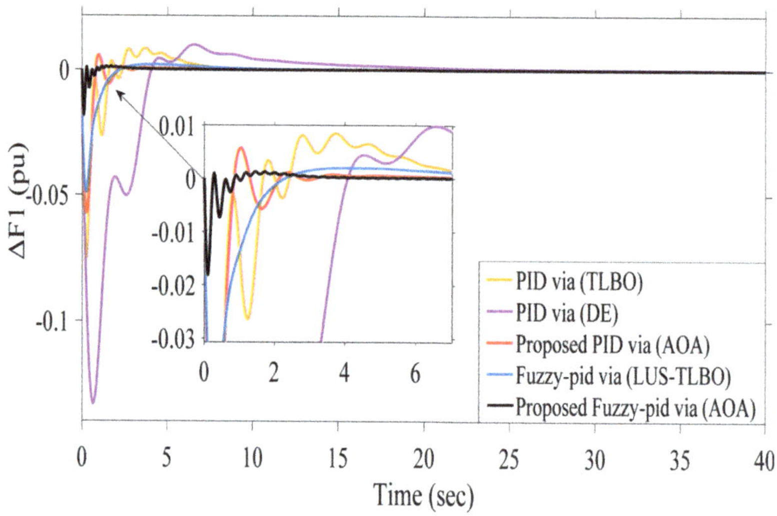

Figure 16. Dynamic response comparison results of Δf_1 for scenario 4.3.

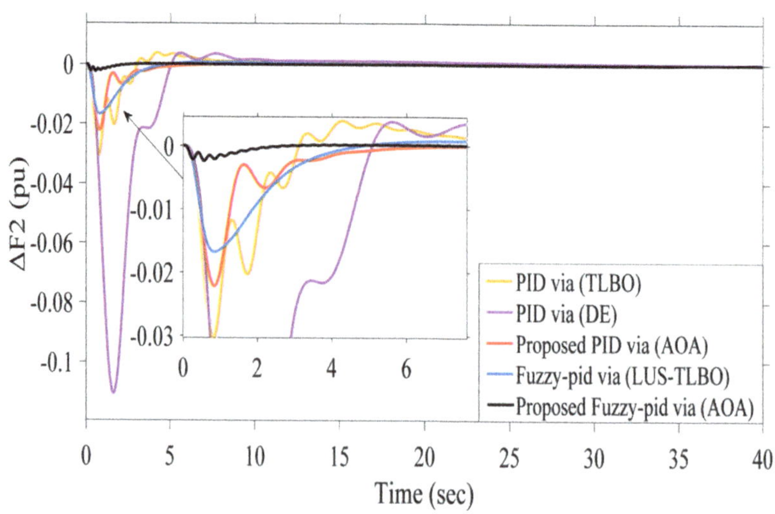

Figure 17. Dynamic response comparison results of Δf_2 for scenario 4.3.

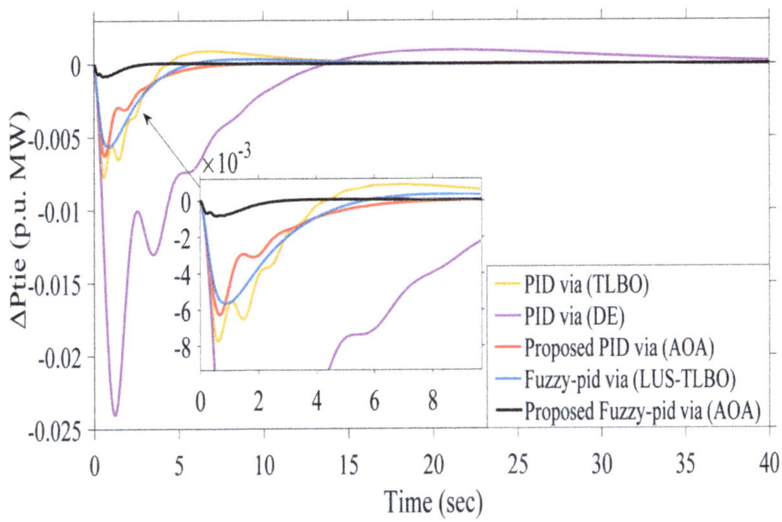

Figure 18. Dynamic response comparison results of Δp_{tie} for scenario 4.3.

Table 11. Performance evaluation of PID and Fuzzy-PID using all mentioned techniques for scenario 4.3.

Different Dynamic Responses	Fuzzy-PID Based LUS-TLBO OS & US × (10^{-3})	Fuzzy-PID Based AOA OS & US × (10^{-3})	PID Based-TLBO OS & US × (10^{-3})	PID Based DE OS & US × (10^{-3})	PID Based AOA OS & US × (10^{-3})
Dynamic response of (ΔF1)	2.063 −48.74	1.28 −18.13	8.552 −74.85	9.956 −133.1	5.795 −57.07
Dynamic response of (ΔF2)	0.8297 −16.77	0.1825 −2.418	3.981 −30.52	3.823 −110.7	0.1048 −22.21
Dynamic response of (ΔPtie)	0.359 −5.6	0.06577 −0.8255	0.9155 −7.719	0.9719 −24.02	0.05535 −6.245

Table 12. Percentage improvement in US and OS for all previous mentioned controllers with different techniques based on PID controller via DE for scenario 4.3.

Controller	Δf_1		Δf_2		ΔP_{tie}	
	U_{sh}	O_{sh}	U_{sh}	O_{sh}	U_{sh}	O_{sh}
Fuzzy-PID (LUS-TLBO)	63.38	79.28	84.85	78.30	76.79	63.06
Fuzzy-PID (AOA)	86.38	87.14	97.82	95.23	96.56	93.23
PID (TLBO)	43.76	14.10	72.43	−4.13	35.33	5.8
PID (AOA)	57.12	41.79	79.94	97.26	74.00	94.30

4.4. Studied Power System Performance Considering the Effect of System Parameters' Variations

The robustness of the studied power system has been tested by making a variation of system parameters such as thermal governor time constant at both areas, thermal turbine time constant at both areas and hydro governor time constant simultaneously in the range of +25% and −25% from their nominal values reported in Table 2. These variations ensure that, the obtained fuzzy-PID controller parameters based AOA that mentioned in Table 4 can efficiently damp and overcome the oscillations and achieve more system stability under system parameters variation. The dynamic performance of both areas frequencies and tie-line power exchange after ±25% system parameters variation is illustrated in Figures 19–21 respectively.

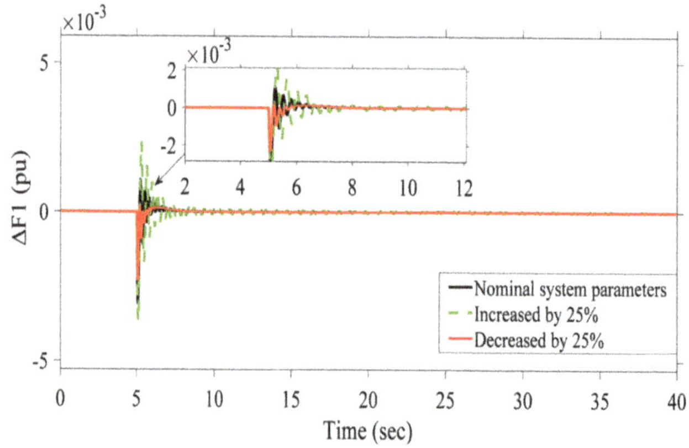

Figure 19. Dynamic response comparison results of Δf_1 for scenario 4.4.

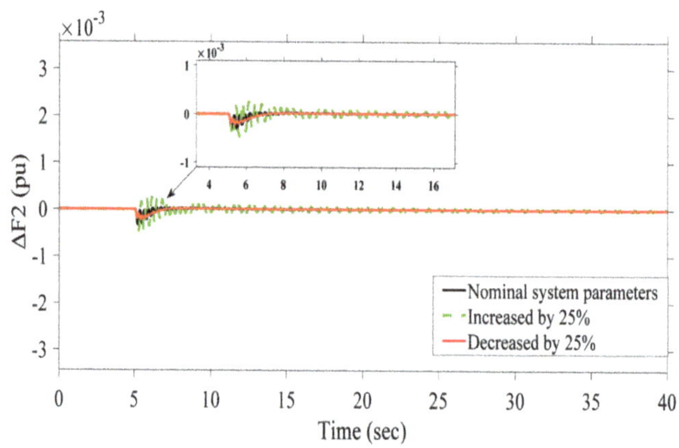

Figure 20. Dynamic response comparison results of Δf_2 for scenario 4.4.

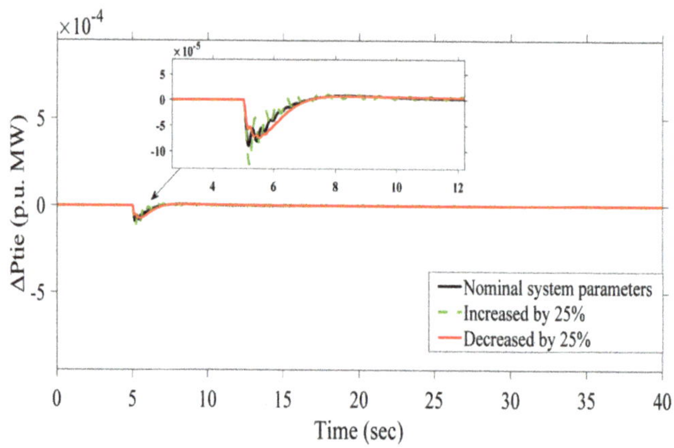

Figure 21. Dynamic response comparison results of Δp_{tie} for scenario 4.4.

4.5. Studied Power System Performance Considering Wind Power Penetration

4.5.1. Case A

The penetration of wind power at both areas of the studied system is tested in this scenario at nominal system parameters. Both wind farms have the same power rating and penetrated the studied system at the same time (t = 0 s) with the instant of 1% load variation. This scenario is applied to ensure the robustness of the proposed fuzzy-PID controller based AOA that mentioned in Table 4 in achieving system stability in existing wind energy. All dynamic system performance (both-area frequencies and tie power) have been plotted in Figures 22–24 respectively. Table 13 illustrates the OSs and USs behavior of both areas frequencies and tie-line power with penetrating of wind energy at the studied power system. Also, Table 14 indicates the percentage performance in US and OS with different controllers.

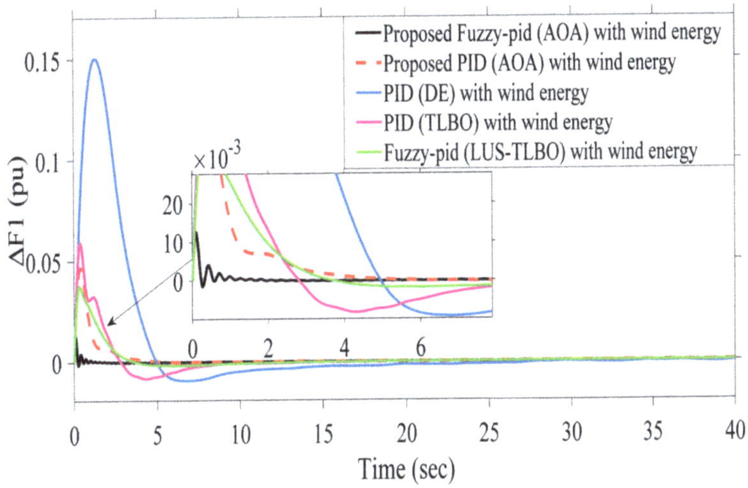

Figure 22. Dynamic response comparison results of Δf_1 for scenario 4.5.1: case A.

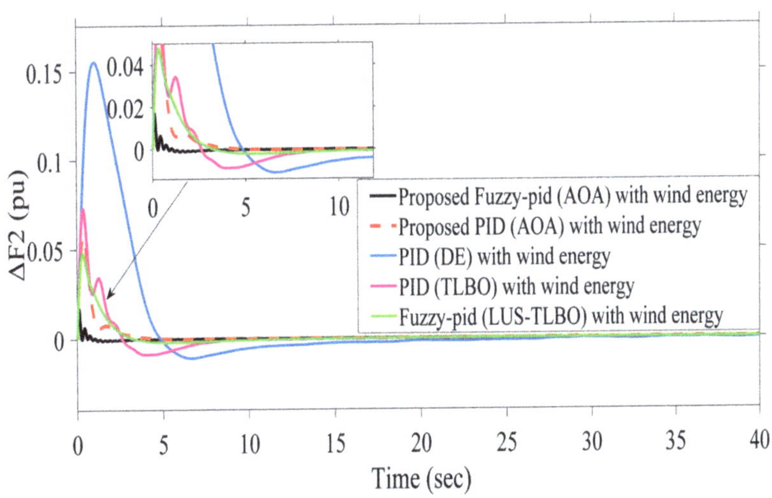

Figure 23. Dynamic response comparison results of Δf_2 for scenario 4.5.1: case A.

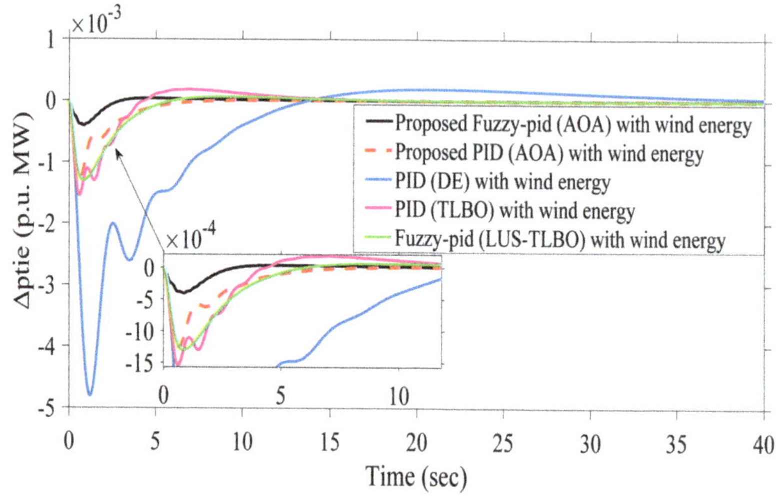

Figure 24. Dynamic response comparison results of Δp_{tie} for scenario 4.5.1: case A.

Table 13. Performance evaluation of PID and Fuzzy-PID using all mentioned techniques for scenario 4.5.1: case A.

Different Dynamic Responses	Fuzzy-PID Based LUS-TLBO OS & US × (10^{-3})	Fuzzy-PID Based AOA OS & US × (10^{-3})	PID Based-TLBO OS & US × (10^{-3})	PID Based DE OS & US × (10^{-3})	PID Based AOA OS & US × (10^{-3})
Dynamic response of (ΔF1)	37.7 −1.88	12.69 −1.526	59.25 −8.257	150.5 −9.28	46.6 −0.2713
Dynamic response of (ΔF2)	47.75 −2.163	17.1 −1.106	73.12 −8.793	155.4 −10.92	56.86 −0.3124
Dynamic response of (ΔPtie)	0.06752 −1.297	0.0361 −0.41	0.1831 −1.543	0.1944 −4.804	0.01107 −1.249

Table 14. Percentage improvement in US and OS for all previous mentioned controllers with different techniques based on PID controller via DE for scenario 4.5.1: case A.

Controller	Δf_1		Δf_2		ΔP_{tie}	
	U_{sh}	O_{sh}	U_{sh}	O_{sh}	U_{sh}	O_{sh}
Fuzzy-PID (LUS-TLBO)	79.74	74.95	80.19	69.27	73.00	65.27
Fuzzy-PID (AOA)	83.56	91.57	89.87	89.00	91.47	81.43
PID (TLBO)	11.02	60.63	19.48	52.95	67.88	5.81
PID (AOA)	97.08	69.04	97.14	63.41	74.00	94.31

4.5.2. Case B

In this case, the studied power system is stabilized until load variation occurred at (t = 40 s) in the first area. Also, the wind farm at the first area shared generated power at (t = 300 s) then the wind energy from the wind farm at the second area enter to feed the system at (t = 600 s). It is observed that the obtained fuzzy-PID controller parameters based AOA that mentioned in Table 5 achieve more system stability than those obtained by LUS-TLBO. Also, the fuzzy-PID controller parameters based on the AOA recover the studied system and back it to nominal operation process faster than PID controller based TLBO, DE, and AOA. Figures 25–27 show the dynamic performance of area frequencies and the exchanged tie-line power when sharing wind energy to the studied system at different instants (t = 300 s, t = 600 s). Table 15 extract the effectiveness of the fuzzy-PID-based AOA compared to other mentioned controllers by showing the OSs and USs of both areas frequencies and the tie-line power. The percentage improvement in US and OS with different controllers via various techniques is mentioned in Table 16.

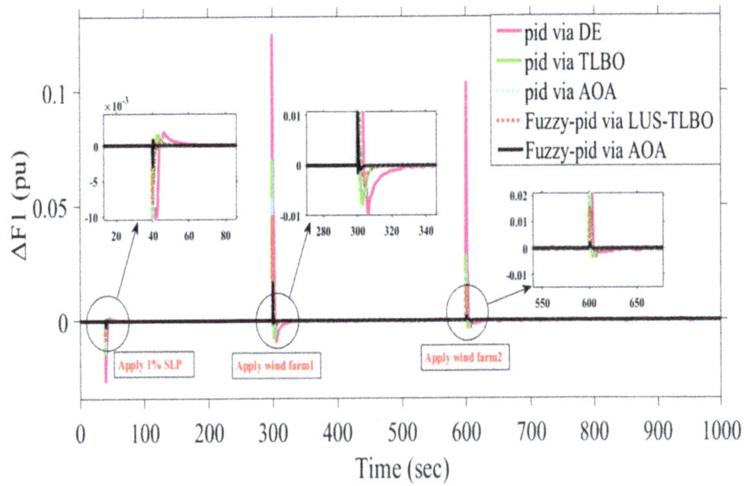

Figure 25. Dynamic response comparison results of Δf_1 for scenario 4.5.2: case B.

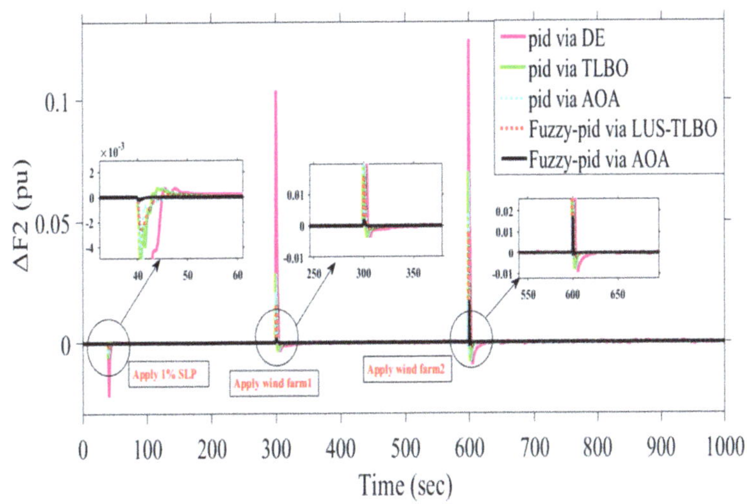

Figure 26. Dynamic response comparison results of Δf_2 for scenario 4.5.2: case B.

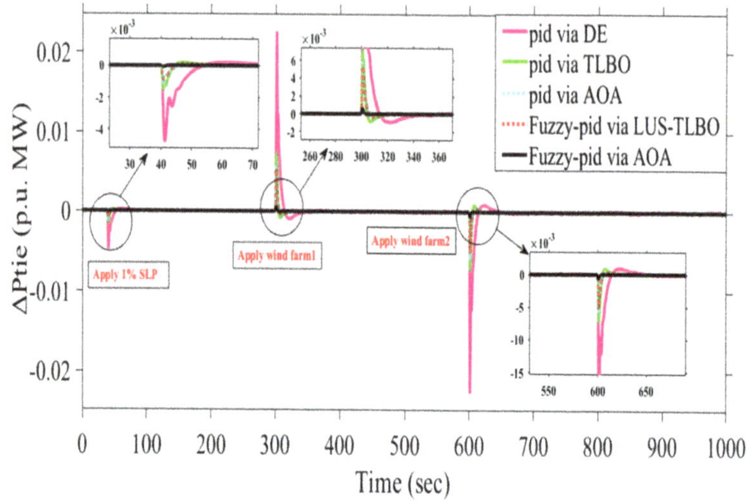

Figure 27. Dynamic response comparison results of Δp_{tie} for scenario 4.5.2: case B.

Table 15. Performance evaluation of PID and Fuzzy-PID using all mentioned techniques for scenario 4.5.2: case B.

Different Dynamic Responses	Fuzzy-PID Based LUS-TLBO OS & US $\times (10^{-3})$	Fuzzy-PID Based AOA OS & US $\times (10^{-3})$	PID Based-TLBO OS & US $\times (10^{-3})$	PID Based DE OS & US $\times (10^{-3})$	PID Based AOA OS & US $\times (10^{-3})$
Dynamic response of (ΔF1)	45.73 −8.18	16.98 −3.058	70.37 −14.98	124.9 −26.6	53.66 −11.43
Dynamic response of (ΔF2)	45.57 −2.586	17.01 −1.738	70.31 −7.94	124.8 −22.13	53.68 −5.42
Dynamic response of (ΔPtie)	5.233 −5.199	0.7004 −0.7178	7.246 −7.257	22.57 −22.60	5.867 −5.874

Table 16. Percentage improvement in US and OS for all previous mentioned controllers with different techniques based on PID controller via DE for scenario 4.5.2: case B.

Controller	Δf_1		Δf_2		ΔP_{tie}	
	U_{sh}	O_{sh}	U_{sh}	O_{sh}	U_{sh}	O_{sh}
Fuzzy-PID (LUS-TLBO)	69.25	63.39	88.31	63.49	77.00	76.81
Fuzzy-PID (AOA)	88.50	86.41	92.15	86.37	96.82	96.90
PID (TLBO)	43.68	43.66	64.12	43.66	67.89	67.89
PID (AOA)	57.03	57.04	75.51	56.99	74.01	74.00

5. Conclusions

In this paper, there are main points that have been included, which can be summarized as follows:

- The proposed fuzzy-PID controller has been implemented on the two-area interconnected multi-source power systems that include thermal, hydro, and gas power plants for tackling the LFC problem.
- The selection the of the proposed controller parameters has been made via a new meta-heuristic optimization technique, which is known as an arithmetic optimization algorithm, to get the optimal solution which leads to stabilizing the system performance. Appling HVDC link in addition to AC links to overcome the demerits of the AC tie-lines.

- Considering several challenges during designing the proposed control parameters such as (i.e., system uncertainties, different load variations, and different levels of wind power penetration).
- Applying different scenarios to validate the robustness of the proposed fuzzy-PID controller than other previous controllers.
- The proposed AOA has tuned the fuzzy-PID controller to achieve a better disturbance rejection ratio than a newly published technique namely a hybrid Local Unimodal Sampling and Teaching Learning Based Optimization using also Fuzzy-PID controller. On the other hand, PID controller-based-AOA gets more system stability than which utilized in previous research work optimized by Differential Evolution, TLBO.
- The system performance has been enhanced by 90.76% by applying the proposed fuzzy-PID controller based on the AOA algorithm in comparison with the fuzzy-PID controller based on the LUS-TLBO algorithm.
- The presence of an HVDC link in parallel with an AC link improved system performance by 95.42% when compared to using only an AC tie-line.
- According to the analysis and simulation results, the Fuzzy-PID controller based on the AOA algorithm gives better results in terms of system stability and security in comparison with other previous control techniques.

There are some points will be taken in consideration in the future work and can be summarized as follows:

- Increasing the penetration level of renewable energy sources in the considered system.
- Applying different types of energy storage devices to study its effect on LFC problem.
- Improving of different recent optimization techniques to achieve the desired control parameters that lead to satisfied performance.

Author Contributions: Conceptualization, A.H.A.E., M.K., G.M. and S.K.; data curation, I.B.M.T.; formal analysis, A.H.A.E., M.K. and G.M.; funding acquisition, I.B.M.T. and S.K.; investigation, A.H.A.E., M.K. and G.M.; methodology, I.B.M.T. and S.K.; project administration, A.H.A.E., M.K. and G.M.; resources, I.B.M.T. and S.K.; supervision, S.K. and I.B.M.T.; validation, A.H.A.E., M.K. and G.M.; visualization, A.H.A.E., M.K. and G.M.; writing—original draft, A.H.A.E., M.K. and G.M.; writing—review and editing, I.B.M.T. and S.K. All authors have read and agreed to the published version of the manuscript.

Funding: Taif University Researchers Supporting Project Number (TURSP-2020/61), Taif University, Taif, Saudi Arabia.

Institutional Review Board Statement: Not applicable.

Informed Consent Statement: Not applicable.

Data Availability Statement: Not applicable.

Acknowledgments: The authors would like to acknowledge the financial support received from Taif University Researchers Supporting Project Number (TURSP-2020/61), Taif University, Taif, Saudi Arabia.

Conflicts of Interest: The authors declare that there is no conflict of interest.

Nomenclature

Symbols	Parameters
FLC	Fuzzy logic control
PID	Proportional-Integral-Derivative
AOA	Arithmetic Optimization Algorithm
LFC	Load Frequency Control
HVDC	High Voltage Direct Current
RESs	Renewable Energy Sources

CESs	Conventional Energy Sources
DE	Differential Evolution
LUS	Local Unimodal Sampling
TLBO	Teaching Learning-based Optimization
OS	overshoot
US	undershoot
SLP	Step load perturbation
ΔP_{wt}	Wind turbine output power
ρ	The air density
A_T	The swept area by the blades of turbine
V_W	The wind speed
C_p	The coefficient of rotor blades
C_1–C_7	The turbine coefficients
β	The pitch angle
r_T	The radius of rotor
ω_T	The rotor speed
λ_T	The optimum tip-speed ratio
λ_i	The intermittent tip-speed ratio
B_1	Frequency bias factor of Area 1
B_2	Frequency bias factor of Area 2
$\Delta f1$	Deviation in frequency waveform in area 1
$\Delta f2$	Deviation in frequency waveform in area 2
ΔP_{tie1-2}	Tie-line power exchange at area 1
ΔP_{tie2-1}	Tie-line power exchange at area 2
T_{12}	Coefficient of synchronizing
R_1	Regulation constant of thermal power plant
R_2	Regulation constant of hydro power plant
R_3	Regulation constant of gas turbine
a_{12}	Control Area Capacity Ratio
K_T	Participation factor for thermal unit
K_H	Participation factor for hydro unit
K_G	Participation factor for gas unit
K_{ps}	Gain constant of power system
T_{ps}	Time constant of power system
T_{sg}	Governor time constant
T_t	Turbine Time Constant
K_r	Gain of reheater steam turbine
T_r	Time Constant of reheater steam turbine
T_{gh}	Speed governor time constant of hydro turbine
T_{rs}	Speed governor reset time of hydro turbine
T_{rh}	Transient droop time constant of hydro turbine speed governor
T_w	Nominal string time of water in penstock
b_g	Gas turbine constant of valve positioner
c_g	Valve positioner of gas turbine
Y_c	Lag time constant of gas turbine speed governor
X_c	Lead time constant of gas turbine speed governor
T_{cr}	Gas turbine combustion reaction time delay
T_{fc}	Gas turbine fuel time constant
T_{cd}	Gas turbine compressor discharge volume-time constant
K_{dc}	Gain of HVDC link
T_{dc}	Time constant of hvdc link
ITAE	Integral time absolute error
ISE	Integral square error
IAE	Integral absolute error
K_1	Input scaling factor
K_2	Derivative input gain
K_3	Proportional output gain
K_4	Integral output gain

NB	Negative big
NS	Negative small
Z	Zero
PB	Positive big
PS	Positive small
UB	Upper boundary value
LB	Lower boundary value

Appendix A

Table A1. Transfer functions included in the studied system.

Control Block	Transfer Functions
Thermal Governor	$\frac{1}{Tsg.s+1}$
Reheater of Thermal Turbine	$\frac{Kr*Tr.s+1}{Tr.s+1}$
Thermal Turbine	$\frac{1}{Tt.s+1}$
Hydro Governor	$\frac{1}{Tgh.s+1}$
Transient Droop Compensation	$\frac{Trs.s+1}{Trh.s+1}$
Hydro Turbine	$\frac{-Tws+1}{0.5*Tw.s+1}$
Valve Positioner of Gas Turbine	$\frac{1}{bg.s+cg}$
Speed Governor of Gas Turbine	$\frac{Xc.s+1}{Yc.s+1}$
Fuel System and Combustor	$\frac{Tcr.s+1}{Tfc.s+1}$
Gas Turbine Dynamics	$\frac{1}{Tcd.s+1}$
Power System 1	$\frac{Kps1}{Tps1.s+1}$
Power System 2	$\frac{Kps2}{Tps2.s+1}$
HVDC 1	$\frac{Kdc1}{Tdc1.s+1}$
HVDC 2	$\frac{Kdc2}{Tdc2.s+1}$

References

1. Mosaad, M.I.; El-Raouf, M.O.A.; Al-Ahmar, M.A.; Bendary, F.M. Optimal PI controller of DVR to enhance the performance of hybrid power system feeding a remote area in Egypt. *Sustain. Cities Soc.* **2019**, *47*, 101469. [CrossRef]
2. Balu, N.J.; Lauby, M.G.; Kundur, P. *Power System Stability and Control*; Electrical Power Research Institute, McGraw-Hill Professional: Washington, DC, USA, 1994.
3. Omer, A.M. Energy, environment and sustainable development. *Renew. Sustain. Energy Rev.* **2008**, *12*, 2265–2300. [CrossRef]
4. Bilgen, S.; Kaygusuz, K.; Sari, A. Renewable Energy for a Clean and Sustainable Future. *Energy Sources* **2004**, *26*, 1119–1129. [CrossRef]
5. Blaabjerg, F.; Teodorescu, R.; Liserre, M.; Timbus, A. Overview of Control and Grid Synchronization for Distributed Power Generation Systems. *IEEE Trans. Ind. Electron.* **2006**, *53*, 1398–1409. [CrossRef]
6. Fang, J.; Li, H.; Tang, Y.; Blaabjerg, F. Distributed Power System Virtual Inertia Implemented by Grid-Connected Power Converters. *IEEE Trans. Power Electron.* **2017**, *33*, 8488–8499. [CrossRef]
7. Bevrani, H. *Robust Power System Frequency Control*; Springer: Berlin/Heidelberg, Germany, 2009.
8. Shahalami, S.H.; Farsi, D. Analysis of Load Frequency Control in a restructured multi-area power system with the Kalman filter and the LQR controller. *AEU-Int. J. Electron. Commun.* **2018**, *86*, 25–46. [CrossRef]
9. Das, S.K.; Rahman, M.; Paul, S.K.; Armin, M.; Roy, P.N.; Paul, N. High-Performance Robust Controller Design of Plug-In Hybrid Electric Vehicle for Frequency Regulation of Smart Grid Using Linear Matrix Inequality Approach. *IEEE Access* **2019**, *7*, 116911–116924. [CrossRef]
10. Liao, K.; Xu, Y. A Robust Load Frequency Control Scheme for Power Systems Based on Second-Order Sliding Mode and Extended Disturbance Observer. *IEEE Trans. Ind. Inform.* **2018**, *14*, 3076–3086. [CrossRef]
11. HBevrani, H.; Feizi, M.R.; Ataee, S. Robust Frequency Control in an Islanded Microgrid: H∞ and μ-Synthesis Approaches. *IEEE Trans. Smart Grid* **2015**, *7*, 706–717. [CrossRef]
12. Wang, Z.-Q.; Sznaier, M. Robust control design for load frequency control using/spl mu/-synthesis. In Proceedings of the SOUTHCON'94, Orlando, FL, USA, 29–31 March 1994; pp. 186–190.
13. Wang, Y.; Zhou, R.; Gao, L. H/sub ∞/controller design for power system load frequency control. In Proceedings of the TENCON'93, IEEE Region 10 International Conference on Computers, Communications and Automation, Beijing, China, 19–21 October 1993.
14. Ma, M.; Liu, X.; Zhang, C. LFC for multi-area interconnected power system concerning wind turbines based on DMPC. *IET Gener. Transm. Distrib.* **2017**, *11*, 2689–2696. [CrossRef]

15. Khooban, M.H.; Gheisarnejad, M. A Novel Deep Reinforcement Learning Controller Based Type-II Fuzzy System: Frequency Regulation in Microgrids. *IEEE Trans. Emerg. Top. Comput. Intell.* **2021**, *5*, 689–699. [CrossRef]
16. Aluko, A.O.; Dorrell, D.G.; Pillay-Carpanen, R.; Ojo, E.E. Frequency Control of Modern Multi-Area Power Systems Using Fuzzy Logic Controller. In Proceedings of the 2019 IEEE PES/IAS PowerAfrica, Abuja, Nigeria, 20–23 August 2019.
17. Yang, D.; Jin, E.; You, J.; Hua, L. Dynamic Frequency Support from a DFIG-Based Wind Turbine Generator via Virtual Inertia Control. *Appl. Sci.* **2020**, *10*, 3376. [CrossRef]
18. Magdy, G.; Shabib, G.; Elbaset, A.A.; Mitani, Y. Renewable power systems dynamic security using a new coordination of frequency control strategy based on virtual synchronous generator and digital frequency protection. *Int. J. Electr. Power Energy Syst.* **2019**, *109*, 351–368. [CrossRef]
19. Mudi, R.K.; Pal, N.R. A self-tuning fuzzy PI controller. *Fuzzy Sets Syst.* **2000**, *115*, 327–338. [CrossRef]
20. Chang, C.; Fu, W. Area load frequency control using fuzzy gain scheduling of PI controllers. *Electr. Power Syst. Res.* **1997**, *42*, 145–152. [CrossRef]
21. Yesil, E.; Güzelkaya, M.; Eksin, I. Self tuning fuzzy PID type load and frequency controller. *Energy Convers. Manag.* **2004**, *45*, 377–390. [CrossRef]
22. Ahmadi, S.; Talami, S.H.; Sahnesaraie, M.A.; Dini, F.; Tahernejadjozam, B.; Ashgevari, Y. FUZZY aided PID controller is optimized by GA algorithm for Load Frequency Control of Multi-Source Power Systems. In Proceedings of the 2020 IEEE 18th World Symposium on Applied Machine Intelligence and Informatics (SAMI), Herlany, Slovakia, 23–25 January 2020.
23. Lal, D.K.; Barisal, A.K. Comparative performances evaluation of FACTS devices on AGC with diverse sources of energy generation and SMES. *Cogent Eng.* **2017**, *4*, 1318466. [CrossRef]
24. Lal, D.K.; Barisal, A.K.; Tripathy, M. Load Frequency Control of Multi Source Multi-Area Nonlinear Power System with DE-PSO Optimized Fuzzy PID Controller in Coordination with SSSC and RFB. *Int. J. Control. Autom.* **2018**, *11*, 61–80. [CrossRef]
25. Lal, D.K.; Barisal, A.K.; Tripathy, M. Load Frequency Control of Multi Area Interconnected Microgrid Power System using Grasshopper Optimization Algorithm Optimized Fuzzy PID Controller. In Proceedings of the 2018 Recent Advances on Engineering, Technology and Computational Sciences (RAETCS), Allahabad, India, 6–8 February 2018.
26. Dhanasekaran, B.; Siddhan, S.; Kaliannan, J. Ant colony optimization technique tuned controller for frequency regulation of single area nuclear power generating system. *Microprocess. Microsyst.* **2020**, *73*, 102953. [CrossRef]
27. Annamraju, A.; Nandiraju, S. Coordinated control of conventional power sources and PHEVs using jaya algorithm optimized PID controller for frequency control of a renewable penetrated power system. *Prot. Control. Mod. Power Syst.* **2019**, *4*, 28. [CrossRef]
28. Magdy, G.; Shabib, G.; Elbaset, A.A.; Kerdphol, T.; Qudaih, Y.; Mitani, Y. Decentralized optimal LFC for a real hybrid power system considering renewable energy sources. *J. Eng. Sci. Technol.* **2019**, *14*, 682–697.
29. Khamari, D.; Kumbhakar, B.; Patra, S.; Laxmi, D.A.; Panigrahi, S. Load Frequency Control of a Single Area Power System using Firefly Algorithm. *Int. J. Eng. Res.* **2020**, *V9*. [CrossRef]
30. Khadanga, R.K.; Kumar, A.; Panda, S. A hybrid shuffled frog-leaping and pattern search algorithm for load frequency controller design of a two-area system composing of PV grid and thermal generator. *Int. J. Numer. Model. Electron. Netw. Devices Fields* **2020**, *33*, e2694. [CrossRef]
31. Elkasem, A.H.A.; Kamel, S.; Rashad, A.; Jurado, F. Optimal Performance of DFIG Integrated with Different Power System Areas Using Multi-Objective Genetic Algorithm. In Proceedings of the 2018 Twentieth International Middle East Power Systems Conference (MEPCON), Cairo, Egypt, 18–20 December 2018; pp. 672–678.
32. Elkasem, A.H.A.; Kamel, S.; Korashy, A.; Nasrat, L. Load Frequency Control Design of Two Area Interconnected Power System Using GWO. In Proceedings of the 2019 IEEE Conference on Power Electronics and Renewable Energy (CPERE), Aswan, Egypt, 23–25 October 2019.
33. Kamel, S.; Elkasem, A.H.A.; Korashy, A.; Ahmed, M.H. Sine Cosine Algorithm for Load Frequency Control Design of Two Area Interconnected Power System with DFIG Based Wind Turbine. In Proceedings of the 2019 International Conference on Computer, Control, Electrical and Electronics Engineering (ICCCEEE), Khartoum, Sudan, 21–23 September 2019.
34. Hamdy, A.; Kamel, S.; Nasrat, L.; Jurado, F. Frequency Stability of Two-Area Interconnected Power System with Doubly Fed Induction Generator Based Wind Turbine. In *Wide Area Power Systems Stability, Protection, and Security*; Springer: Berlin/Heidelberg, Germany, 2021; pp. 293–324.
35. Khamies, M.; Magdy, G.; Hussein, M.E.; Banakhr, F.A.; Kamel, S. An Efficient Control Strategy for Enhancing Frequency Stability of Multi-Area Power System Considering High Wind Energy Penetration. *IEEE Access* **2020**, *8*, 140062–140078. [CrossRef]
36. Khamies, M.; Magdy, G.; Ebeed, M.; Kamel, S. A robust PID controller based on linear quadratic gaussian approach for improving frequency stability of power systems considering renewables. *ISA Trans.* **2021**, *117*, 118–138. [CrossRef] [PubMed]
37. Gürses, D.; Bureerat, S.; Sait, S.M.; Yıldız, A.R. Comparison of the arithmetic optimization algorithm, the slime mold optimization algorithm, the marine predators algorithm, the salp swarm algorithm for real-world engineering applications. *Mater. Test.* **2021**, *63*, 448–452. [CrossRef]
38. Abualigah, L.; Diabat, A.; Sumari, P.; Gandomi, A. A Novel Evolutionary Arithmetic Optimization Algorithm for Multilevel Thresholding Segmentation of COVID-19 CT Images. *Processes* **2021**, *9*, 1155. [CrossRef]
39. Dash, P.; Saikia, L.C.; Sinha, N. Flower Pollination Algorithm Optimized PI-PD Cascade Controller in Automatic Generation Control of a Multi-area Power System. *Int. J. Electr. Power Energy Syst.* **2016**, *82*, 19–28. [CrossRef]

40. Guha, D.; Roy, P.; Banerjee, S. Load frequency control of interconnected power system using grey wolf optimization. *Swarm Evol. Comput.* **2016**, *27*, 97–115. [CrossRef]
41. Ibraheem; Nizamuddin; Bhatti, T.S. AGC of two area power system interconnected by AC/DC links with diverse sources in each area. *Int. J. Electr. Power Energy Syst.* **2014**, *55*, 297–304. [CrossRef]
42. Neshat, M.; Nezhad, M.M.; Abbasnejad, E.; Mirjalili, S.; Tjernberg, L.B.; Garcia, D.A.; Alexander, B.; Wagner, M. A deep learning-based evolutionary model for short-term wind speed forecasting. *Energy Convers. Manag.* **2021**, *236*, 114002. [CrossRef]
43. Li, D.; Jiang, F.; Chen, M.; Qian, T. Multi-step-ahead wind speed forecasting based on a hybrid decomposition method and temporal convolutional networks. *Energy* **2022**, *238*, 121981. [CrossRef]
44. Emeksiz, C.; Tan, M. Multi-step wind speed forecasting and Hurst analysis using novel hybrid secondary decomposition approach. *Energy* **2022**, *238*, 121764. [CrossRef]
45. Jiang, P.; Liu, Z.; Niu, X.; Zhang, L. A combined forecasting system based on statistical method, artificial neural networks, and deep learning methods for short-term wind speed forecasting. *Energy* **2021**, *217*, 119361. [CrossRef]
46. Deveci, M.; Özcan, E.; John, R.; Pamucar, D.; Karaman, H. Offshore wind farm site selection using interval rough numbers based Best-Worst Method and MARCOS. *Appl. Soft Comput.* **2021**, *109*, 107532. [CrossRef]
47. Deveci, M.; Erdogan, N.; Cali, U.; Stekli, J.; Zhong, S. Type-2 neutrosophic number based multi-attributive border approximation area comparison (MABAC) approach for offshore wind farm site selection in USA. *Eng. Appl. Artif. Intell.* **2021**, *103*, 104311. [CrossRef]
48. Jena, T.; Debnath, M.K.; Sanyal, S.K. Optimal fuzzy-PID controller with derivative filter for load frequency control including UPFC and SMES. *IJECE* **2019**, *9*, 2813. [CrossRef]
49. Mohanty, B.; Panda, S.; Hota, P.K. Controller parameters tuning of differential evolution algorithm and its application to load frequency control of multi-source power system. *Int. J. Electr. Power Energy Syst.* **2014**, *54*, 77–85. [CrossRef]
50. Sahu, B.K.; Pati, T.K.; Nayak, J.R.; Panda, S.; Kar, S.K. A novel hybrid LUS–TLBO optimized fuzzy-PID controller for load frequency control of multi-source power system. *Int. J. Electr. Power Energy Syst.* **2016**, *74*, 58–69. [CrossRef]
51. Parmar, K.S.; Majhi, S.; Kothari, D. Load frequency control of a realistic power system with multi-source power generation. *Int. J. Electr. Power Energy Syst.* **2012**, *42*, 426–433. [CrossRef]
52. Magdy, G.; Mohamed, E.A.; Shabib, G.; Elbaset, A.A.; Mitani, Y. SMES based a new PID controller for frequency stability of a real hybrid power system considering high wind power penetration. *IET Renew. Power Gener.* **2018**, *12*, 1304–1313. [CrossRef]
53. Kickert, W.; Lemke, H.V.N. Application of a fuzzy controller in a warm water plant. *Automatica* **1976**, *12*, 301–308. [CrossRef]
54. Sahu, B.K.; Pati, S.; Panda, S. Hybrid differential evolution particle swarm optimisation optimised fuzzy proportional–integral derivative controller for automatic generation control of interconnected power system. *IET Gener. Transm. Distrib.* **2014**, *8*, 1789–1800. [CrossRef]
55. Zangeneh, M.; Aghajari, E.; Forouzanfar, M. A survey: Fuzzify parameters and membership function in electrical applications. *Int. J. Dyn. Control* **2020**, *8*, 1040–1051. [CrossRef]
56. Sadollah, A. Introductory Chapter: Which Membership Function is Appropriate in Fuzzy System? In *Fuzzy Logic Based in Optimization Methods and Control Systems and its Applications*; InTech Open: Rijeka, Croatia, 2018.
57. Abualigah, L.; Diabat, A.; Mirjalili, S.; Elaziz, M.A.; Gandomi, A.H. The Arithmetic Optimization Algorithm. *Comput. Methods Appl. Mech. Eng.* **2021**, *376*, 113609. [CrossRef]

sustainability

MDPI

Article

An Improved Heap-Based Optimizer for Optimal Design of a Hybrid Microgrid Considering Reliability and Availability Constraints

Mohammed Kharrich [1], Salah Kamel [2,*], Mohamed H. Hassan [2], Salah K. ElSayed [3] and Ibrahim B. M. Taha [3]

1 Department of Electrical Engineering, Mohammadia School of Engineers, Mohammed V University, Rabat 10090, Morocco; mohammedkharrich@research.emi.ac.ma
2 Department of Electrical Engineering, Faculty of Engineering, Aswan University, Aswan 81542, Egypt; mohamed.hosny@moere.gov.eg
3 Department of Electrical Engineering, College of Engineering, Taif University, Taif 21944, Saudi Arabia; sabdelhamid@tu.edu.sa (S.K.E.); i.taha@tu.edu.sa (I.B.M.T.)
* Correspondence: skamel@aswu.edu.eg

Abstract: The hybrid microgrid system is considered one of the best solution methods for many problems, such as the electricity problem in regions without electricity, to minimize pollution and the depletion of fossil sources. This study aims to propose and implement a new algorithm called improved heap-based optimizer (IHBO). The objective of minimizing the microgrid system cost is to reduce the net present cost while respecting the reliability, power availability, and renewable fraction factors of the microgrid system. The results show that the PV/diesel/battery hybrid renewable energy system (HRES) gives the best solution, with a net present cost of MAD 120463, equivalent to the energy cost of MAD 0.1384/kWh. The reliability is about 3.89%, the renewable fraction is about 95%, and the power availability is near to 99%. The optimal size considered is represented as 167.3864 m^2 of PV area, which is equivalent to 44.2582 kW and 3.8860 kW of diesel capacity. The study results show that the proposed optimization algorithm of IHBO is better than the artificial electric field algorithm, the grey wolf optimizer, Harris hawks optimization, and the original HBO algorithm. A comparison of the net present cost with a different fuel price is carried out, in which it is observed that the net present cost is reduced even though its quantity used is mediocre.

Keywords: HRES; microgrid design and sizing; optimization algorithm; HBO algorithm; reliability

Citation: Kharrich, M.; Kamel, S.; Hassan, M.H.; ElSayed, S.K.; Taha, I.B.M. An Improved Heap-Based Optimizer for Optimal Design of a Hybrid Microgrid Considering Reliability and Availability Constraints. *Sustainability* **2021**, *13*, 10419. https://doi.org/10.3390/su131810419

Academic Editors: Nicu Bizon, Mamadou Baïlo Camara and Bhargav Appasani

Received: 9 August 2021
Accepted: 14 September 2021
Published: 18 September 2021

1. Introduction

The implementation of hybrid microgrids is necessary due to their advantages. Many projects and studies have proven their essential ecological and economic effects. The literature has assessed the microgrid from all directions, including design, operation, optimization, control, and others. Literature reviews have provided more comprehensive studies. In [1], a comprehensive study on the optimization of microgrid operations has been presented. In [2], a review of AC and DC microgrid protection has been presented. Reference [3] presented a D.C. microgrid protection comprehensive review. Reference [4] presented a review on optimization and control techniques of the hybrid AC/DC microgrid, as well as the integration challenges. Reference [5] presented a comprehensive review of the planning, the operation, and the control of a DC microgrid. Reference [6] presented a review of microgrid sizing, design, and energy management.

The design and operation optimization of microgrids, considered the main objective of this work, has been presented in many papers. Reference [7] presented a design and assessment of the microgrid using a statistical methodology that calculates the effect of energy reliability and variability on microgrid performance. The paper used a REopt platform to explore the cost savings and revenue streams. In [8,9], the microgrid design has been investigated using several algorithms and configurations. In [10], a hybrid simulated annealing particle swarm (SAPS) algorithm has been presented to determine the microgrid optimal size that is subject to the economic and reliable operation constraints and to subsequently boost power supply security and stability. The paper [11] presented a new compromise method based on the Six Sigma approach to compare several multi-objective algorithms. The new approach has been applied to microgrid sizing and design based on PV, wind turbine, diesel, and battery systems. Reference [12] presented a graph-theoretic algorithm known as P-graph which allows the identification of optimal and near-optimal solutions for practical decision making. This study proposed a multi-period P-graph optimization framework for optimizing photovoltaic-based microgrids with battery-hydrogen energy storage. The proposed approach is demonstrated through two case studies. Reference [13] proposed a novel cash-flow model for Li-ion battery storage used in the energy system; the study considers the Li-ion battery degradation characteristic.

Optimization techniques are more competent in solving non-linear optimization problems, such as optimal reactive power dispatch (ORPD) [14], economic emission dispatch [15], intelligent energy management [16], and parameter estimation of photovoltaic models [17]. Reference [18] used an experimental validation of a lab-scale microgrid. Reference [19] concerns the undervoltage in smart distribution systems. The optimal power flow from attackers has been presented in [20].

The development of tools to design microgrids has become an important research area; the development of meta-heuristic algorithms begins a trend. In the literature, many papers presented different algorithms which have been applied to design a hybrid microgrid. In [21], an improved two-archive many-objective evolutionary algorithm (TA-MaEA) based on fuzzy decisions has been used to solve the sizing optimization problem for HRES. The simulation considered the following objective function: costs, probability of loss of power supply, pollutant emissions, and power balance. Reference [22] proposes an HRES of PV and fuel cells with an optimal total annual cost; the study used a new, improved metaheuristic called the amended water strider algorithm (AWSA). The reliability is considered, and the sensitivity analysis is applied. Reference [23] presents a microgrid design composed of PV, wind, an inverter, a rectifier, an electrolyzer, and a fuel cell. The paper used a modified seagull optimization technique to find the best cost of the optimal sizing. The proposed algorithm is compared with the original seagull optimization algorithm (SOA) and modified farmland fertility algorithm (MFFA). Reference [24] presents a new hybrid algorithm called IWO/BSA to resolve the microgrid design of any configuration, including PV/wind turbine (WT)/biomass/battery, PV/biomass, PV/diesel/battery, and WT/diesel/battery systems. The study's objective is to obtain the best system with optimal cost, pollution, availability, and reliability. Reference [25] presents an adaptive version of the marine predators algorithm (AMPA) to design a PV/diesel/battery microgrid system. The objective function minimizes the annualized cost, respecting the ecologic and reliability factors of the system. The results are compared with PSO and HOMER. Reference [26] proposed an improved version of the bonobo optimizer (BO) based on the quasi-oppositional technique to resolve the design problem of the HRES considering the PV, wind turbines, battery, and diesel. A comparison between the traditional BO, the new QOBO, and other optimization techniques is investigated to prove the efficacy of QOBO. Reference [27] proposed a deterministic approach to size a PV, battery, anaerobic digestion, and biogas power plant to meet a demand load in Kenya. The levelized cost of energy (LCOE) is considered the objective function, while the energy imbalance between generation and demand is considered.

The present paper proposes a new tool consisting of platforms using an improved version of the HBO algorithm called IHBO. The improvement of the HBO algorithm depends on enhancing the performance of the HBO algorithm using the velocity equation from the particle swarm optimization (PSO) algorithm. This equation improves the convergence capability behavior and enables different diversified solutions in the search space, which is necessary for such an algorithm and achieves the fitness function's optimal value. The proposed platforms design hybrid microgrid systems composed of PV, wind, diesel, and batteries. Two configurations are presented, and four algorithms are used in the comparison. In summary, the paper addresses the following points:

- An improved version of the conventional HBO algorithm is proposed with the aim of improving its performance;
- The conventional HBO and proposed IHBO algorithms are applied for optimal design of a hybrid microgrid system including RES (photovoltaic panels, wind turbines, and batteries) with diesel generators;
- In the designed microgrid, the reliability, availability and the renewable fraction constraints are considered;
- The proposed IHBO algorithm's efficiency and performance are evaluated on different benchmark functions, including the statistical measurement;
- The impact of the fuel price variation on the project investment is analyzed.

The paper is organized as follows: the introduction occurs in Section 1; the modeling of HRES components is contained in Section 2; Section 3 presents the objective functions and constraints; Section 4 presents the new, improved algorithm, namely IHBO; the results and discussion are presented in Section 5; and the conclusion is presented in Section 6.

2. HRES Components Modeling

2.1. PV Panel Modeling

The PV output power is calculated as follows [28,29]:

$$P_{pv} = I\langle t\rangle \times \eta_{pv} t \times A_{pv} \tag{1}$$

where I represents the irradiation, η_{pv} represents the efficiency of PV, and A_{pv} is the area of PV. The efficiency of PV can be calculated based on reference efficiency (η_r), the efficiency of MPPT (η_t), temperature coefficient (β), ambient temperature (T_a), PV cell reference temperature (T_r) and nominal operating cell temperature (NOCT), as follows:

$$\eta_{pv}(t) = \eta_r \times \eta_t \times \left[1 - \beta \times (T_a\langle t\rangle - T_r) - \beta \times I\langle t\rangle \times \left(\frac{NOCT - 20}{800}\right) \times (1 - \eta_r \times \eta_t)\right] \tag{2}$$

2.2. Wind System Modeling

The wind turbine output power can be calculated following these conditions [30]:

$$P_{wind} = \begin{cases} 0, & v\langle t\rangle \le v_{ci},\ v\langle t\rangle \ge v_{co} \\ a \times V\langle t\rangle^3 - b \times P_r, & v_{ci} < v\langle t\rangle < v_r \\ P_r, & v_r \le v\langle t\rangle < v_{co} \end{cases} \tag{3}$$

where V represents the wind velocity, P_r is rated power, v_{ci} is cut-in, v_{co} represents cut-out, and v_r is the rated wind. a and b are constant values that expressed as:

$$\begin{aligned} a &= P_r / ({v_r}^3 - {v_{ci}}^3) \\ b &= {v_{ci}}^3 / ({v_r}^3 - {v_{ci}}^3) \end{aligned} \tag{4}$$

The rated power of wind turbine can be calculated as:

$$P_r = \frac{1}{2} \times \rho \times A_{wind} \times C_p \times {v_r}^3 \tag{5}$$

where ρ is the air density, A_{wind} represents the swept area of the wind turbine, and C_p is the maximum power coefficient (from 0.25 to 0.45).

2.3. Diesel System Modeling

The diesel rated power can be calculated as [31]:

$$P_{dg} = \frac{F_{dg}\langle t\rangle - A_g \times P_{dg,out}}{B_g} \tag{6}$$

where F_{dg} represents the fuel consumption, $P_{dg,out}$ is the output power of the diesel generator, and A_g and B_g are two constant values represent the fuel linear consumption.

2.4. Battery System Modeling

The battery capacity of the battery can be calculated as [31]:

$$C_{BESS} = \frac{E_l \times AD}{DOD \times \eta_i \times \eta_b} \tag{7}$$

where E_l is the load demand, AD is the autonomy of the battery which can lead power to the load on rainy days, DOD represents the depth of discharge, and η_i and η_b represent the inverter and battery efficiency, respectively.

3. Objective Function and Constraints

3.1. Net Present Cost

The NPC represents an economic factor, which is considered the objective function in this study. The goal of the paper is to minimize the NPC, which is the sum of all costs during the project lifetime. The NPC is calculated as [32,33]:

$$\text{NPC} = C + OM + R + FC_{dg} \tag{8}$$

where C represent the capital cost, OM is the operation and maintenance costs, R is the replacement cost, and FC_{dg} is the fuel cost.

3.2. LCOE Index

The LCOE represents the price of energy and is a critical factor which is calculated as [31]:

$$\text{LCOE} = \frac{NPC \times CRF}{\sum_{t=1}^{8760} P_{load} t} \tag{9}$$

where CRF represents the capital recovery factor (obtained by converting the initial cost to annual capital cost), and P_{load} represents the power load. The CRF is calculated as:

$$\text{CRF}(ir, N) = \frac{i_r \times (1 + i_r)^N}{(1 + i_r)^R - 1} \tag{10}$$

3.3. LPSP Index

The loss of power supply probability (LPSP) is a technical index that ranges from 0 to 1. It is used to indicate the reliability of the microgrid system. The LPSP is calculated as follows [31]:

$$\text{LPSP} = \frac{\sum_{t=1}^{8760} \left(P_{load}\langle t\rangle - P_{pv}\langle t\rangle - P_{wind}\langle t\rangle + P_{dg,out}\langle t\rangle + E_{bmin}\right)}{\sum_{t=1}^{8760} P_{load}\langle t\rangle} \tag{11}$$

3.4. Renewable Energy Index

Renewable energy (RF) is calculated to determine the renewable energy percent that is penetrated into the microgrid system. The RF is expressed as [31]:

$$\mathrm{RF} = \left(1 - \frac{\sum_{t=1}^{8760} P_{dg,out}\langle t \rangle}{\sum_{t=1}^{8760} P_{re}\langle t \rangle}\right) \times 100 \tag{12}$$

where P_{re} represents the sum of renewable energy powers.

3.5. Availability Index

The availability factor (Av) is assumed as an index of the customer's satisfaction; it measures the ability of the microgrid to convert the total energy to load charge. The availability is calculated as [33]:

$$\mathrm{Av} = 1 - \frac{DMN}{\sum_{t=1}^{8760} P_{load}\langle t \rangle} \tag{13}$$

$$DMN = P_{bmin}\langle t \rangle - P_b\langle t \rangle - \left(P_{pv}\langle t \rangle + P_{wind}\langle t \rangle + P_{dg,out}\langle t \rangle - P_{load}\langle t \rangle\right) \times u\langle t \rangle \tag{14}$$

where P_{bmin} represents the battery min state, P_b represents the battery power, and u is a fixed value which equals 1 when the load is not satisfied and which equals 0 otherwise.

3.6. Constraints

Constraints are introduced to tune the microgrid system factors and help to improve the microgrid service quality. In this work, the constraints proposed are:

$$\begin{cases} 0 \le A_{pv} \le A_{pv}^{max} \\ 0 \le A_{wind} \le A_{wind}^{max} \\ 0 \le P_{dg} \le P_{dg}^{max} \\ 0 \le C_{BESS} \le C_{BESS}^{max} \\ LPSP \le \mathrm{LPSP}^{max} \\ \mathrm{RF}^{min} \le RF \\ Av^{min} \le Av \\ \mathrm{AD}^{min} \le AD \end{cases} \tag{15}$$

where $\mathrm{LPSP}^{max} = 0.05$, $\mathrm{RF}^{min} = 70\%$, $\mathrm{Av}^{min} = 90\%$, and $\mathrm{AD}^{min} = 1$ day. The sizing limit is different from configuration to the others. All other parameters are shown in Table A1.

4. Proposed Algorithm

4.1. Heap-Based Optimizer (HBO)

The heap-based optimizer algorithm (HBO) is inspired by the social behavior of human beings [34]. One sort of social interaction between human beings can be observed in organizations where people in teams are arranged in a hierarchy for achieving a specific target; this is known as corporate rank hierarchy (CRH). CRH is presented in Figure 1a. The HBO algorithm is based on CRH in a very distinctive manner. In this regard, the concept of CRH is to arrange the search agents based on their suitability in this hierarchy using a heap tree-based data structure to enact the implementation of priority queues. Figure 1b shows an example of 3 degrees (3-ary) of min-heap. Three types of employees' behaviors were used in the HBO algorithm. These types are: (i) the interaction of subordinates with their immediate head; (ii) the interaction between co-workers; and (iii) the self-contribution of individuals.

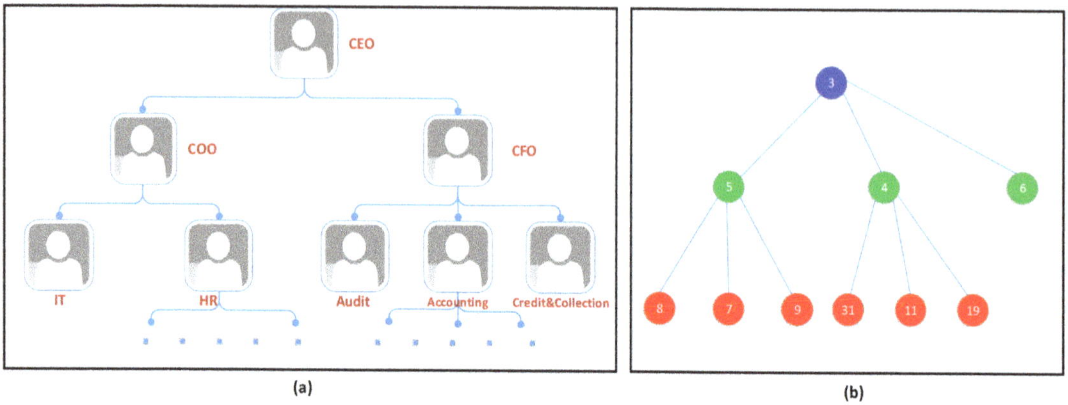

Figure 1. Partial examples of corporate rank hierarchy (**a**) and 3-ary min-heap (**b**).

The mapping of the heap concept is divided into four steps:

A. *Modeling the corporate rank hierarchy*

Figure 2 displays the procedure of CRH modeling through a heap data structure, wherein xi is the ith search agent of the population. The curve in the objective space describes the shape of the supposed objective function, and the search agents are drawn on the fitness shape according to their convenience.

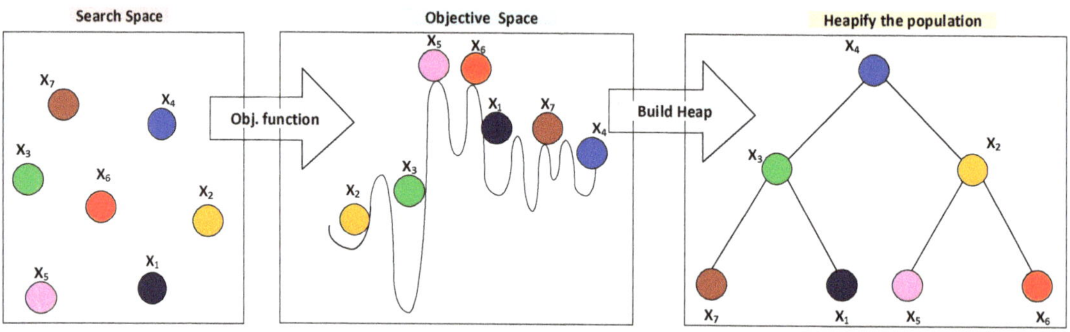

Figure 2. An illustration of the modeling of the CRH with min-heap.

B. *Mathematically modeling the collaboration with the boss*

In a centralized organizational structure, the regulations and policies are enforced from the upper levels, and subordinates must follow their direct manager.

This can be mathematically described by updating the agent position of each search as follows:

$$x_i^k(t+1) = B^k + \gamma\lambda^k\left|B^k - x_i^k(t)\right| \tag{16}$$

$$\gamma = \left|2 - \frac{\left(t \bmod \frac{T}{C}\right)}{\frac{T}{4C}}\right| \tag{17}$$

$$\lambda^k = (2r - 1) \tag{18}$$

where t is the current iteration, k is the kth component of a vector, B denotes the parent node, r is a random number from the range [0, 1], T is the maximum number of iterations, and C represents a user-defined parameter.

C. *Mathematically modeling the interaction between the colleagues*

Colleagues cooperate and perform official tasks. It is assumed in a heap that the nodes at the same level are colleagues, and each search agent x_i updates its location based on its randomly selected colleague S_r as follows:

$$x_i^k(t+1) = \begin{cases} s_r^k + \gamma\lambda^k \left| s_r^k - x_i^k(t) \right|, & f(S_r) < f(x_i(t)) \\ x_i^k + \gamma\lambda^k \left| s_r^k - x_i^k(t) \right|, & f(S_r) \geq f(x_i(t)) \end{cases} \tag{19}$$

D. *Self-contribution of an employee to accomplish a task*

In this phase, the self-contribution of a worker is mapped as follows:

$$x_i^k(t+1) = x_i^k(t) \tag{20}$$

The following part explains how exploration can be controlled with this equation.

E. *putting all together*

The principal challenge is determining the selection probabilities for the three equations to balance exploration and exploitation. The purpose of the roulette wheel is to achieve a balance of possibilities. The roulette wheel is divided into three parts: p_1, p_2, and p_3. The value of p_1 makes a population changes their position, and it is calculated from the following equation:

$$p_1 = 1 - \frac{t}{T} \tag{21}$$

The selection of p_2 is computed from the following equation:

$$p_2 = p_1 - \frac{1 - p_1}{2} \tag{22}$$

Finally, the selection of p_3 is calculated as follows:

$$p_3 = p_2 - \frac{1 - p_1}{2} = 1 \tag{23}$$

Accordingly, a general position-updating mechanism of the HBO algorithm is mathematically represented as follows:

$$x_i^k(t+1) = \begin{cases} x_i^k(t), p \leq p_1 \\ B^k + \gamma\lambda^k \left| B^k - x_i^k(t) \right|, p > p_1 \; and \; p \leq p_2 \\ s_r^k + \gamma\lambda^k \left| s_r^k - x_i^k(t) \right|, p > p_2 \; and \; p \leq p_3 \; and \; f(S_r) < f(x_i(t)) \\ x_i^k + \gamma\lambda^k \left| s_r^k - x_i^k(t) \right|, p > p_2 \; and \; p \leq p_3 \; and \; f(S_r) \geq f(x_i(t)) \end{cases} \tag{24}$$

where p is a random number in the range (0, 1).

4.2. Improved Heap-Based Optimizer(IHBO)

In order to enhance the strength of the proposed IHBO algorithm for many high-dimensional optimization problems, core aspects of one of the most used meta-heuristic algorithms, PSO, are utilized. The PSO algorithm is introduced by [35]. The velocity equation from the PSO algorithm is used in the proposed IHBO algorithm. This modification leads to the improvement of the ability of the global search and enhances the local search capabilities of the improved algorithm. This core equation is as follows:

$$V_i^k(t+1) = w \cdot V_i^k(t) + C_1 \cdot r_1 \times (p_{best} - x_i^k(t)) + C_2 \cdot r_2 \times \left(g_{best} - x_i^k(t)\right) \tag{25}$$

$$x_i^k(t+1) = x_i^k(t) + V_i^k(t+1) \tag{26}$$

where $C_1 = C_2 = 0.5$, as these values gave the best solution in [36]; $w = 0.7$; r_1 and r_2 are a random number in the range (0, 1); p_{best} is the best solution of an individual population, and g_{best} is the best solution so far.

The flow chart of the proposed IHBO algorithm is shown in Figure 3.

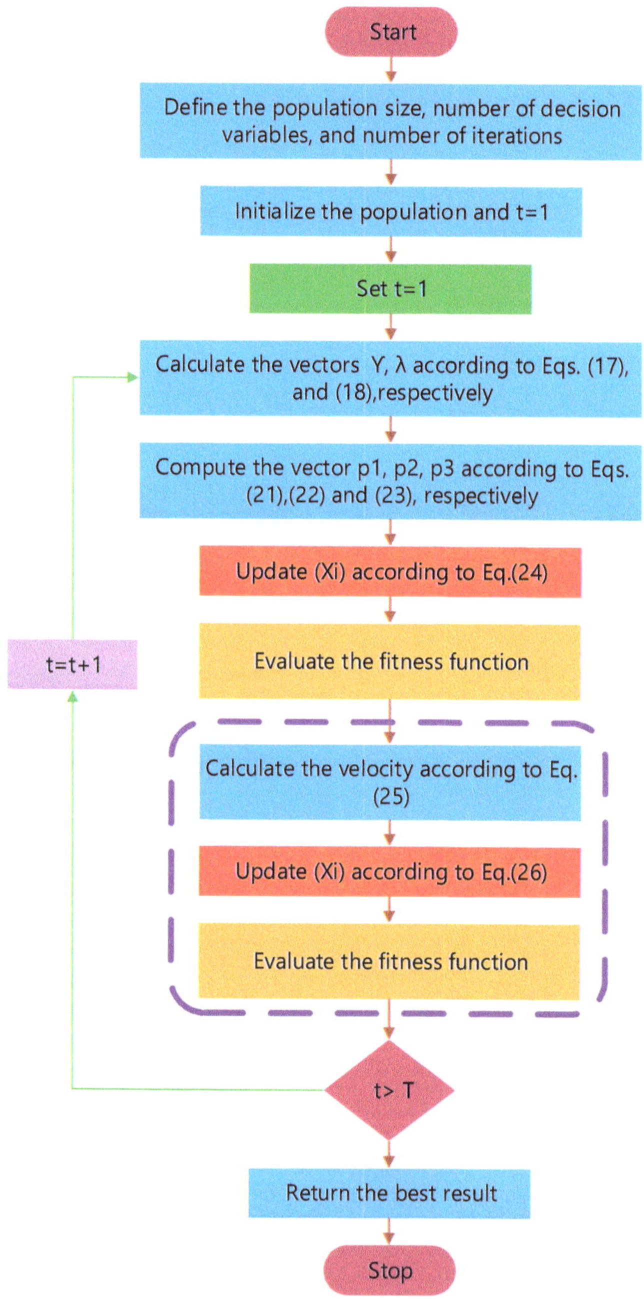

Figure 3. Flowchart of the proposed IHBO technique.

Performance of the Proposed IHBO Algorithm

The proposed IHBO algorithm's efficiency and performance are evaluated on different benchmark functions, including statistical measurements, such as minimum values, mean values, maximum values, and standard deviation (STD) for best solutions obtained by the proposed IHBO algorithm and the other recent optimization algorithms. The results obtained with the proposed IHBO technique is compared with three well-known optimization algorithms, including the sine cosine algorithm (SCA) [37], salp swarm algorithm (SSA) [38], movable damped wave algorithm (MDWA) [39], and the original heap-based optimizer (HBO). Table 1 shows the parameters of all compared algorithms (SSA, MDWA, SCA, IHBO, and HBO). Qualitative metrics on F1, F4, F7, F9, F11, F12, F15, and F18, including 2D views of the functions, search history, average fitness history, and convergence curve, are presented in Figure 4.

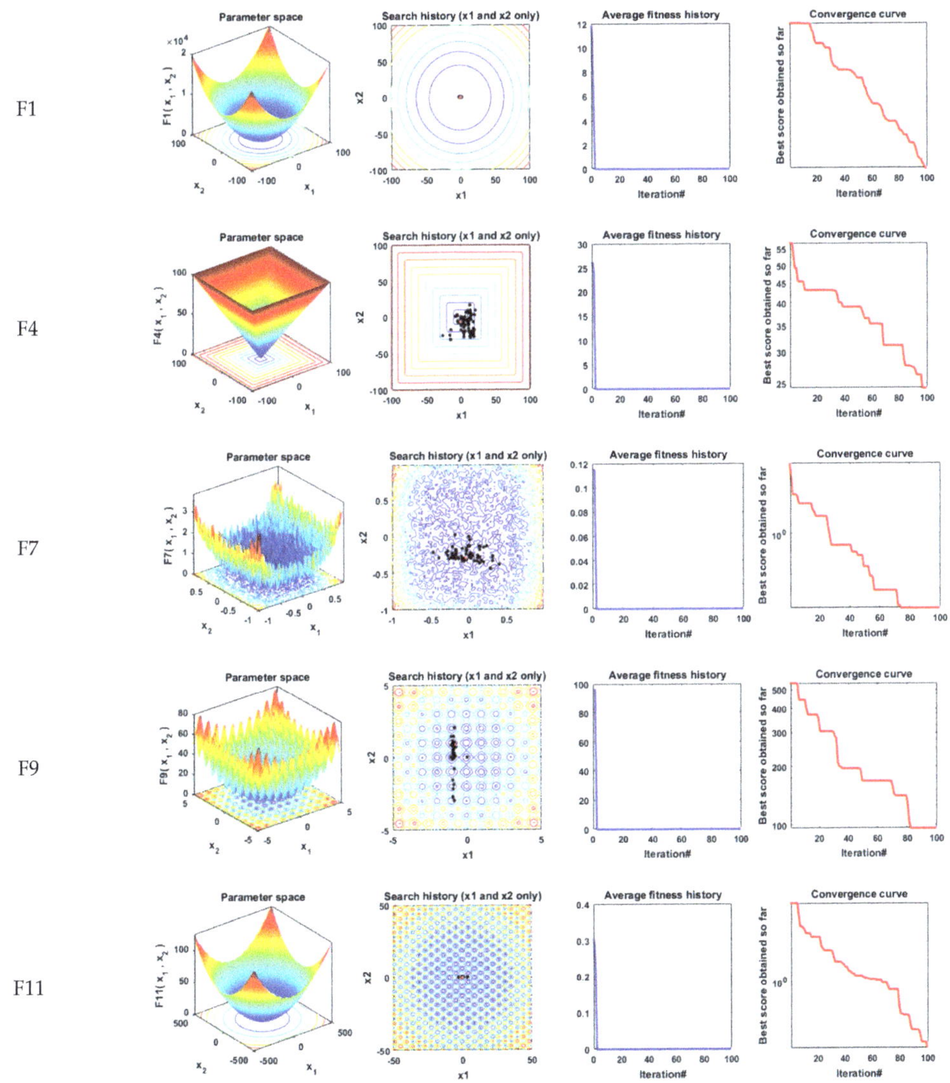

Figure 4. *Cont.*

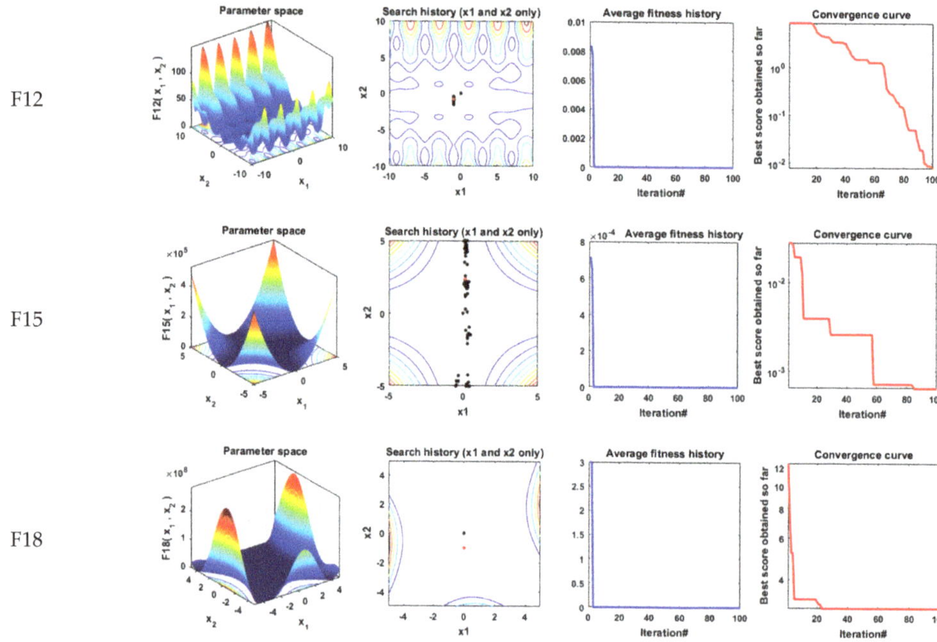

Figure 4. Qualitative metrics on F1, F4, F7, F9, F11, F12, F15 and F18: 2D views of the functions, search history, average fitness history, and convergence curve.

Table 1. Parameter settings of the selected techniques.

Algorithms	Parameter Settings
Common settings	Population size: nPop = 50 Maximum iterations: Max_iter = 1000 Number of independent runs: 20
SSA	c_2 = rand; c_3 = rand
MDWA	amax = 1; amin = 0
SCA	A = 2
IHBO	sv = 100; degree = 3; w = 0.7; C_1 = 0.5; C_2 = 0.5; r_1 = rand; r_2 = rand
HBO	sv = 100; degree = 3

Tables 2–4 tabulate the statistical results of the proposed IHBO algorithm and other well-known algorithms when applied for unimodal benchmark functions, named F1 to F7, multimodal benchmark functions, named F8 to F13, and composite benchmark functions, named F14 to F23, respectively. The best values, shown in bold, were achieved with the proposed IHBO algorithm, as well as MDWA and SCA, but the proposed IHBO technique achieves the best results for most of the benchmark functions. The convergence curves of all algorithms for the unimodal benchmark functions are shown in Figure 5 while Figure 6 shows the boxplots of each algorithm for these unimodal benchmark functions. Figure 7 displays the convergence characteristics curves of all algorithms for the multimodal benchmark functions. The boxplots for each algorithm for these types of benchmark functions are presented in Figure 8. The convergence curves of all algorithms for the composite benchmark functions are displayed in Figure 9 while Figure 10 illustrates the boxplots for each algorithm for these benchmark functions. The proposed algorithm reached a stable point for all functions. Also, the boxplots of the proposed IHBO technique are very narrow for most functions compared to the other algorithms.

Table 2. Results of unimodal benchmark functions.

Function		HBO	IHBO	SCA	MDWA	SSA
F1	Best	8.81×10^{-65}	$\mathbf{3.11 \times 10^{-86}}$	5.61×10^{-41}	1.34×10^{-44}	2.21×10^{-10}
	Worst	7.28×10^{-59}	$\mathbf{3.81 \times 10^{-81}}$	1.24×10^{-28}	3×10^{-39}	8.48×10^{-10}
	Mean	4×10^{-60}	$\mathbf{3.72 \times 10^{-82}}$	8.72×10^{-30}	3.01×10^{-40}	5.63×10^{-10}
	std	1.63×10^{-59}	$\mathbf{8.56 \times 10^{-82}}$	2.92×10^{-29}	8.19×10^{-40}	1.99×10^{-10}
F2	Best	1.24×10^{-39}	$\mathbf{4.59 \times 10^{-53}}$	6.81×10^{-27}	1.14×10^{-22}	2.69×10^{-06}
	Worst	4.86×10^{-37}	$\mathbf{1.2 \times 10^{-49}}$	1.84×10^{-19}	2.86×10^{-22}	1.54×10^{-05}
	Mean	5.65×10^{-38}	$\mathbf{1.99 \times 10^{-50}}$	9.67×10^{-21}	1.89×10^{-22}	6.7×10^{-06}
	std	1.12×10^{-37}	$\mathbf{3.32 \times 10^{-50}}$	4.11×10^{-20}	4.64×10^{-23}	2.62×10^{-06}
F3	Best	2.24×10^{-09}	3×10^{-16}	1.42×10^{-19}	$\mathbf{5.59 \times 10^{-22}}$	3.12×10^{-10}
	Worst	2.29×10^{-05}	1.8×10^{-11}	4.44×10^{-12}	$\mathbf{5.44 \times 10^{-13}}$	1.84×10^{-09}
	Mean	1.78×10^{-06}	1.45×10^{-12}	2.57×10^{-13}	$\mathbf{3.52 \times 10^{-14}}$	1.14×10^{-09}
	std	5.22×10^{-06}	4.09×10^{-12}	9.86×10^{-13}	$\mathbf{1.21 \times 10^{-13}}$	4.43×10^{-10}
F4	Best	8.73×10^{-14}	$\mathbf{2.76 \times 10^{-16}}$	1.04×10^{-13}	1.67×10^{-15}	7.74×10^{-06}
	Worst	1.94×10^{-10}	$\mathbf{1.5 \times 10^{-13}}$	3.01×10^{-08}	1.72×10^{-11}	1.73×10^{-05}
	Mean	2.54×10^{-11}	$\mathbf{2.05 \times 10^{-14}}$	2×10^{-09}	2.27×10^{-12}	1.25×10^{-05}
	std	4.6×10^{-11}	$\mathbf{3.92 \times 10^{-14}}$	6.71×10^{-09}	4.03×10^{-12}	2.36×10^{-06}
F5	Best	0.072812	**0.010245**	6.364761	0.068609	4.046853
	Worst	8.553338	**4.229361**	8.078748	4.466474	600.9202
	Mean	1.999632	**1.430528**	6.967484	0.879403	73.87534
	std	2.507143	**1.504423**	0.50586	1.21085	150.4287
F6	Best	**0.00**	**0.00**	7.99×10^{-02}	6.83×10^{-06}	2.94×10^{-10}
	Worst	1.23×10^{-32}	**0.00**	5.71×10^{-01}	1.11×10^{-04}	8.89×10^{-10}
	Mean	1.23×10^{-33}	**0.00**	3.12×10^{-01}	2.93×10^{-05}	5.74×10^{-10}
	std	3.79×10^{-33}	**0.00**	1.35×10^{-01}	2.67×10^{-05}	1.60×10^{-10}
F7	Best	7.43×10^{-04}	3.54×10^{-04}	$\mathbf{5.78 \times 10^{-05}}$	2.88×10^{-04}	6.27×10^{-04}
	Worst	3.22×10^{-03}	3.30×10^{-03}	$\mathbf{1.82 \times 10^{-03}}$	4.33×10^{-03}	1.49×10^{-02}
	Mean	1.99×10^{-03}	1.50×10^{-03}	$\mathbf{6.88 \times 10^{-04}}$	1.46×10^{-03}	4.60×10^{-03}
	std	7.46×10^{-04}	7.90×10^{-04}	$\mathbf{5.20 \times 10^{-04}}$	1.23×10^{-03}	3.22×10^{-03}

The best values obtained are in bold.

Table 3. Results of multimodal benchmark functions.

Function		HBO	IHBO	SCA	MDWA	SSA
F8	Best	**−4189.83**	**−4189.83**	−2724.77	−2752.03	−3262.03
	Worst	**−4189.83**	**−4189.83**	−1976.68	−1566.57	−2423.89
	Mean	**−4189.83**	**−4189.83**	−2297.91	−2116.51	−2851.97
	std	$\mathbf{1.87 \times 10^{-12}}$	$\mathbf{1.87 \times 10^{-12}}$	187.9469	338.86	263.4202
F9	Best	**0.00**	**0.00**	**0.00**	**0.00**	4.974795
	Worst	**0.00**	0.994959	19.68757	**0.00**	24.87393
	Mean	**0.00**	0.049748	0.984379	**0.00**	13.82991
	std	**0.00**	0.22248	4.402275	**0.00**	5.448648
F10	Best	4.44×10^{-15}	4.44×10^{-15}	4.44×10^{-15}	$\mathbf{8.88 \times 10^{-16}}$	6.91×10^{-06}
	Worst	$\mathbf{4.44 \times 10^{-15}}$	$\mathbf{4.44 \times 10^{-15}}$	2.43×10^{-12}	7.99×10^{-15}	$2.01 \times 10^{+00}$
	Mean	$\mathbf{4.44 \times 10^{-15}}$	$\mathbf{4.44 \times 10^{-15}}$	1.26×10^{-13}	4.44×10^{-15}	1.58×10^{-01}
	std	**0.00**	**0.00**	5.43×10^{-13}	1.15×10^{-15}	5.07×10^{-01}
F11	Best	**0.00**	**0.00**	**0.00**	**0.00**	7.38×10^{-02}
	Worst	1.06×10^{-09}	2.46×10^{-02}	1.31×10^{-01}	**0.00**	6.13×10^{-01}
	Mean	5.32×10^{-11}	2.59×10^{-03}	8.38×10^{-03}	**0.00**	2.77×10^{-01}
	std	2.38×10^{-10}	6.16×10^{-03}	2.96×10^{-02}	**0.00**	1.51×10^{-01}
F12	Best	$\mathbf{4.71 \times 10^{-32}}$	$\mathbf{4.71 \times 10^{-32}}$	1.85×10^{-02}	9.67×10^{-07}	3.46×10^{-12}
	Worst	$\mathbf{4.81 \times 10^{-32}}$	$\mathbf{4.71 \times 10^{-32}}$	9.96×10^{-02}	1.31×10^{-04}	$3.12 \times 10^{+00}$
	Mean	$\mathbf{4.72 \times 10^{-32}}$	$\mathbf{4.71 \times 10^{-32}}$	6.12×10^{-02}	2.06×10^{-05}	3.30×10^{-01}
	std	2.16×10^{-34}	$\mathbf{5.62 \times 10^{-48}}$	2.05×10^{-02}	3.56×10^{-05}	7.87×10^{-01}

Table 3. *Cont.*

Function		HBO	IHBO	SCA	MDWA	SSA
F13	Best	**1.35×10^{-32}**	**1.35×10^{-32}**	5.71×10^{-02}	2.48×10^{-06}	1.51×10^{-11}
	Worst	1.84×10^{-32}	**1.35×10^{-32}**	3.59×10^{-01}	3.70×10^{-04}	1.10×10^{-02}
	Mean	1.37×10^{-32}	**1.35×10^{-32}**	2.16×10^{-01}	4.53×10^{-05}	1.65×10^{-03}
	std	1.10×10^{-33}	**2.81×10^{-48}**	7.48×10^{-02}	8.66×10^{-05}	4.03×10^{-03}

The best values obtained are in bold.

Table 4. Results of composite benchmark functions.

Function		HBO	IHBO	SCA	MDWA	SSA
F14	Best	**9.98×10^{-01}**	**9.98×10^{-01}**	**9.98×10^{-01}**	**9.98×10^{-01}**	**9.98×10^{-01}**
	Worst	**9.98×10^{-01}**	**9.98×10^{-01}**	$2.98 \times 10^{+00}$	$6.90 \times 10^{+00}$	**9.98×10^{-01}**
	Mean	**9.98×10^{-01}**	**9.98×10^{-01}**	$1.49 \times 10^{+00}$	$4.12 \times 10^{+00}$	**9.98×10^{-01}**
	std	**0.00**	**0.00**	8.81×10^{-01}	$2.51 \times 10^{+00}$	1.25×10^{-16}
F15	Best	3.15×10^{-04}	**3.07×10^{-04}**	3.46×10^{-04}	3.10×10^{-04}	**3.07×10^{-04}**
	Worst	7.59×10^{-04}	**3.56×10^{-04}**	1.50×10^{-03}	1.66×10^{-03}	1.27×10^{-03}
	Mean	5.77×10^{-04}	**3.10×10^{-04}**	8.01×10^{-04}	5.92×10^{-04}	9.01×10^{-04}
	std	1.55×10^{-04}	**1.08×10^{-05}**	3.78×10^{-04}	3.77×10^{-04}	3.21×10^{-04}
F16	Best	**−1.03163**	**−1.03163**	**−1.03163**	**−1.03163**	**−1.03163**
	Worst	**−1.03163**	**−1.03163**	**−1.03159**	**−1.03163**	**−1.03163**
	Mean	**−1.03163**	**−1.03163**	**−1.03161**	**−1.03163**	**−1.03163**
	std	**2.28×10^{-16}**	**2.28×10^{-16}**	1.2×10^{-05}	4.18×10^{-07}	4.92×10^{-15}
F17	Best	**0.397887**	**0.397887**	0.397907	**0.397887**	**0.397887**
	Worst	**0.397887**	**0.397887**	0.401488	0.397999	**0.397887**
	Mean	**0.397887**	**0.397887**	0.398743	0.397896	**0.397887**
	std	**0.00**	**0.00**	0.000945	2.48×10^{-05}	1.1×10^{-14}
F18	Best	**3**	**3**	**3**	**3**	**3**
	Worst	**3**	**3**	3.000052	3.000228	**3**
	Mean	**3**	**3**	3.000007	3.000026	**3**
	std	**6.03×10^{-16}**	1.11×10^{-15}	1.2×10^{-05}	5.53×10^{-05}	5.2×10^{-14}
F19	Best	**−3.86278**	**−3.86278**	−3.86221	−3.86278	**−3.86278**
	Worst	**−3.86278**	**−3.86278**	−3.85312	−3.86276	**−3.86278**
	Mean	**−3.86278**	**−3.86278**	−3.85612	−3.86278	**−3.86278**
	std	**2.28×10^{-15}**	**2.28×10^{-15}**	0.003147	5.9×10^{-06}	1.22×10^{-14}
F20	Best	**−3.322**	**−3.322**	−3.18286	−3.32199	−3.322
	Worst	**−3.322**	**−3.322**	−1.92056	−3.20299	−3.1952
	Mean	**−3.322**	**−3.322**	−2.9198	−3.21496	−3.22015
	std	**4.56×10^{-16}**	5.19×10^{-16}	0.37441	0.036596	0.043929
F21	Best	**−10.1532**	**−10.1532**	−5.86842	**−10.1532**	**−10.1532**
	Worst	**−10.1532**	**−10.1532**	−0.49729	−2.63044	−2.63047
	Mean	**−10.1532**	**−10.1532**	−2.22823	−6.14446	−8.6434
	std	**3.43×10^{-15}**	3.65×10^{-15}	1.885656	3.476701	2.751568
F22	Best	**−10.4029**	**−10.4029**	−9.10162	**−10.4029**	**−10.4029**
	Worst	**−10.4029**	**−10.4029**	−0.90756	−2.75188	−2.7659
	Mean	**−10.4029**	**−10.4029**	−4.00724	−6.90966	−9.37553
	std	3.21×10^{-15}	**2.51×10^{-15}**	2.111063	3.308424	2.548
F23	Best	**−10.5364**	**−10.5364**	−7.64188	**−10.5363**	**−10.5364**
	Worst	**−10.5364**	**−10.5364**	−3.70826	−2.42157	−2.42734
	Mean	**−10.5364**	**−10.5364**	−5.51857	−6.16316	−9.86292
	std	**1.78×10^{-15}**	1.95×10^{-15}	0.95525	3.420207	2.120365

The best values obtained are in bold.

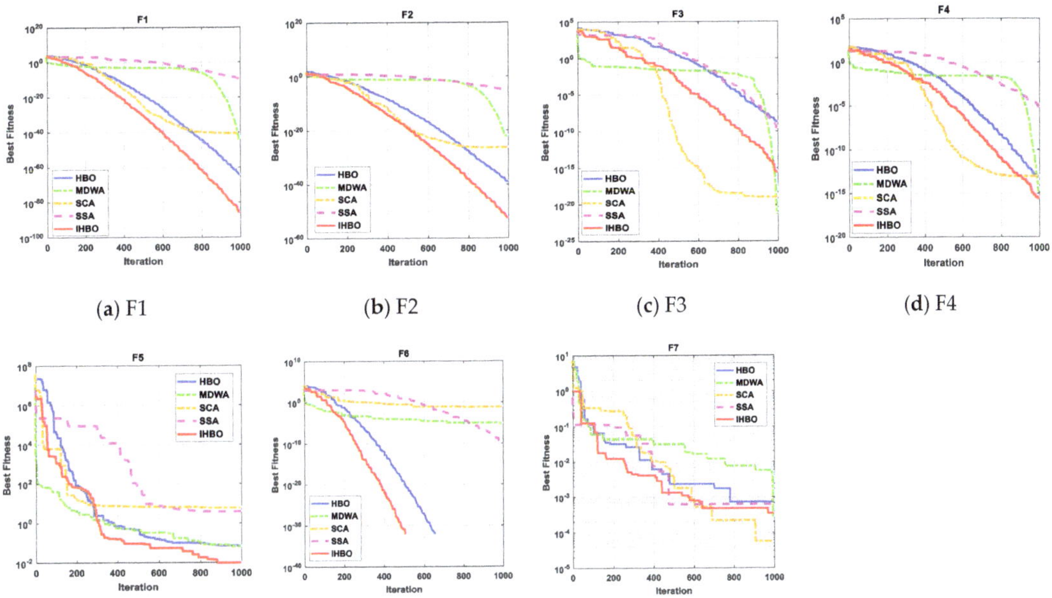

(**a**) F1 (**b**) F2 (**c**) F3 (**d**) F4

(**e**) F5 (**f**) F6 (**g**) F7

Figure 5. The convergence curves of all algorithms for unimodal benchmark functions (**a**) F1, (**b**) F2, (**c**) F3, (**d**) F4, (**e**) F5, (**f**) F6, and (**g**) F7.

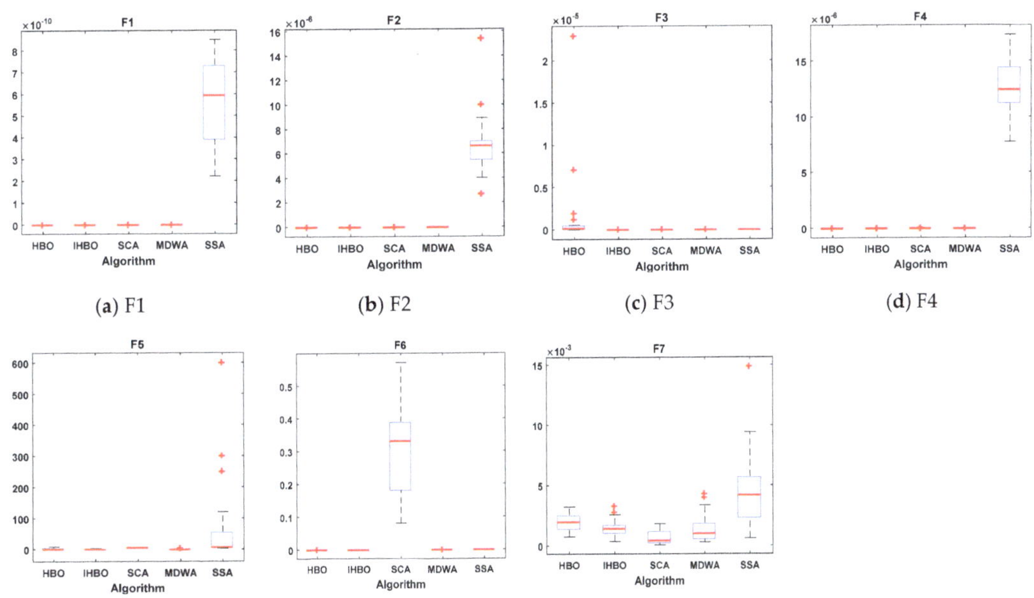

(**a**) F1 (**b**) F2 (**c**) F3 (**d**) F4

(**e**) F5 (**f**) F6 (**g**) F7

Figure 6. Boxplots for all algorithms for unimodal benchmark functions (**a**) F1, (**b**) F2, (**c**) F3, (**d**) F4, (**e**) F5, (**f**) F6, and (**g**) F7.

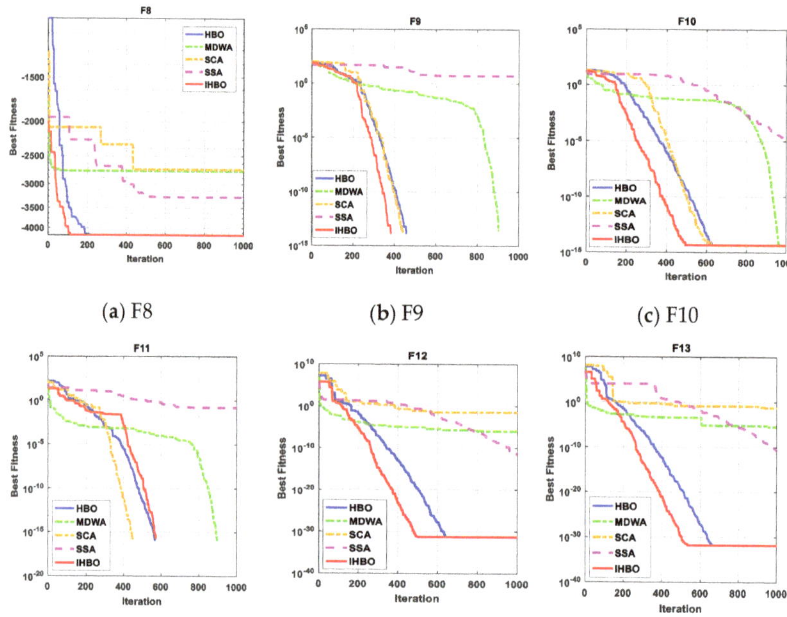

(**a**) F8 (**b**) F9 (**c**) F10

(**d**) F11 (**e**) F12 (**f**) F13

Figure 7. The convergence curves of all algorithms for multi-modal benchmark functions (**a**) F8, (**b**) F9, (**c**) F10, (**d**) F11, (**e**) F12 and (**f**) F13.

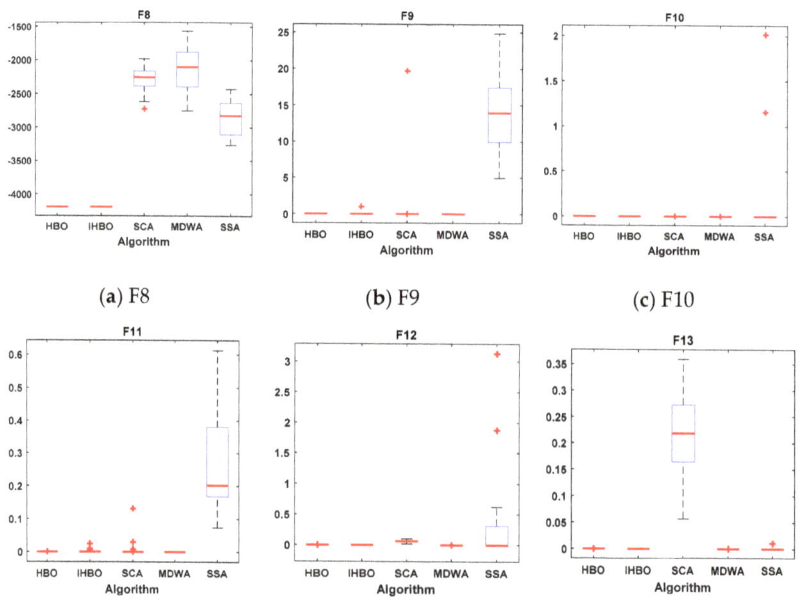

(**a**) F8 (**b**) F9 (**c**) F10

(**d**) F11 (**e**) F12 (**f**) F13

Figure 8. Boxplots for all algorithms for multi-modal benchmark functions (**a**) F8, (**b**) F9, (**c**) F10, (**d**) F11, (**e**) F12 and (**f**) F13.

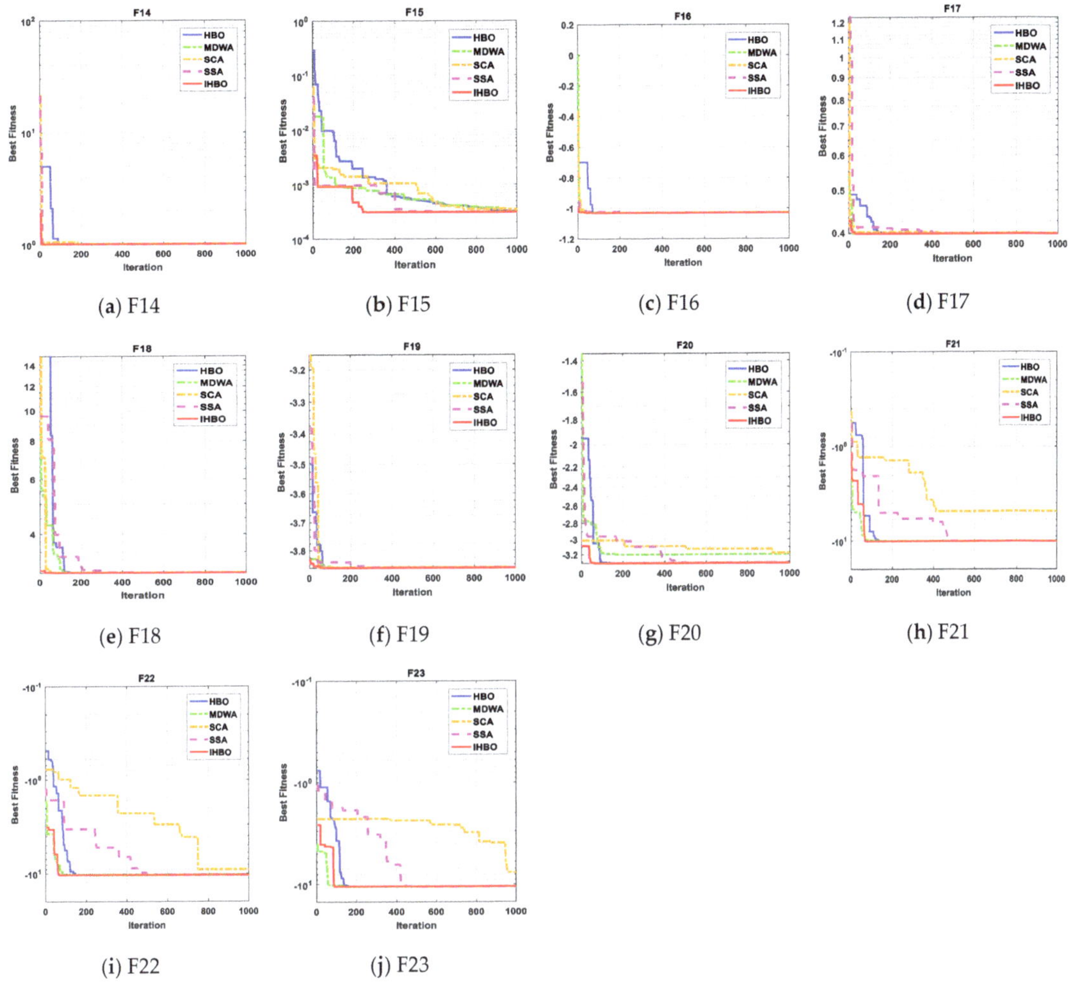

(**a**) F14 (**b**) F15 (**c**) F16 (**d**) F17

(**e**) F18 (**f**) F19 (**g**) F20 (**h**) F21

(**i**) F22 (**j**) F23

Figure 9. The convergence curves of all algorithms for composite benchmark functions (**a**) F14, (**b**) F15, (**c**) F16, (**d**) F17, (**e**) F18, (**f**) F19, (**g**) F20, (**h**) F21, (**i**) F22, and (**j**) F23.

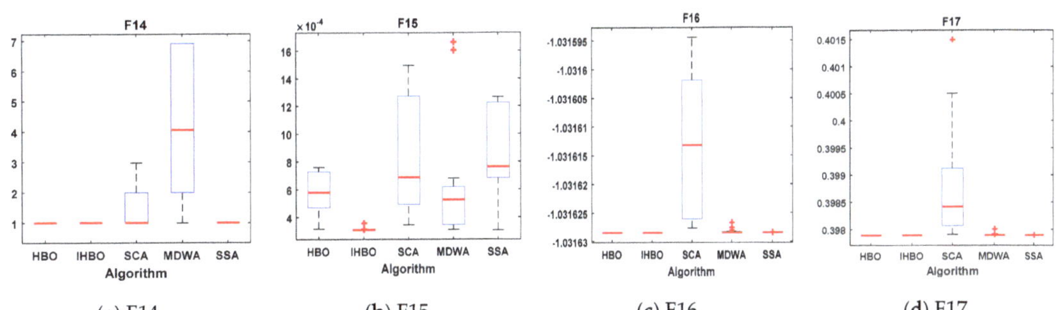

(**a**) F14 (**b**) F15 (**c**) F16 (**d**) F17

Figure 10. *Cont.*

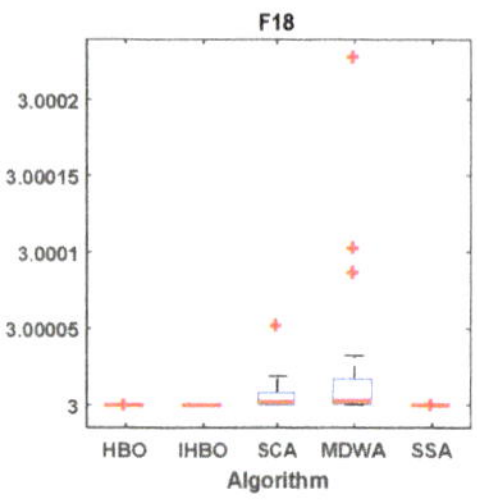

(**e**) F18

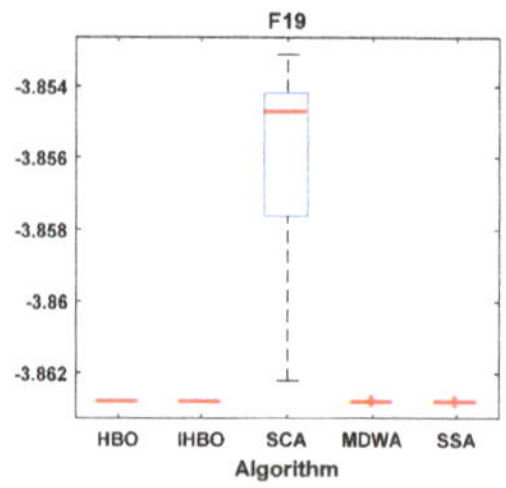

(**f**) F19

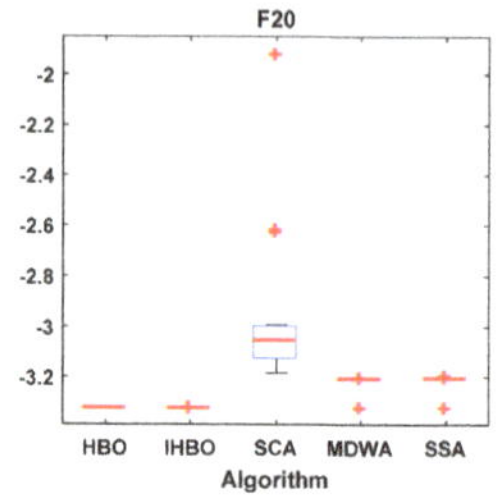

(**g**) F20

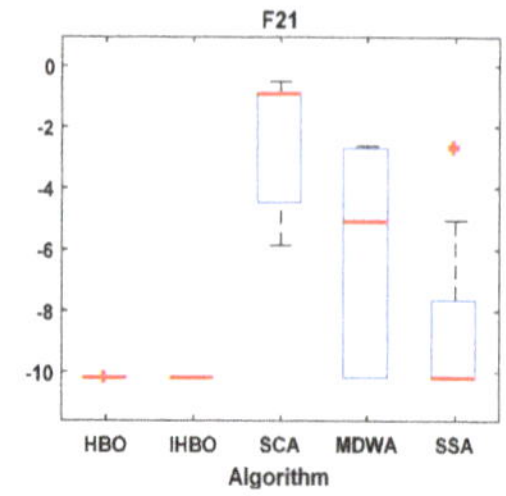

(**h**) F21

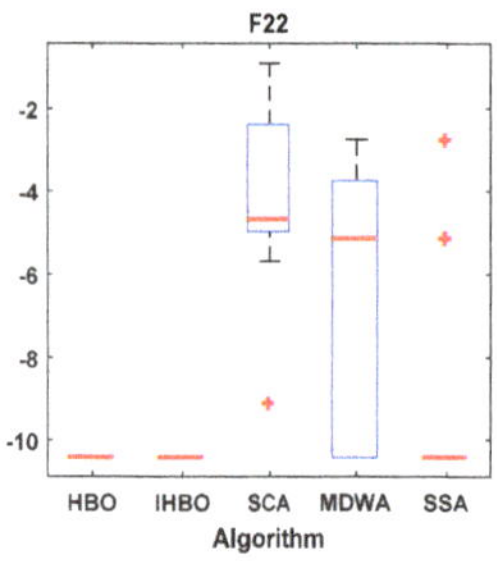

(**i**) F22

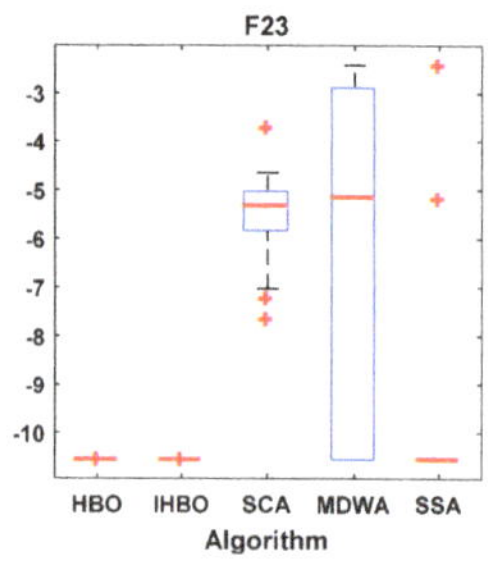

(**j**) F23

Figure 10. Boxplots for all algorithms for composite benchmark functions (**a**) F14, (**b**) F15, (**c**) F16, (**d**) F17, (**e**) F18, (**f**) F19, (**g**) F20, (**h**) F21, (**i**) F22, and (**j**) F23.

5. Project Implementation Location

The project was implemented in a small region in the west of Morocco called Terfaya, at coordinating latitude 27.932 and longitude −12.935.

6. Results and Discussion

In this paper, the Terfaya region of Morocco is selected as the case study to implement an HRES platform based on an improved optimization algorithm called IHBO. The maps for the project location, the load charge, the annual ambient radiation, temperature, wind speed, and pressure are presented in Figures 11–15, respectively.

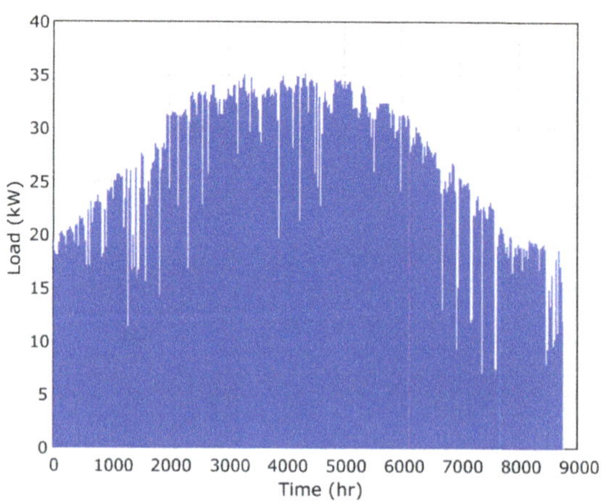

Figure 11. Load power.

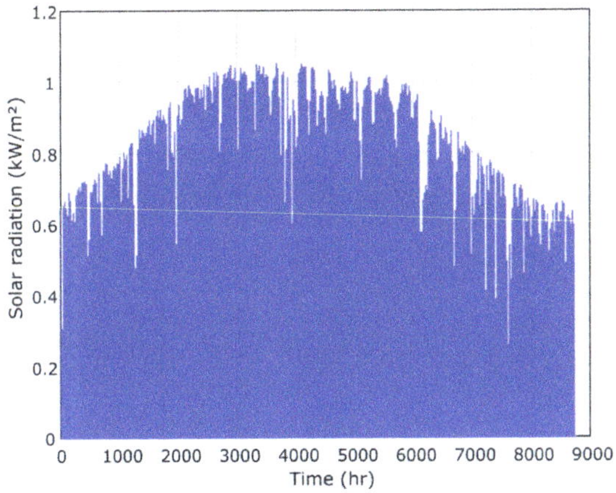

Figure 12. Solar radiation.

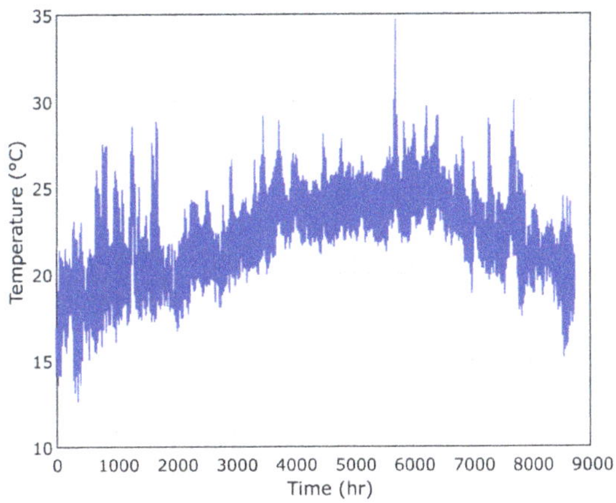

Figure 13. Temperature.

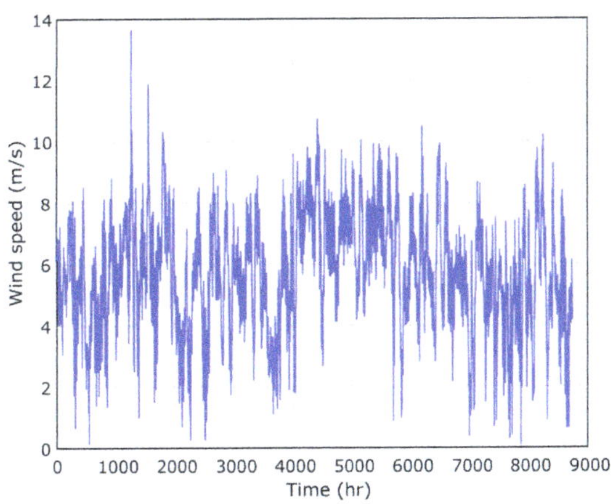

Figure 14. Wind speed.

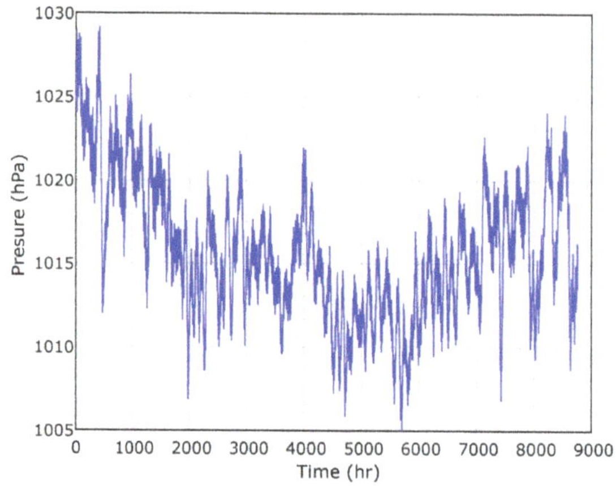

Figure 15. Pressure.

The proposed HRES includes two renewable sources (PV and wind turbines), a diesel generator, and a battery storage system. According to the mathematical modeling of the mentioned systems, the PV output can be affected by the solar radiation data; otherwise, the output power of the wind is influenced by the wind speed data. The decision variables in this study are dedicated to the size of the HRES where: x(1) is the PV area (A_{pv},), x(2) is the wind swept area (A_{wind}), x(3) represents the battery capacity (C_{BESS}) and x(4) is the rated power of the diesel generator (P_{dg}). In this paper, an analysis of fuel price variation is carried out.

6.1. Optimal HRES Design of PV/Diesel/Battery and PV/Wind/Diesel/Battery

6.1.1. PV/Diesel/Battery HRES

The results of the optimal HRES design for the case study concerning the PV/diesel/battery HRES are summarized in Table 5. The table presents all used algorithms concerning the predefined constraints, including the LPSP, RF, and the availability. The algorithms are arranged as GWO, HBO, AEFA, HHO, and IHBO, with a net present cost of MAD 191,661, MAD 175,321, MAD 169,142, MAD 147,527, and MAD 120,463, respectively. The optimal system needs MAD 120,463, equivalent to an LCOE of MAD 0.13/kWh. The system designed respected the constraints very well, with a reliability (LPSP) of 3%, a renewable fraction of 95%, and power availability of 98%. Table 6 presents the optimal size of each algorithm; the best solution is then obtained by IHBO, with 1,673,864 m^2 and 38,860 kW of diesel generator capacity. Table 7 presents the convergence time of all simulations.

Table 5. Results of the PV/diesel/battery HRES.

Algorithm	NPC (MAD)	LCOE (MAD/kWh)	LPSP	RF (%)	Availability (%)
AEFA	169,142	0.1361	0.0360	99.7905	98.5686
GWO	191,661	0.1543	0.0386	99.6961	98.6947
HHO	147,527	0.1187	0.0269	99.7289	98.5725
HBO	175,321	0.1411	0.0383	95.5051	98.9342
IHBO	120,463	0.1384	0.0389	95.3802	98.8665

Table 6. Sizing results of the PV/diesel/battery HRES.

Algorithm	PV (m^2)	Battery (kWh)	Diesel (kW)
AEFA	306.3	4.38	0.65
GWO	334	5.16	1.1
HHO	263.4	1.85	0.65
HBO	170.6	0.63	4
IHBO	167.4	0	3.88

Table 7. Convergence time of algorithms.

Algorithm	Convergence Time (s)
AEFA	5421
GWO	5959
HHO	7283
HBO	333
IHBO	14,017

The convergence curve results for all scenarios are presented in Figure 16, in which the IHBO proves its efficacy to reach the optimal solution.

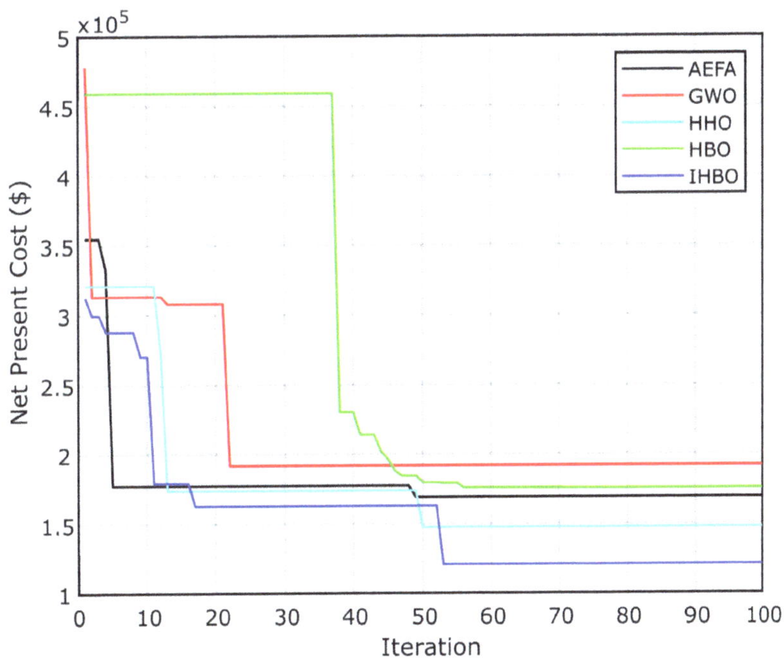

Figure 16. PV/diesel/battery.

6.1.2. PV/Wind/Diesel/Battery HRES

The second configuration used in this paper concerns the PV/wind/diesel/battery HRES. From Table 8, the results respect the constraints; then, the best algorithms results converge as HBO, GWO, AEFA, HHO, and IHBO, with an investment cost of MAD 461,233, MAD 226,559, MAD 221,694, MAD 215,371, and MAD 100,337, respectively. The best cost needs MAD 100,337, equivalent to MAD 0.08/kWh; in this situation, the LPSP is about 4%, the renewable fraction is near 100%, and the power availability is more than 99%. Table 9 presents the size results, which show that the best project needs 261.3031 m^2 of PV area, 102.7114 m^2 of swept area of the wind turbines, 23.2177 kWh of battery, and 1.0762 kW of diesel. Table 10 presents the convergence time of all simulations.

Table 8. Results of the PV/wind/diesel/battery HRES.

Algorithm	NPC (MAD)	LCOE (MAD/kWh)	LPSP	RF (%)	Availability (%)
AEFA	221,694	0.1784	0.0440	99.4362	98.7671
GWO	226,559	0.1823	0.0233	98.6033	99.5884
HHO	215,371	0.1733	0.0076	99.9540	99.6287
HBO	461,233	0.3712	0.0030	99.8870	99.9540
IHBO	165,999	0.1336	0.0445	99.9133	99.0391

Table 9. Sizing results of the PV/wind/diesel/battery HRES.

Algorithm	PV (m^2)	Wind (m^2)	Battery (kWh)	Diesel (kW)
AEFA	92.9	659.8	0.14	4.83
GWO	143.9	174	14.64	7.24
HHO	317.8	232.6	3.59	1.47
HBO	403.9	652.8	2.35	10.6
IHBO	261.3	102.7	23.2	1

Table 10. Convergence time of the algorithms.

Algorithm	Convergence Time (s)
AEFA	1176
GWO	1079
HHO	3931
HBO	680
IHBO	4942

Figure 17 presents the convergence curve of the NPC for the PV/wind/diesel/battery HRES; the curve shows that the IHBO algorithm gives better convergence results.

6.2. *Impact of Fuel Price Variation*

In the paper, if we suppose that the price of fuel is about MAD 0.41/L, then we can compare the total investment cost with the previous study that used the actual price, which is MAD 0.96/L.

From Table 11, it is clearly shown that the NPC of the HRES is reduced strongly while it is passed from MAD 120,463. Table 12 presents the optimal HRES size using all optimization algorithms. Figure 18 presents the convergence curve of the NPC for the PV/diesel/battery HRES, with a fuel price of MAD 0.54/L. This figure shows that the IHBO algorithm gives the better convergence results.

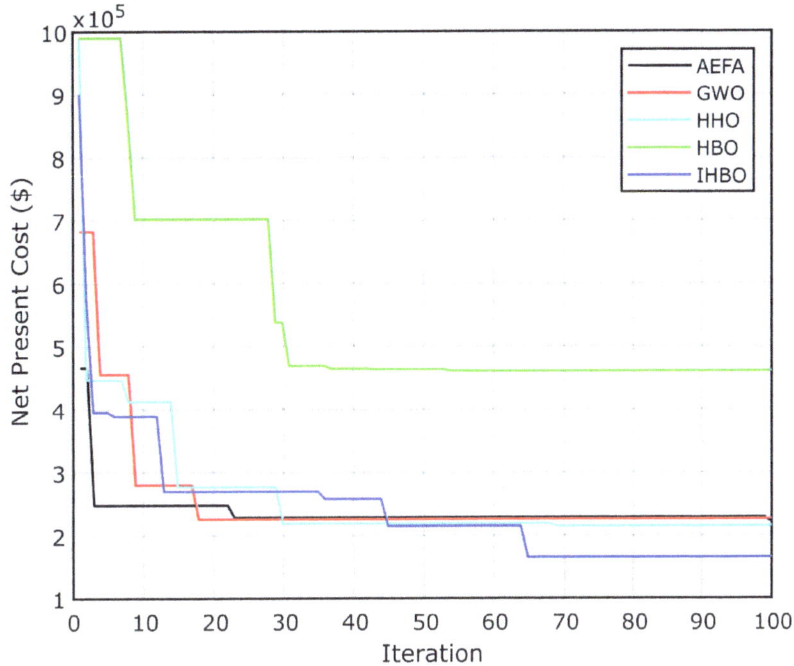

Figure 17. Convergence curve for the PV/wind/diesel/battery HRES.

Table 11. Results of the PV/diesel/battery HRES with fuel prices.

Algorithm	NPC (MAD)	LCOE (MAD/kWh)	LPSP	RF (%)	Availability (%)
AEFA	166,303	0.1339	0.0324	96.1598	99.4268
GWO	107,532	0.0865	0.0483	97.0846	98.0032
HHO	87,394	0.0703	0.0496	99.6090	96.1823
HBO	125,791	0.1012	0.0296	97.7468	98.8978
IHBO	68,121	0.0548	0.1119	99.9999	88.8055

Table 12. Sizing results of the PV/diesel/battery HRES.

Algorithm	PV (m^2)	Battery (kWh)	Diesel (kW)
AEFA	216.7	3.7	6
GWO	163.7	0.4	2.24
HHO	161.8	0.7	0.29
HBO	194	0	2.79
IHBO	124.9	0	0

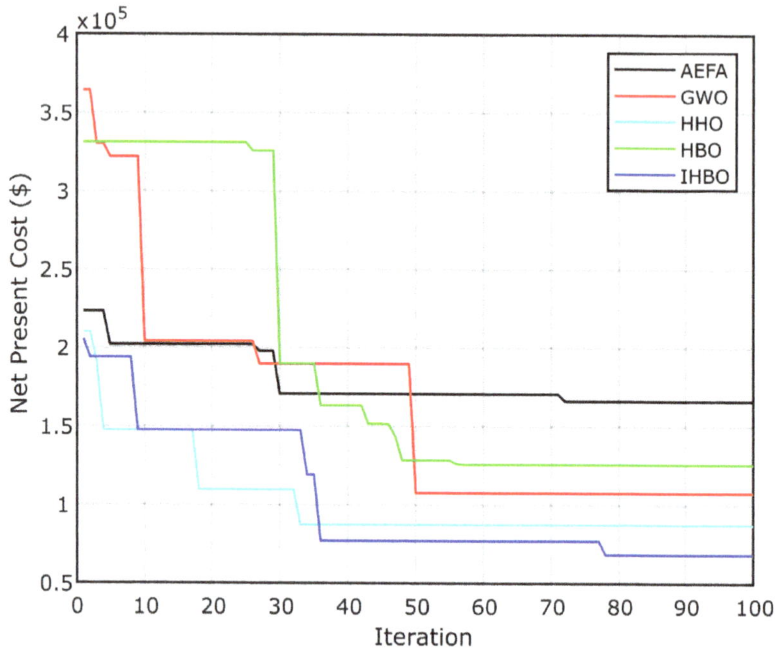

Figure 18. Convergence PV/diesel/battery with a fuel price of 0.54.

7. Conclusions

This paper proposed a platform to design an HRES microgrid system based on two configurations, PV/diesel/battery, and PV/wind/diesel/battery. The platform is based on modeling, power management, and a cost optimization study using an improved IHBO algorithm. The proposed IHBO algorithm proved its efficacy in finding the optimal solution compared with many algorithms, including AEFA, GWO, HHO, and the original HBO. In the paper, we discussed the case of reducing fuel prices and its impact on the investment cost. The results show that the NPC is highly reduced when the use of diesel is small. Several systems, such as hydrogen storage and biomass systems, can be integrated in the microgrid. Future work will focus on developing configurations considering the degradation of battery characteristics.

Author Contributions: Conceptualization, M.K. and S.K.; methodology, S.K.E., M.H.H. and I.B.M.T.; software, M.K., S.K. and M.H.H.; validation, S.K.E. and I.B.M.T.; formal analysis, M.K., S.K. and M.H.H.; investigation, S.K.E. and I.B.M.T.; resources, M.K., S.K. and M.H.H.; data curation, S.K.E. and I.B.M.T.; writing original draft preparation, M.K., S.K. and M.H.H.; writing on characteristic, S.K.E. and I.B.M.T.; visualization, S.K.E. and I.B.M.T.; supervision, S.K.; project administration, M.K. All authors have read and agreed to the published version of the manuscript.

Funding: This study was funded by Taif University Researchers Supporting Project Number (TURSP-2020/61), Taif University, Taif, Saudi Arabia.

Institutional Review Board Statement: Not applicable.

Informed Consent Statement: Not applicable.

Data Availability Statement: Not applicable.

Acknowledgments: The authors would like to acknowledge the financial support received from Taif University Researchers Supporting Project Number (TURSP-2020/61), Taif University, Taif, Saudi Arabia.

Conflicts of Interest: The authors declare no conflict of interest.

Nomenclature

AEFA	Artificial electric field algorithm
BESS	Battery energy storage system
BO	Bonobo optimizer
BSA	Backtracking aearch algorithm
COE	Cost of energy
DG	Diesel generator
GWO	Grey wolf optimizer
CRH	Corporate rank hierarchy
HBO	Heap-based optimizer
HHO	Harris hawks optimizer
HRES	Hybrid renewable energy system
IWO	Invasive weed optimization
MPPT	Maximum power point tracker
NPC	Net present cost
PSO	Particle swarm optimization
QOBO	Quasi-oppositional bonobo optimizer
RESSOC	Renewable energy sourcesState of charge
WT	Wind turbine
Symbols	
A_g, B_g	Constants of the linear consumption of the fuel (L/kW)
A_{pv}	PV area (m^2)
A_{wind}	Swept area of the wind turbine (m^2)
C_{BESS}	Capacity of BESS (kWh)
C_p	Maximum power coefficient (%)
E_{bmin}	Min battery energy in discharge (kWh)
E_l	Energy Load (kWh)
FC_{dg}	Fuel cost (MAD)
F_{dg}	Fuel consumption (L/h)
$P_{dg,out}$	Output power of diesel generator (kW)
P_{dg}	Rated power of diesel generator (kW)
P_{load}	Load power (kW)
P_{pv}	Output power of PV (kW)
P_{re}	Output power of renewable energy sources (kW)
P_{wind}	Output wind power (kW)
T_a	Ambient temperature (°C),
T_r	Photovoltaic cell reference temperature (°C).
i_r	Interest rate (%)
Max_iter	Maximum iteration
v_{ci}	Cut-in speed (m/s)
v_{co}	Cut-out speed (m/s)
v_r	Rated wind speed (m/s)
η_b	Efficiency of the battery (%)
η_i	Efficiency of the inverter (%)
η_{pv}	Efficiency of the PV (%)
η_r	Reference efficiency
η_t	Efficiency of the MPPT equipment
STD	Standard deviation
Av	Availability index (%)
AD	Autonomy daily of the battery (day)
AD^{min}	Minimum allowed autonomy daily of the battery (day)
C	Capital cost (MAD)
CRF	Capital recovery factor
DOD	Depth of discharge (%)
I	Solar irradiation (kW/m^2)

LCOE	Levelized cost of energy (MAD/kWh)
LPSP	Loss of power supply probability (%)
NOCT	Nominal operating cell temperature (°C),
NPC	Net present cost (MAD)
OM	Operation and maintenance cost (MAD)
R	Replacement cost (MAD)
RF	Renewable fraction (%)
v	Wind velocity (m/s)
ρ	Air density (Kg/m^3)

Appendix A

Table A1. Summary of the HRES parameters.

Symbol	Quantity	Conversion
N	Project lifetime	20 year
η_r	Reference efficiency of the PV	25%
η_t	Efficiency of MPPT	100%
T_r	PV cell reference temperature	25 °C
β	Temperature coefficient	0.005 °C
$NOCT$	Nominal operating cell temperature	47 °C
N_{pv}	PV system lifetime	20 year
C_p	Maximum power coefficient	48%
V_{ci}	Cut-in wind speed	2.6 m/s
V_{co}	Cut-out wind speed	25 m/s
V_r	Rated wind speed	9.5 m/s
N_{wind}	Wind system lifetime	20 year
p_f	Fuel price in Morocco	MAD 0.96/L
N_{diesel}	Diesel system lifetime	7 year
DOD	Depth of discharge	80%
η_b	Battery efficiency	97%
N_{bat}	Battery system lifetime	5 year
η_{inv}	Inverter efficiency	97%

References

1. Gao, K.; Wang, T.; Han, C.; Xie, J.; Ma, Y.; Peng, R. A Review of Optimization of Microgrid Operation. *Energies* **2021**, *14*, 2842. [CrossRef]
2. Dagar, A.; Gupta, P.; Niranjan, V. Microgrid protection: A comprehensive review. *Renew. Sustain. Energy Rev.* **2021**, *149*, 111401. [CrossRef]
3. Beheshtaein, S.; Cuzner, R.M.; Forouzesh, M.; Savaghebi, M.; Guerrero, J.M. DC Microgrid Protection: A Comprehensive Review. *IEEE J. Emerg. Sel. Top. Power Electron.* **2019**. [CrossRef]
4. Azeem, O.; Ali, M.; Abbas, G.; Uzair, M.; Qahmash, A.; Algarni, A.; Hussain, M.R. A Comprehensive Review on Integration Challenges, Optimization Techniques and Control Strategies of Hybrid AC/DC Microgrid. *Appl. Sci.* **2021**, *11*, 6242. [CrossRef]
5. Al-Ismail, F.S. DC Microgrid Planning, Operation, and Control: A Comprehensive Review. *IEEE Access* **2021**, *9*, 36154–36172. [CrossRef]
6. Roy, A.; Auger, F.; Olivier, J.-C.; Schaeffer, E.; Auvity, B. Design, Sizing, and Energy Management of Microgrids in Harbor Areas: A Review. *Energies* **2020**, *13*, 5314. [CrossRef]
7. Marqusee, J.; Becker, W.; Ericson, S. Resilience and economics of microgrids with PV, battery storage, and networked diesel generators. *Adv. Appl. Energy* **2021**, *3*, 100049. [CrossRef]
8. Kharrich, M.; Kamel, S.; Abdeen, M.; Mohammed, O.H.; Akherraz, M.; Khurshaid, T.; Rhee, S.-B. Developed approach Based on Equilibrium Optimizer for Optimal Design of Hybrid PV/Wind/Diesel/Battery Microgrid in Dakhla, Morocco. *IEEE Access* **2021**, *9*, 13655–13670. [CrossRef]
9. Fathy, A.; Kaaniche, K.; Alanazi, T.M. Recent Approach Based Social Spider Optimizer for Optimal Sizing of Hybrid PV/Wind/Battery/Diesel Integrated Microgrid in Aljouf Region. *IEEE Access* **2020**, *8*, 57630–57645. [CrossRef]
10. Hafez, A.A.; Abdelaziz, A.Y.; Hendy, M.A.; Ali, A.F.M. Optimal sizing of off-line microgrid via hybrid multi-objective simulated annealing particle swarm optimizer. *Comput. Electr. Eng.* **2021**, *94*, 107294. [CrossRef]
11. Kharrich, M.; Mohammed, O.H.; Alshammari, N.; Akherraz, M. Multi-objective optimization and the effect of the economic factors on the design of the microgrid hybrid system. *Sustain. Cities Soc.* **2021**, *65*, 102646. [CrossRef]
12. Mah, A.X.Y.; Ho, W.S.; Hassim, M.H.; Hashim, H.; Ling, G.H.T.; Ho, C.S.; Muis, Z.A. Optimization of photovoltaic-based microgrid with hybrid energy storage: A P-graph approach. *Energy* **2021**, *233*, 121088. [CrossRef]
13. Lai, C.S.; Locatelli, G.; Pimm, A.; Tao, Y.; Li, X.; Lai, L.L. A financial model for lithium-ion storage in a photovoltaic and biogas energy system. *Appl. Energy* **2019**, *251*, 113179. [CrossRef]

14. Naderi, E.; Narimani, H.; Pourakbari-Kasmaei, M.; Cerna, F.V.; Marzband, M.; Lehtonen, M. State-of-the-Art of Optimal Active and Reactive Power Flow: A Comprehensive Review from Various Standpoints. *Processes* **2021**, *9*, 1319. [CrossRef]
15. Hassan, M.H.; Kamel, S.; Abualigah, L.; Eid, A. Development and application of slime mould algorithm for optimal economic emission dispatch. *Expert Syst. Appl.* **2021**, *182*, 115205. [CrossRef]
16. Cerna, F.V.; Pourakbari-Kasmaei, M.; Pinheiro, L.S.; Naderi, E.; Lehtonen, M.; Contreras, J. Intelligent Energy Management in a Prosumer Community Considering the Load Factor Enhancement. *Energies* **2021**, *14*, 3624. [CrossRef]
17. Ramadan, A.; Kamel, S.; Hassan, M.H.; Khurshaid, T.; Rahmann, C. An Improved Bald Eagle Search Algorithm for Parameter Estimation of Different Photovoltaic Models. *Processes* **2021**, *9*, 1127. [CrossRef]
18. Naderi, E.; Asrari, A. Hardware-in-the-loop experimental validation for a lab-scale microgrid targeted by cyberattacks. In Proceedings of the 9th International Conference on Smart Grid, Setubal, Portugal, 29 June–1 July 2021; Volume 29.
19. Naderi, E.; Pazouki, S.; Asrari, A. A Region-based Framework for Cyberattacks Leading to Undervoltage in Smart Distribution Systems. In Proceedings of the 2021 IEEE Power and Energy Conference at Illinois (PECI), Urbana, IL, USA, 1–2 April 2021.
20. Naderi, E.; Asrari, A. Approaching Optimal Power Flow from Attacker's Standpoint to Launch False Data Injection Cyberattack. In Proceedings of the 2020 IEEE Green Energy and Smart Systems Conference (IGESSC), Long Beach, CA, USA, 2–3 November 2020.
21. Cao, B.; Dong, W.; Lv, Z.; Gu, Y.; Singh, S.; Kumar, P. Hybrid Microgrid Many-Objective Sizing Optimization with Fuzzy Decision. *IEEE Trans. Fuzzy Syst.* **2020**, *28*, 2702–2710. [CrossRef]
22. Xu, Y.-P.; Ouyang, P.; Xing, S.-M.; Qi, L.-Y.; Khayatnezhad, M.; Jafari, H. Optimal structure design of a PV/FC HRES using amended Water Strider Algorithm. *Energy Rep.* **2021**, *7*, 2057–2067. [CrossRef]
23. Lei, G.; Song, H.; Rodriguez, D. Power generation cost minimization of the grid-connected hybrid renewable energy system through optimal sizing using the modified seagull optimization technique. *Energy Rep.* **2020**, *6*, 3365–3376. [CrossRef]
24. Kharrich, M.; Kamel, S.; Ellaia, R.; Akherraz, M.; Alghamdi, A.S.; Abdel-Akher, M.; Eid, A.; Mosaad, M.I. Economic and Ecological Design of Hybrid Renewable Energy Systems Based on a Developed IWO/BSA Algorithm. *Electronics* **2021**, *10*, 687. [CrossRef]
25. Yu, G.; Meng, Z.; Ma, H.; Liu, L. An adaptive Marine Predators Algorithm for optimizing a hybrid PV/DG/Battery System for a remote area in China. *Energy Rep.* **2021**, *7*, 398–412. [CrossRef]
26. Kharrich, M.; Mohammed, O.H.; Kamel, S.; Selim, A.; Sultan, H.M.; Akherraz, M.; Jurado, F. Development and Implementation of a Novel Optimization Algorithm for Reliable and Economic Grid-Independent Hybrid Power System. *Appl. Sci.* **2020**, *10*, 6604. [CrossRef]
27. Lai, C.S.; McCulloch, M.D. Sizing of Stand-Alone Solar PV and Storage System with Anaerobic Digestion Biogas Power Plants. *IEEE Trans. Ind. Electron.* **2017**, *64*, 2112–2121. [CrossRef]
28. Heydari, A.; Askarzadeh, A. Optimization of a biomass-based photovoltaic power plant for an off-grid application subject to loss of power supply probability concept. *Appl. Energy* **2016**, *165*, 601–611. [CrossRef]
29. Tabak, A.; Kayabasi, E.; Guneser, M.T.; Ozkaymak, M. Grey wolf optimization for optimum sizing and controlling of a PV/WT/BM hybrid energy system considering TNPC, LPSP, and LCOE concepts. Energy Sources. *Part A Recovery Util. Environ. Eff.* **2019**, 1–21. [CrossRef]
30. Guangqian, D.; Bekhrad, K.; Azarikhah, P.; Maleki, A.A. hybrid algorithm based optimization on modeling of grid independent biodiesel-based hybrid solar/wind systems. *Renew. Energy* **2018**, *122*, 551–560. [CrossRef]
31. Ramli, M.A.M.; Bouchekara, H.R.E.H.; Alghamdi, A.S. Optimal sizing of PV/wind/diesel hybrid microgrid system using multi-objective self-adaptive differential evolution algorithm. *Renew. Energy* **2018**, *121*, 400–411. [CrossRef]
32. Movahediyan, Z.; Askarzadeh, A. Multi-objective optimization framework of a photovoltaic-diesel generator hybrid energy system considering operating reserve. *Sustain. Cities Soc.* **2018**, *41*, 1–12. [CrossRef]
33. Ghiasi, M. Detailed study, multi-objective optimization, and design of an AC-DC smart microgrid with hybrid renewable energy resources. *Energy* **2019**, *169*, 496–507. [CrossRef]
34. Askari, Qamar, Mehreen Saeed, and Irfan Younas. Heap-based optimizer inspired by corporate rank hierarchy for global optimization. *Expert Syst. Appl.* **2020**, *161*, 113702. [CrossRef]
35. Eberhart, R.; Kennedy, J. A new optimizer using particle swarm theory. MHS'95. In Proceedings of the Sixth International Symposium on Micro Machine and Human Science, Nagoya, Japan, 4–6 October 1995.
36. Pervez, I.; Malick, I.H.; Tariq, M.; Sarwar, A.; Zaid, M. A maximum power point tracking method using a hybrid PSO and grey wolf optimization algorithm. In Proceedings of the 2019 2nd International Conference on Power Energy, Environment and Intelligent Control (PEEIC), Greater Noida, India, 18–19 October 2019; pp. 565–569.
37. Mirjalili, S. SCA: A sine cosine algorithm for solving optimization problems. *Knowl.-Based Syst.* **2016**, *96*, 120–133. [CrossRef]
38. Mirjalili, S.; Gandomi, A.; Mirjalili, S.Z.; Saremi, S.; Faris, H.; Mirjalili, S.M. Salp Swarm Algorithm: A bio-inspired optimizer for engineering design problems. *Adv. Eng. Softw.* **2017**, *114*, 163–191. [CrossRef]
39. Rizk-Allah, R.M.; Hassanien, A.E. A movable damped wave algorithm for solving global optimization problems. *Evol. Intell.* **2018**, *12*, 49–72. [CrossRef]

sustainability

MDPI

Article

Design, Modeling, and Differential Flatness Based Control of Permanent Magnet-Assisted Synchronous Reluctance Motor for e-Vehicle Applications

Songklod Sriprang [1,2], Nitchamon Poonnoy [2,*], Damien Guilbert [1], Babak Nahid-Mobarakeh [3], Noureddine Takorabet [1], Nicu Bizon [4] and Phatiphat Thounthong [2,*]

1. Groupe de Recherche en Energie Electrique de Nancy (GREEN), Université de Lorraine, F-54000 Nancy, France; songklod.sriprang@univ-lorraine.fr (S.S.); damien.guilbert@univ-lorraine.fr (D.G.); noureddine.takorabet@univ-lorraine.fr (N.T.)
2. Renewable Energy Research Centre (RERC), Department of Teacher Training in Electrical Engineering, Faculty of Technical Education, King Mongkut's University of Technology North Bangkok, Bangkok 10800, Thailand
3. Department of Electrical and Computer Engineering, McMaster University, Hamilton, ON L8S 4L8, Canada; babak.nahid@mcmaster.ca
4. Faculty of Electronics, Communications and Computers, University of Pitesti, 110040 Pitesti, Romania; nicu.bizon@upit.ro

* Correspondence: nitchamon.p@fte.kmutnb.ac.th (N.P.); phatiphat.t@fte.kmutnb.ac.th (P.T.)

Citation: Sriprang, S.; Poonnoy, N.; Guilbert, D.; Nahid-Mobarakeh, B.; Takorabet, N.; Bizon, N.; Thounthong, P. Design, Modeling, and Differential Flatness Based Control of Permanent Magnet-Assisted Synchronous Reluctance Motor for e-Vehicle Applications. *Sustainability* **2021**, *13*, 9502. https://doi.org/10.3390/su13179502

Academic Editors: Marc A. Rosen and Lin Li

Received: 29 June 2021
Accepted: 18 August 2021
Published: 24 August 2021

Publisher's Note: MDPI stays neutral with regard to jurisdictional claims in published maps and institutional affiliations.

Abstract: This paper presents the utilization of differential flatness techniques from nonlinear control theory to permanent magnet assisted (PMa) synchronous reluctance motor (SynRM). The significant advantage of the proposed control approach is the potentiality to establish the behavior of the state variable system during the steady-state and transients operations as well. The mathematical models of PMa-SynRM are initially proved by the nonlinear case to show the flatness property. Then, the intelligent proportional-integral (iPI) is utilized as a control law to deal with some inevitable modeling errors and uncertainties for the torque and speed of the motor. Finally, a MicroLab Box dSPACE has been employed to implement the proposed control scheme. A small-scale test bench 1-KW relying on the PMa-SynRM has been designed and developed in the laboratory to approve the proposed control algorithm. The experimental results reflect that the proposed control effectively performs high performance during dynamic operating conditions for the inner torque loop control and outer speed loop control of the motor drive compared to the traditional PI control.

Keywords: electric vehicle; inverter; permanent magnet assisted synchronous reluctance motor; differential flatness-based control; parameter observers; traction drive

1. Introduction

Permanent magnet synchronous motors (PMSMs) are the most widespread motor technologies in transportation applications including more electric aircraft (MEA), electric vehicles (EVs or e-vehicle), and hybrid electric vehicles (HEVs) [1–5]. Indeed, this technology enables offering high torque, power density, and high efficiency; while providing an extensive speed range. Besides, due to their design, they are extremely versatile and can also be employed for low-power applications, offering high performance [1]. On the other side, these motors require the use of rare-earth metals to make permanent magnets (PMs) such as Nd-Fe-B (neodymium-iron-boron), located on the rotor. Due to the growing development of electric vehicles, the interest in rare-earth metals has been increasing; leading up consequently to high cost and environmental consequences for the extraction and refining of rare-earth elements.

As a result, to cope with these important issues, a new permanent magnet-assisted (PMa) synchronous reluctance motor (SynRM) has been conceived to reduce the size of PMs and to increase the use of ferrite magnet materials in the rotor part. Besides, the modern PMa-SynRM is more advantageous than the classic SynRM [6–8]. Its cost is reduced compared to the usual PMSM since ferrite magnets are cost-effective over rare-earth PMs. In summary, the PMa-SynRM is an emerging and attractive motor for the dissemination of the next generations of electric cars. The constitution of the proposed four-pole PMa-SynRM prototype relying on ferrite magnets, and the d-axis and q-axis are shown in Figure 1. It can be noted that the permanent magnets are positioned in the flux barriers of the rotor part. Hence, magnetization takes place along the negative q-axis. Given that the PMa-SynRM is relatively new, the achievement of high performance for large functioning conditions by controlling it remains a technical barrier. Indeed, its nonlinear features, parameter mismatch, and also parametric uncertainty make its control challenging.

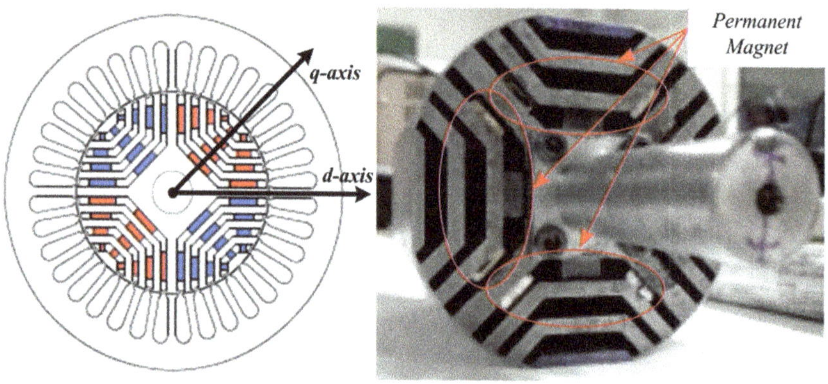

Figure 1. The rotor's structure of the proposed PMa-SynRM.

To face this technical barrier from the control point of view, various control approaches have been reported and analyzed recently in the literature. First and foremost, Ion Boldea et al. [9] have studied a direct torque and flux control with space vector modulation (DTFC-SVM) drive control of PM-assisted reluctance synchronous motor/generator employed for mild hybrid vehicles applications. Peyman Niazi et al. [10] have conceived a maximum torque per ampere (MTPA) control strategy coupled with a parameter observer applied to a PMa-SynRM to face the change of motor parameters (inductances and PM flux density) and saturation effect due to the internal temperature. However, the tests have been performed when the PMa-SynRM operates under a constant torque region. By comparison, Elena Trancho et al. [11] have designed a robust torque-based control scheme addressed to the PM-Assisted Synchronous Reluctance Machine in EVs and HEVs. In this work, the authors have employed an arrangement of a second-order-based sliding mode control for inner current regulation and a look-up table/voltage restriction pursuit-based hybrid field weakening operation to cope with parameters deviation during operation. Despite these relevant works introducing and designing robust controllers, the achievement of high performance is still a challenging barrier. As emphasized in these works, this barrier is due to the variation of machine electrical parameters and the nonlinear properties. To meet these technical issues for the control of the PMa-SynRM, a nonlinear control named "Differential Flatness" is elaborated and has recently been suggested to reach the expected performances in controlling PMSM [12]. It has been demonstrated that differential flatness control presents higher performance than traditional control systems. Besides, adding the nonlinear observation for motor parameters estimating makes the control system more robust. This novel control approach has been applied in various nonlinear systems; for instance, for the energy management of a hybrid power plant (including a fuel cell and supercapacitors) [13], and the command of double fed inductor motor [14].

This paper presents the differential flatness control to control PMa-SynRM. As a result of this introduction reviewing the main issues for the control of PMa-SynRM and the current state-of-the-art, the mathematical models of PMa-SynRM are developed to prove the differential flat property in Section 2. Then, in Section 3, the intelligent proportional-integral (*i*PI) [15,16] controller is conceived as a control law to compensate for the torque and speed of the motor. A nonlinear estimator is introduced to estimate external torque disturbance. Finally, in Section 4, the comparison between differential flatness and conventional PI control is discussed to demonstrate the benefits of the proposed control algorithm. A small-scale test bench 1-KW relying on the PMa-SynRM with ferrite magnets has been realized to attest to the performance of the designed control scheme in the laboratory [17].

2. FEM-Based Magnetic Model

Given that the inductances L_d and L_q play major roles in the overload capacity, the field-weakening operation has been enhanced, and an accurate regulation has been determined to make their calculation easier. The inductances in the d–q axis have been computed considering a nonlinear case where the saturation of the stator teeth and rotor ribs have been taken into account. Due to the effect of the internal temperature, the inductances can be saturated as emphasized in [10]. Hence, the saturation effects of the inductances L_d and L_q can be computed relying on the link between the flux linkage change and the small rise in the current of the d–q axis, as provided by the Equations (1)–(3) [18,19]. Furthermore, the cross-coupling effects due to saturation have been investigated as well for the PMa-SynRM. The parameters have been assessed in the d–q axis. The operating constraints of the current supply have been reproduced, and the flux linkages have been assessed by incorporating the magnetic vector potential. The d-q flux linkages linked to the d- and q-axes currents are illustrated in Figure 2. The d- and q-flux linkages related to based on FEM analysis are represented in Figure 2. Figure 2a depicts the lookup table of d-axis flux linkage $\Psi_d(i_d,i_q)$ (LUT1); while Figure 2b exhibits the lookup q-axis flux linkage $\Psi_q(i_d,i_q)$ (LUT2). In addition, Table 1 shows the given flux linkages of the PMs and d–q axes inductances of the control network design.

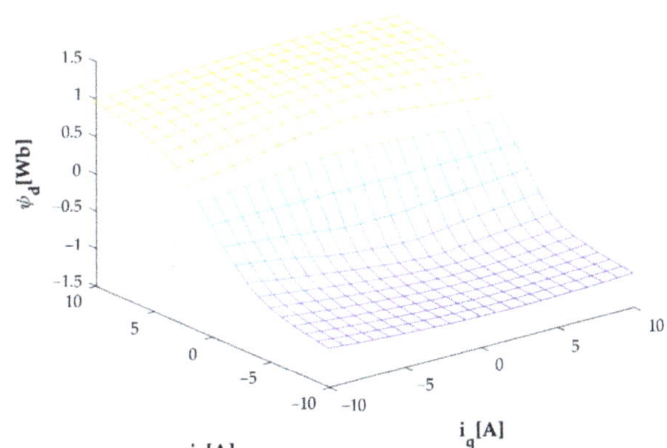

(a)

Figure 2. *Cont.*

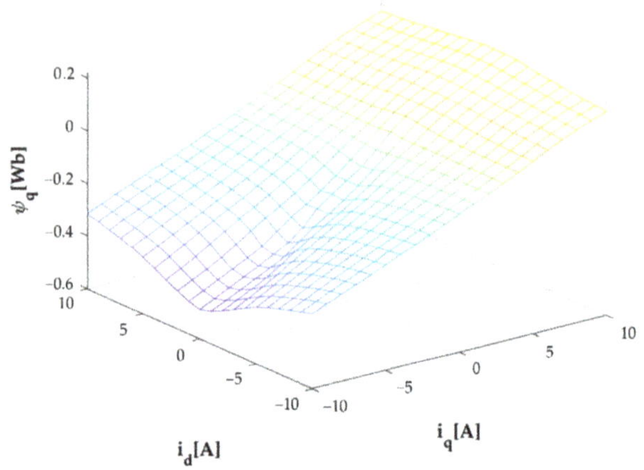

(b)

Figure 2. The flux linkages Ψ_d and Ψ_q in the function of i_d and i_q. (**a**) $\Psi_d(i_d,i_q)$. (**b**) $\Psi_q(i_d,i_q)$.

$$L_d = \frac{\partial \Psi_d(i_d, i_q)}{\partial i_d} = \left.\frac{\Delta \Psi_d(i_d, i_q)}{\Delta i_d}\right|_{i_q=\text{constant}} \tag{1}$$

$$L_q = \frac{\partial \Psi_q(i_d, i_q)}{\partial i_q} = \left.\frac{\Delta \Psi_q(i_d, i_q)}{\Delta i_q}\right|_{i_d=\text{constant}} \tag{2}$$

$$L_{dq} = \frac{\partial \Psi_d(i_d, i_q)}{\partial i_q} = \left.\frac{\Delta \Psi_d(i_d, i_q)}{\Delta i_q}\right|_{i_d=\text{constant}} \tag{3}$$

Table 1. Properties of the PMa-SynRM parameters relying on FEM analysis.

Symbol	Quantity	Value
Ψ_m	Permanent magnet flux	0.138 Wb
L_d	Normal d-axis self-inductance	288 mH
L_q	Normal q-axis self-inductance	38 mH
L_{dq}	Mutual inductance	4 mH

3. A Shot Briefly Differential Flatness Control and Control Law

3.1. Differential Flatness Briefly

The differential flatness-based control approach is crucial to control different types of systems [2,13,14]. A summary of differential flatness control theory is provided below. Considering a nonlinear system expressed by the following state-variable:

$$\dot{x} = f(x, u). \tag{4}$$

The system (4) is said to be "differentially flat" if a set of flat output equal to the number of inputs can be found. More precisely, the control output variable must be written as the function of the flat output and their derivatives as follow:

$$x = \phi(y, \dot{y}, \ldots, y^{(\beta)}), \tag{5a}$$

$$u = \psi(y, \dot{y}, \ldots, y^{(\beta+1)}), \tag{5b}$$

where β is the finite number of derivatives.

3.2. Control Law

The control law's block diagram is provided in Figure 3. The trajectory planning, the feedback control (relying on two controller gains K_i and K_p), the controller output λ, and the inverse dynamic equations are detailed below.

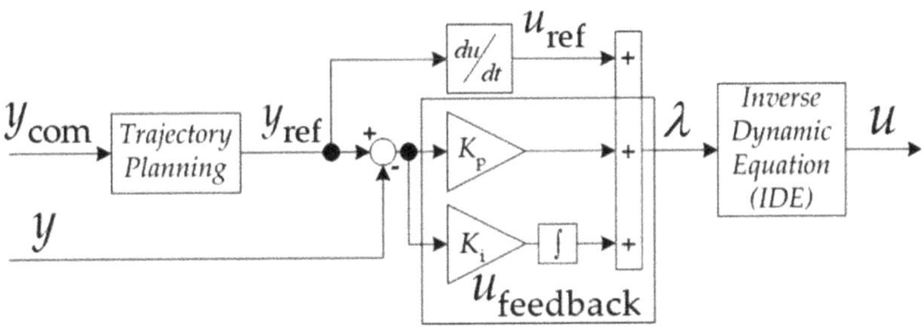

Figure 3. Control law's block diagram.

As shown in Figure 3, a controller output, λ can be defined as follow:

$$\lambda = u_{\text{ref}} + u_{\text{feedback}}(\varepsilon), \quad (6)$$

with

$$u_{ref} = \psi(y_{\text{ref}}, \dot{y}_{\text{ref}}, \ddot{y}_{\text{ref}}, \ldots, y_{\text{ref}}^{(\beta+1)}), \quad (7)$$

$$\varepsilon = y_{ref} - y. \quad (8)$$

As will be seen later, thank to "Inverse Dynamic Equation" (IDE), we will obtain.

$$\dot{y} = \lambda. \quad (9)$$

According to the control law's block diagram, combining (6)–(8) yields.

$$\dot{y} = \dot{y}_{ref} + K_p\varepsilon + K_i \int \varepsilon dt = 0. \quad (10)$$

Taking time derivative (10) obtains.

$$\ddot{y}_{ref} + K_p\dot{\varepsilon} + K_i\varepsilon = 0. \quad (11)$$

By comparing to the standard second-order equation, parameters K_p and K_i can define as follow:

$$\ddot{q}(s) + 2\zeta\omega_n\dot{q}(s) + \omega_n^2 q(s) = 0, \quad (12)$$

Consequently, the controller gains define as follow:

$$\begin{aligned} K_p &= 2\zeta\omega_n \\ K_i &= \omega_n^2 \end{aligned} \quad (13)$$

4. PMa-SynRM Modeling and Development of the Proposed Control Scheme

4.1. Mathematic Model of PMa-SynRM/Inverter

The inverter shown in Figure 4 provides a symmetric sinusoidal three-phase AC voltage source for supplying to PMa-SynRM. In Figure 4, V_{BUS}, i_{BUS}, and i_A, i_C are respectively the input DC grid voltage, the inverter current, and the load motor phase current. According to Figure 2a,b, The electrical modeling equations of PMa-SynRM are discussed by the nonlinear case. In Figure 2, the flux linkage of direct and quadrature axes may be defined according to d- and q-axes current id and iq as in the following equations.

$$\Psi_{\rm d} = \Psi_{\rm d}(i_{\rm d}, i_{\rm q}), \tag{14}$$

$$\Psi_{\rm q} = \Psi_{\rm q}(i_{\rm d}, i_{\rm q}). \tag{15}$$

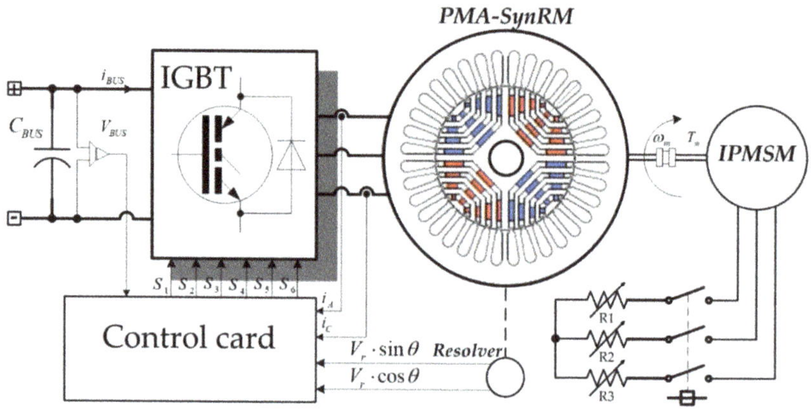

Figure 4. A three-phase inverter to control the PMa-SynRM prototype.

By considering Equations (14) and (15) as well, as mentioned above, the rotating electrical modeling of PMa-SynRM is given by the following equations [10,17]:

$$v_{\rm d} = R_{\rm s} \cdot i_{\rm d} + \frac{d\Psi_{\rm d}(i_{\rm d}, i_{\rm q})}{dt} - \omega_{\rm e} \cdot \Psi_{\rm q}(i_{\rm d}, i_{\rm q}), \tag{16}$$

$$v_{\rm q} = R_{\rm s} \cdot i_{\rm q} + \frac{d\Psi_{\rm q}(i_{\rm d}, i_{\rm q})}{dt} + \omega_{\rm e} \cdot \Psi_{\rm d}(i_{\rm d}, i_{\rm q}), \tag{17}$$

where

$$\omega_{\rm e} = n_p \cdot \omega_m, \tag{18}$$

where v_d is the d-axis voltage, v_q is the q-axis voltage, $\omega_{\rm e}$ is the electrical angular frequency, n_p is the number of pole pairs, and ω_m is the mechanical angular frequency. The important electromagnetic torque of the machine composed of torque produced with the interactivity between the magnet and the reluctance torque is expressed as follows.

$$T_e = n_p(\Psi_d i_q - \Psi_q i_d). \tag{19}$$

The mechanical equation in the rotation moving is:

$$J\frac{d\omega_{\rm m}}{dt} = T_{\rm e} - B_{\rm f}\omega_{\rm m} - T_{\rm L}, \tag{20}$$

where J is the moment of inertia, $B_{\rm f}$ is the viscosity, and $T_{\rm L}$ is the load torque.

4.2. Differential Flatness Control of Current (or Torque) Loop Development

By referring to Equations (5), (16) and (17), the Ψ_d and Ψ_q are determined as the state variables (x). The v_d and v_q are control variables (u). The flat output (y) candidates are the measured parameters, which are i_d and i_q. The systems can be seen as differentially flat if the control output variable must be noted according to the flat output, which are

$$v_d = u_1 = R_s y_1 - \omega_e \Psi_q(y_1, y_2) + \frac{\partial \Psi_d(y_1, y_2)}{\partial y_1} \cdot \dot{y}_1 + \frac{\partial \Psi_d(y_1, y_2)}{\partial y_1} \cdot \dot{y}_2 = \psi_1(y_1, y_2, \dot{y}_1, \dot{y}_2), \tag{21}$$

$$v_q = u_2 = R_s y_2 + \omega_e \Psi_d(y_1, y_2) + \frac{\partial \Psi_q(y_1, y_2)}{\partial y_1} \cdot \dot{y}_1 + \frac{\partial \Psi_q(y_1, y_2)}{\partial y_2} \cdot \dot{y}_2 = \psi_2(y_1, y_2, \dot{y}_1, \dot{y}_2). \tag{22}$$

The control scheme mentioned in Section 3.2 is applied to deal with some inevitable modeling errors and uncertainties. By referring to Equation (6), the control laws of current control can express as follow:

$$\dot{y}_1 = \dot{\Psi}_{dref} + K_p \varepsilon_d + K_i \int \varepsilon_d dt, \tag{23}$$

$$\dot{y}_2 = \dot{\Psi}_{qref} + K_p \varepsilon_q + K_i \int \varepsilon_q dt. \tag{24}$$

Consequently, the output control variables yield as follow:

$$u_1 = v_d = \dot{y}_1 + IDE_d, \tag{25}$$

$$u_2 = v_q = \dot{y}_2 + IDE_q, \tag{26}$$

where the Inverse Dynamic Equations (IDEs) are

$$IDE_d = -R_s i_d + \omega_e \Psi_q, \tag{27}$$

$$IDE_q = -R_s i_q + \omega_e \Psi_d. \tag{28}$$

The controller parameters are

$$K_{pd} = K_{pq} = 2\zeta_1 \omega_{n1}, \tag{29}$$

and

$$K_{id} = K_{iq} = \omega_{n1}^2, \tag{30}$$

where ζ_1 and ω_{n1} are respectively the desirated governing damping ratio and natural frequency.

4.3. Differential Flatness Control of Speed Control Loop Development

Figure 5 shows the proposed control diagram. The outer speed loop enables evaluating the torque reference value of the MPTA, which generates the current command for the inner current loop.

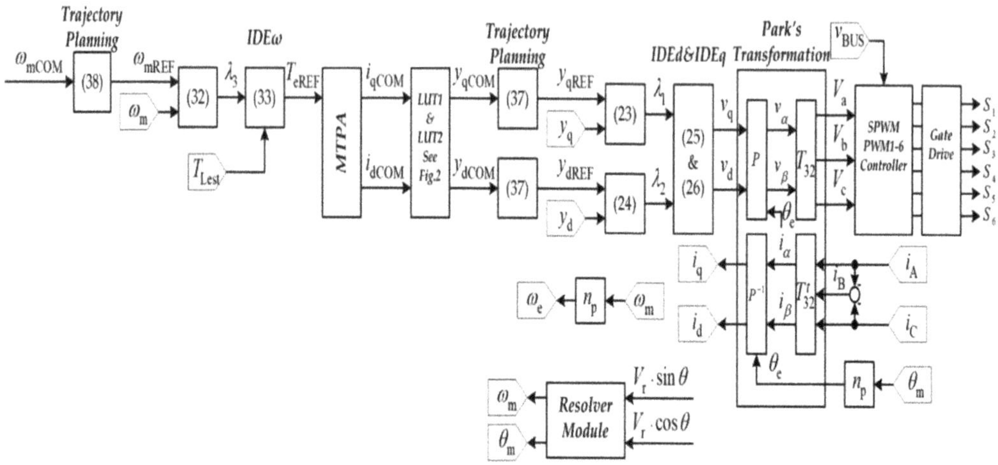

Figure 5. Schematic drawing of the designed control scheme.

For the MTPA algorithm, it has been proposed in [20]. So, the T_e is defined as a command variable u_3, and the flat output y_3 is ω_{m} (or measured angular speed). The system is flat if the control variable is a function of flat output that is

$$u_3 = T_e = J \cdot \dot{y}_3 + B_f y_3 + T_L = \psi_3(y_3, \dot{y}_3). \tag{31}$$

The control strategy of the speed control loop is

$$\dot{y}_3 = \dot{\omega}_{ref} + K_{p\omega}\varepsilon_\omega + K_{i\omega}\int \varepsilon_\omega dt. \tag{32}$$

The control variable of the speed loop can be express as follow:

$$u_3 = J \cdot \dot{y}_3 + J \cdot IDE_\omega, \tag{33}$$

where the inverse dynamic equation of the speed control loop (IDEω) is

$$IDE_\omega = \frac{1}{J}(B_{\mathrm{f}} \cdot y_3 + T_{\mathrm{L}}). \tag{34}$$

The controller parameters are defined as the following equation.

$$K_{p\omega} = 2\zeta_2\omega_{n2}, \tag{35}$$

and

$$K_{i\omega} = \omega_{n2}^2, \tag{36}$$

where ζ_2 and ω_{n2} are, respectively, the desirated governing damping ratio and the natural frequency of the outer speed regulation loop.

Based on the current and speed control law development, the natural frequency setting of the designed controller is depicted in Figure 6. The switching frequency f_s of the inverter shown in Figure 4 is equal to 16 kHz (ω_s = 10^5 rad.s^{-1}) and it is reported in Figure 6. According to Figure 5, the speed control loop must be faster than the current control loop given that the outer speed loop enables assessing the torque reference value, and consequently the current. Considering the Nyquist-Shannon Theorem, the natural frequency ω_{n1} must be chosen lower than a frequency equal to 10^2 rad.s^{-1} (namely two times lower than the switching frequency ω_s). Therefore, the natural frequency ω_{n1} for the current control loop has been tuned at 2000 rad.s^{-1}; while for the speed control loop, a natural frequency ω_{n2} has been set at 20 rad.s^{-1} (100 times lower than ω_{n1}). Both values have been reported in Figure 6, allowing defining the stable region included between these two values; whereas the unstable region is outside the natural frequency ω_{n1}. Regarding the tuning of the damping ratios ζ_1 and ζ_2, to guarantee underdamped transient behaviors with low overshoot and fast response, both parameters have been set at 0.7.

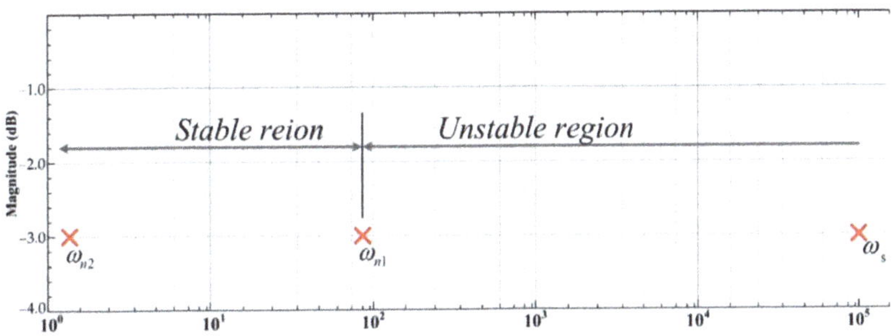

Figure 6. The natural frequency setting of the designed controller.

Note that the stability and response of the differential flatness-based control are easy to set compared to the traditional PI controller. By defining and selecting the governing damping and natural frequency [21,22], as shown in Figure 6, the controller parameters of current and speed loop control may be calculated by Equations (29), (30), (35) and (36).

4.4. Trajectory Planning

The trajectory planning enables restricting the derivative terms. The reference inputs have been defined by trajectory planning utilized by the second-order low-pass filter. The trajectory planning of the current control loops are

$$\frac{y_{1REF}(s)}{y_{1COM}(s)} = \frac{y_{2REF}(s)}{y_{2COM}(s)} = 1 \bigg/ \left(\left(\frac{s}{\omega_{n3}} \right)^2 + \frac{2\zeta_3}{\omega_{n3}} + 1 \right). \tag{37}$$

In the speed control loop, the trajectory planning has been determined by the following equation.

$$\frac{y_{3REF}(s)}{y_{3COM}(s)} = 1 \bigg/ \left(\left(\frac{s}{\omega_{n4}} \right)^2 + \frac{2\zeta_4}{\omega_{n4}} + 1 \right), \tag{38}$$

where ζ_3, ω_3, ζ_4, and ω_4 are the governing damping and natural frequencies of the second-order low-pass filters, respectively.

5. Simulation and Experimental Validation

5.1. Experimental Setup

A small-scale test bench 1-KW relying on the PMa-SynRM has been conceived in the laboratory, as shown in Figure 7. Table 2 sums up the principal parameters of the studied machine. Table 3 outlines the controller parameters. The motor is supplied by a 3-kW 3-phase inverter (DC/AC) operating at a switching frequency of 16 kHz. Besides, the input DC grid voltage of the inverter is fed by a 3-phase variable power supply combined with a 3-phase diode rectifier. The PMa-SynRM is mechanically coupled with an IPMSM (interior permanent magnet synchronous motor) feeding a resistive load (see Figures 4 and 7). Regarding the measurements both for the speed and rotor angle, they have been acquired by a resolver placed on the rotor shaft. The developed control scheme (see Figure 5) relying on the differential flatness controller has been modeled in the Matlab/Simulink software, and then it has been incorporated in the dSPACE 1202 MicroLabBox real-time interface to generate the gate control signals applied to the VSI.

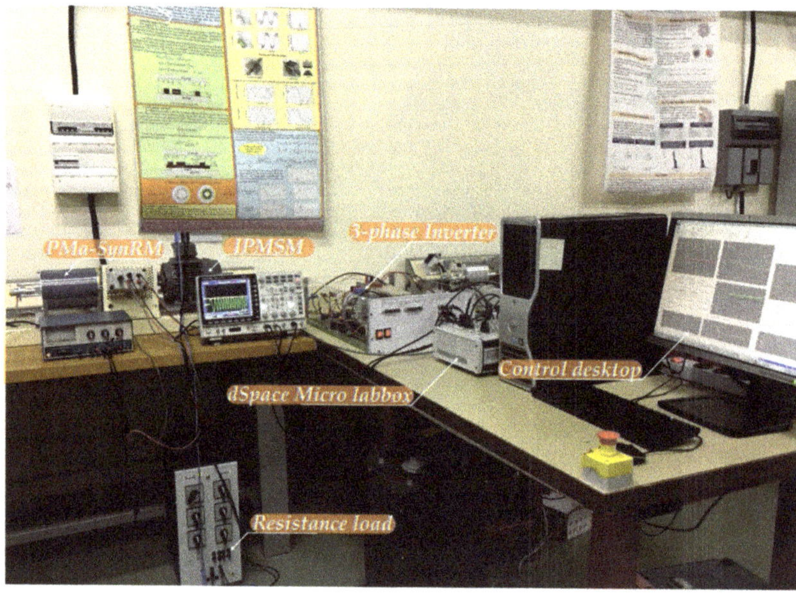

Figure 7. The experimental setup.

Table 2. Specification and parameters of the motor/inverter.

Symbol	Quantity	Value
P_{rated}	Rated power	1 kW
n_{rated}	Rated speed	1350 rpm
T_{rated}	Rated torque	7.07 Nm
n_p	Number of pole pairs	2
PF.	Power factor	0.80
R_s	Resistance (motor + inverter)	3.2 Ω
L_d	Nominal d-axis Inductance	288 mH
L_q	Nominal q-axis Inductance	38 mH
J	Equivalent inertia	0.017 kg m^2
B_f	Viscous friction coefficient	0.008 Nm s/rad
Ψ_m	PMs flux linkage	0.138 Wb
f_s	Switching frequency	16 kHz
V_{dc}	DC bus voltage	400 V

Table 3. Current/torque and speed regulation parameters.

Symbol	Quantity	Value
ζ_1	Governing Damping ratio 1	0.7 pu.
ω_{n1}	Natural frequency 1	2000 Rad.s^{-1}
ζ_2	Governing Damping ratio 2	0.7 pu.
ω_{n2}	Natural frequency 2	20 Rad.s^{-1}
ζ_3	Governing Damping ratio 3	1 pu.
ω_{n3}	Natural frequency 3	200 Rad.s^{-1}
ζ_4	Governing Damping ratio 4	1 pu.
ω_{n4}	Natural frequency 4	20 Rad.s^{-1}
T_{emax}	Maximum Torque	+10 Nm
T_{emin}	Minimum Torque	−10 Nm

5.2. Simulation and Test-Bench Results of the Speed Reversal Employing the Differential Flatness Controller

For the first scenario, Figure 8 reports the obtained simulation results; whereas, Figure 9 exhibits the performed experimental tests to assess the dynamic performance of the designed controller when forcing the motor to reverse direction. In Figure 8, Ch1–10 are the command signal of the speed n_{COM}, the reference signal of the speed n_{REF}, the measured speed n, the command of q-axis current i_{qCOM}, the reference of q-axis current i_{qREF}, the q-axis current i_{q}, the command of d-axis current i_{dCOM}, the reference of d-axis current i_{dREF}, and the d-axis current i_{d}, respectively. In comparison, in Figure 9, Ch1–8 are the speed command n_{COM}, the speed reference n_{REF}, the measured speed n, the current i_{q}, the current reference i_{dREF}, the current i_{d}, and the current reference i_{qREF}, respectively. Firstly, the PMa-SynRM model has been tested by using Matlab/Simulink to support that the elaborated control system is appropriately conceived. Simulations and experimental tests have demonstrated that both simulation and experimental results are similar. Thus, the PMa-SynRM model is fit, and the controller parameters are suitably designed by choosing desired parameters. The experimental results indicate that the PMa-SynRM behaves in a good way when operating under the regenerative mode up to the speed the reference gets positive. Furthermore, it can be emphasized that the measured speed through the resolver enables tracking adequately the speed reference value. Afterward, the operation of the PMa-SynRM is shifted to the motoring mode up to the rotor speed comes to the speed command. The *d*-and *q*-axes currents reveal an appropriate behavior without surpassing the imposed limits. The dominant parameters of the PMa-SynRM enable being ensured, and the elaborated control offers worthwhile dynamic performance.

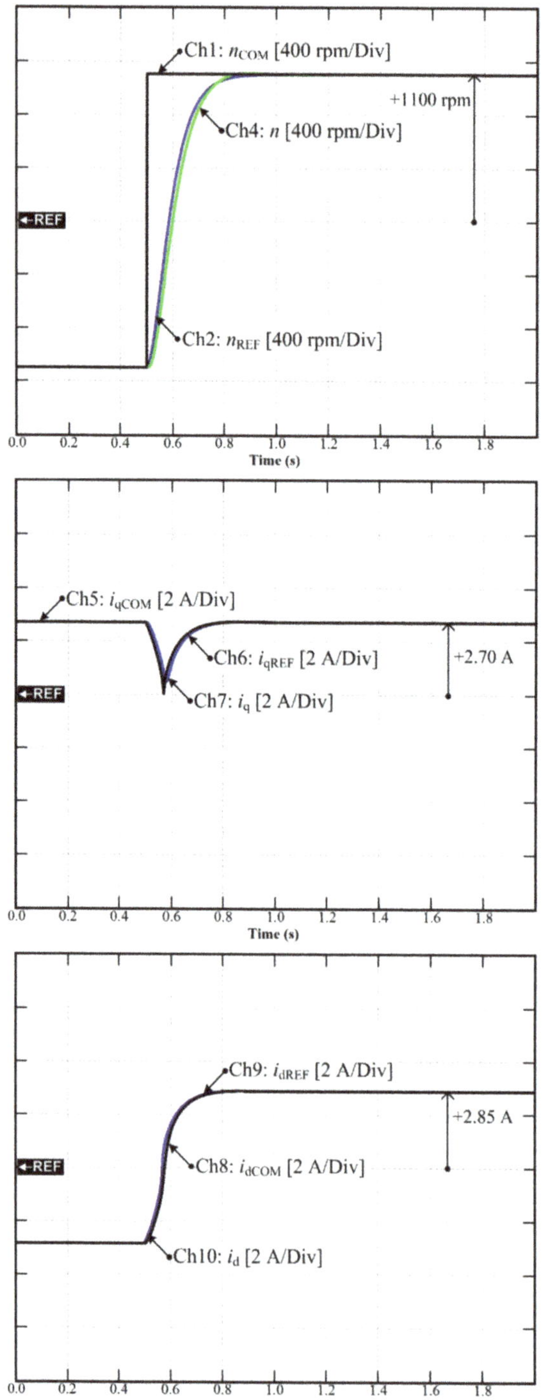

Figure 8. Simulation results: motor speed reversal.

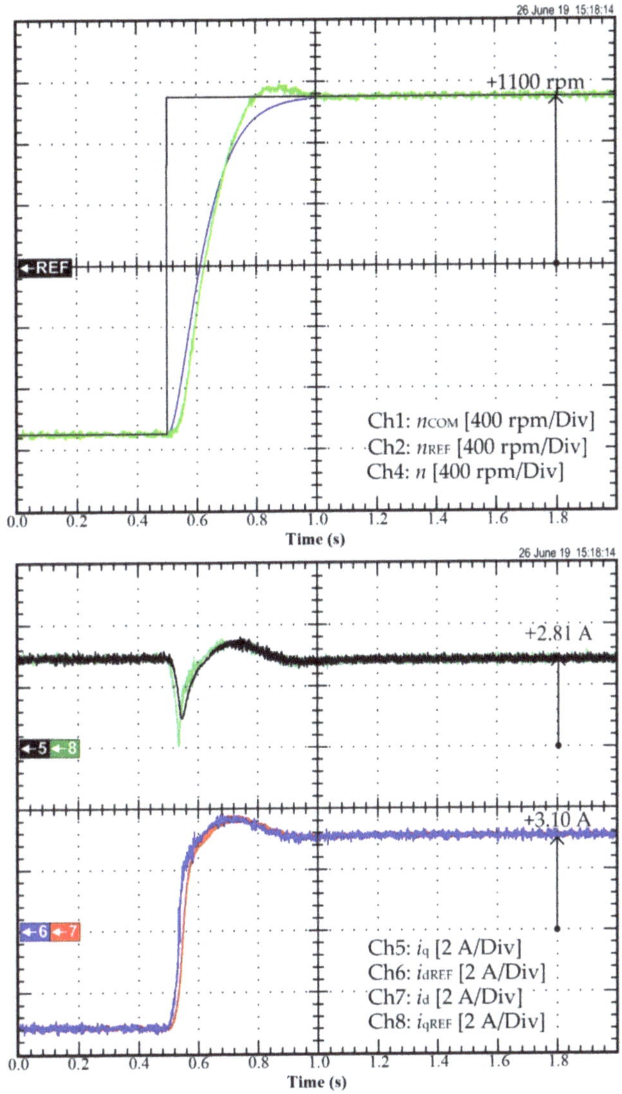

Figure 9. Experimental results: motor speed reversal.

5.3. Experimental Results of the Comparison between the Conventional PI Control and Differential Flatness Control

For the second scenario, the performance of the system when the torque/current loop employs traditional PI control and differential flatness is compared to assess the benefits of the elaborated control scheme. Figure 10a represents the experimental results of the conventional PI control, and Figure 10b illustrates the experimental results of differential flatness control. In Figure 10a, Ch1 is the current i_{dCOM}, Ch3 is the measured current i_d, Ch4 is the measured current i_q, and Ch5 is the measured speed n. In Figure 10b, Ch1 is the current i_{dCOM}, Ch3 is the measured current i_d, Ch4 is the measured current i_q, and Ch5 is the measured speed n. As shown in Figure 10a,b, in a transitory operation, the i_d of PI control exhibits a small overshoot, compared to the differential flatness controller, and the i_q of the PI control shows oscillations.

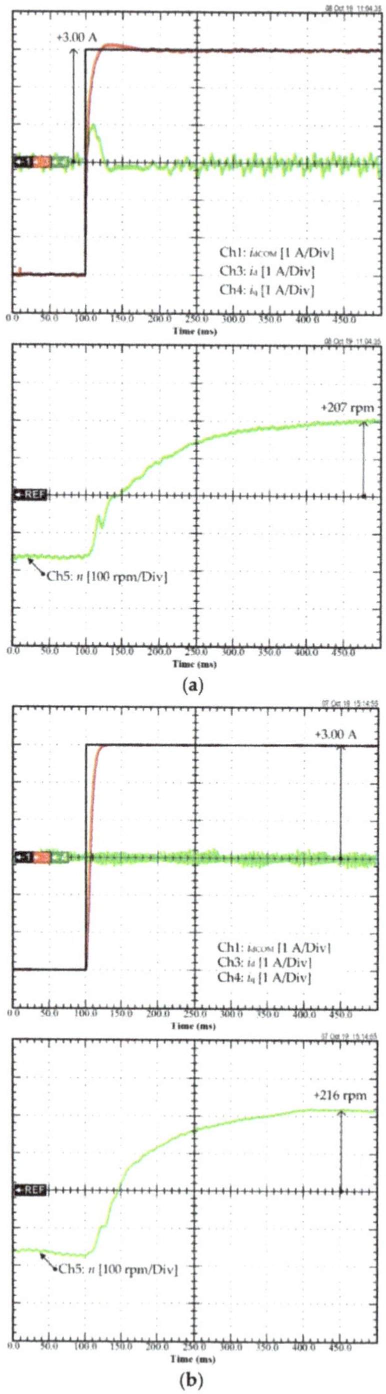

Figure 10. Experimental results: (**a**) the traditional PI control, (**b**) the differential flatness control.

Although the PI controller has used decoupling and back-emf compensation, it demonstrates that the proposed nonlinear controller has a better transient current performance than the traditional linear controller. Furthermore, the speed response (Ch5 of Figure 10a) of the linear cascaded PI controller includes fluctuations, unlike the proposed nonlinear controller (Ch5 of Figure 10b).

For the third scenario, Figure 11a confirms the experimental results of the conventional PI controller, and Figure 11b indicates the experimental validation results of the differential flatness controller when the motor is forced to reverse direction from −1000 rpm to 1000 rpm. In Figure 11a, Ch1 is the speed command n_{COM}, Ch2 is the speed n, Ch3 is the current command i_{qCOM}, Ch4 is the measured current i_q, Ch5 is the current command i_{dCOM}, and Ch6 is the measured current i_d. In Figure 11b, Ch1 is the speed command n_{COM}, and Ch2 is the acquired speed n, Ch3 is the current command i_{qCOM}, Ch4 is the measured current i_q, Ch5 is the current command i_{dCOM}, and Ch6 is the measured current i_d. On one hand, as demonstrated in Figure 11a, during a transient process, the acquired speed n of the PI controller shows an overshoot, and the settling time is approximately 0.45 s. On the other hand, the differential flatness controller (Figure 11b), approximate time is around 0.15 s, as well as the measured i_q of the PI controller, which fluctuates sharply (Ch4 of Figure 11a) in a transient process. The experimental results reflect that differential flatness has a better dynamic speed performance than the traditional PI controller.

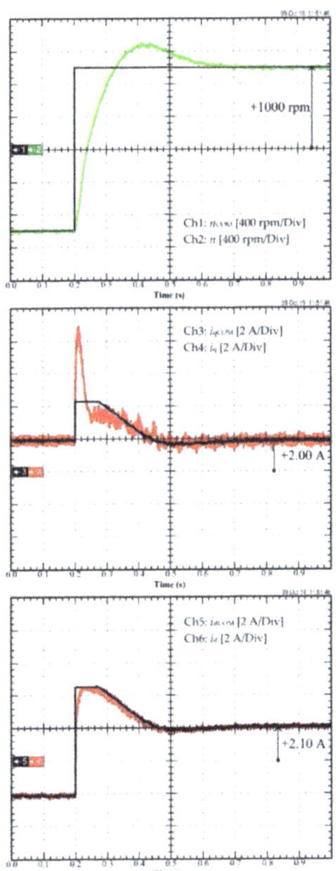

(a)

Figure 11. *Cont.*

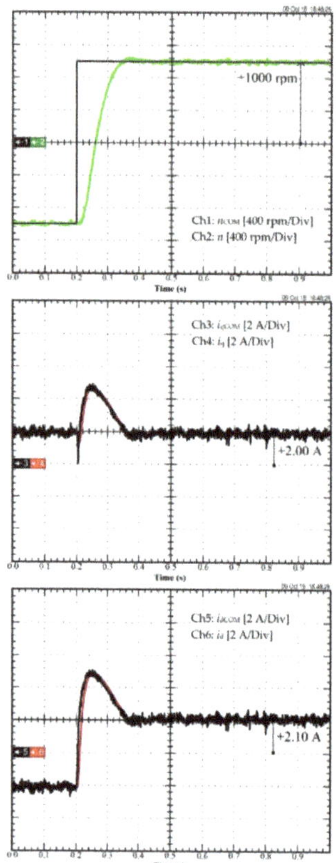

(**b**)

Figure 11. Experimental results: (**a**) the traditional PI control, (**b**) the differential flatness control.

For the fourth scenario, Figure 12a shows the test-bench results of the conventional PI controller; while, Figure 12b depicts the preliminary results of the differential flatness control when suddenly adding an external torque disturbance. In Figure 12a, Ch2 is the measured d-axis current i_d, Ch3 is the measured q-axis current i_q, Ch4 is the measured speed n, Ch5 is the measured phase-A current i_a, and the trajectories of the transient stator current. In Figure 12b, Ch2 is the measured d-axis current i_d, Ch3 is the measured q-axis current i_q, Ch4 is the measured speed n, Ch5 is the phase-A current i_a, and the path of the transient stator current. The experimental results are shown in Figure 12b, validating that differential flatness speed oscillation is roughly 113 rpm; whereas that for the PI controller is 221 rpm. The recuperation time of speed with the elaborated controller is also shorter than that with conventional PI control. These results corroborate that differential flatness control has better dynamic performance both for the torque/current and speed loop control and the external rejection ability.

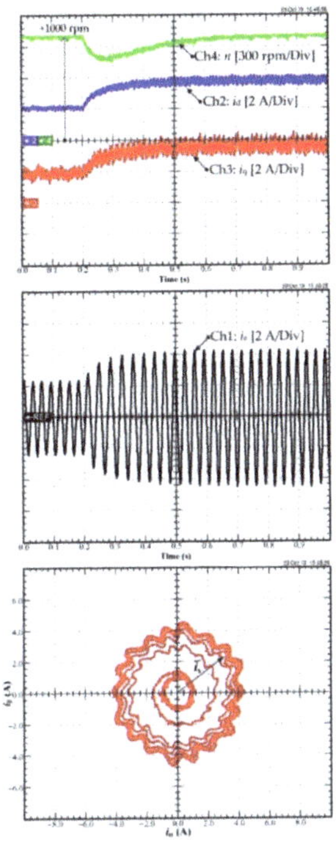

(a)

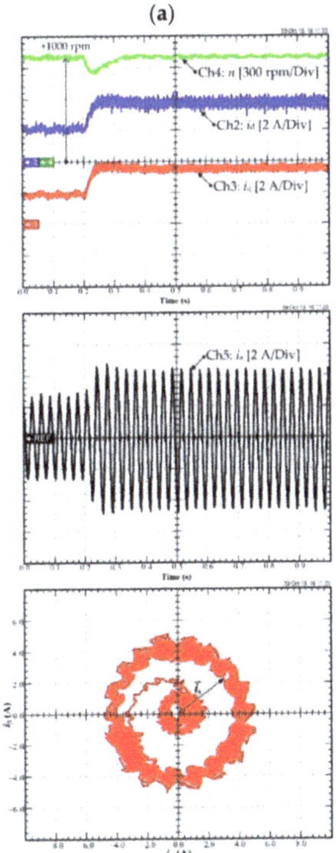

(b)

Figure 12. Experimental results: (**a**) the traditional PI control (**b**) the differential flatness control.

6. Conclusions

To cope with the control issues met in a permanent magnet-assisted synchronous reluctance motor (PMa-SynRM), various control approaches have been previously investigated. Nonetheless, the achievement of high performance is still a challenging barrier due to the nonlinear characteristics and parameter uncertainty conditions of this motor. In this work, a differential flatness control law has been elaborated and designed to control both current/torque and the speed of the PMa-SynRM. Furthermore, an intelligent proportional-integral (*i*PI) has been combined with the nonlinear differential flatness controller to face unavoidable modeling errors and uncertainties for the torque and speed of the motor. This model-based approach requires an accurate model. In case the model is not perfectly known, the estimation of the unknown part is necessary to achieve the expected high performance. Through simulations and experimental tests performed on a small-scale test bench 1-KW including the PMa-SynRM, the dynamic performances of the system have been validated; while demonstrating the performance superiority of the differential flatness controller over the conventional PI controller from the overshoot and oscillation point of view. Furthermore, the results reflect that the dynamic recovery time response is faster using intelligent PI control than the field-oriented control (FOC) based on PI controller with approximately 0.15 s.

In the future work, another control approach will be tentatively applied to the control of PMa-SynRM. This approach, called the model-free control, does not require an accurate model. Indeed, only very limited knowledge of the controlled system is enough to regenerate the control action. Advantages and drawbacks of this controller will be discussed and its performance will be compared to the flatness-based controller in the next work.

Author Contributions: Conceptualization, B.N.-M. and N.T.; methodology, S.S., N.P., D.G. and P.T.; validation, S.S., N.P., D.G. and P.T.; formal analysis, N.B.; writing—original draft preparation, P.T.; writing—review and editing, S.S., D.G. and N.B.; visualization, N.B.; supervision, B.N.-M. and N.T. All authors have read and agreed to the published version of the manuscript.

Funding: This work was partially supported by the International Research Partnerships: Electrical Engineering Thai-French Research Center (EE-TFRC) between Université de Lorraine (UL) and King Mongkut's University of Technology North Bangkok (KMUTNB) and Framework Agreement between the University of Pitesti and King Mongkut's University of Technology North Bangkok through the Research Program Cooperation under Grant KMUTNB–61–GOV–01–67. Besides, this work was supported partly by the French PIA project «Lorraine Université d'Excellence», reference ANR-15-IDEX-04-LUE.

Institutional Review Board Statement: Not applicable.

Informed Consent Statement: Not applicable.

Data Availability Statement: The data presented in this study are available on request from the corresponding authors. The data are not publicly available due to their current utilization for future works involving the authors of this paper.

Acknowledgments: The authors would like to express their gratitude to the GREEN laboratory at the University of Lorraine and King Mongkut's University of Technology North Bangkok (KMUTNB) for their constant support in boosting collaborations between France and Thailand.

Conflicts of Interest: The authors declare no conflict of interest.

References

1. Grasso, E.; Palmieri, M.; Corti, F.; Nienhaus, M.; Cupertino, F.; Grasso, F. Detection of stator turns short-circuit during sensorless operation by means of the Direct Flux Control technique. In Proceedings of the 2020 AEIT International Annual Conference (AEIT), Catania, Italy, 23–25 September 2020.
2. Lim, C.; Rahim, N.; Hew, W.; Levi, E. Model Predictive Control of a Two-Motor Drive with Five-Leg-Inverter Supply. *IEEE Trans. Ind. Electron.* **2013**, *60*, 54–65. [CrossRef]
3. Wang, L.; Jatskevich, J.; Dommel, H. Re-examination of Synchronous Machine Modeling Techniques for Electromagnetic Transient Simulations. *IEEE Trans. Power Syst.* **2007**, *22*, 1221–1230. [CrossRef]

4. Fan, Y.; Zhang, Q.; Wang, W.; Zhou, X. Speed Regulation System of a Flux-Modulated Permanent-Magnet In-Wheel Motor Based on Sliding Mode Control and Adaptive Notch Filter. *IEEE Trans. Energy Convers.* **2018**, *33*, 2183–2190. [CrossRef]
5. Erazo, D.; Wallscheid, O.; Bocker, J. Improved Fusion of Permanent Magnet Temperature Estimation Techniques for Synchronous Motors Using a Kalman Filter. *IEEE Trans. Ind. Electron.* **2020**, *67*, 1708–1717. [CrossRef]
6. Morimoto, S.; Ooi, S.; Inoue, Y.; Sanada, M. Experimental Evaluation of a Rare-Earth-Free PMASynRM With Ferrite Magnets for Automotive Applications. *IEEE Trans. Ind. Electron.* **2014**, *61*, 5749–5756. [CrossRef]
7. Morimoto, S.; Sanada, M.; Takeda, Y. Performance of PM-assisted synchronous reluctance motor for high-efficiency and wide constant-power operation. *IEEE Trans. Ind. Appl.* **2001**, *37*, 1234–1240. [CrossRef]
8. Park, G.; Kim, J.; Son, B.; Jung, S. Optimal Design of PMa-synRM for an Electric Propulsion System Considering Wide Operation Range and Demagnetization. *IEEE Trans. Appl. Supercond.* **2018**, *28*, 1–4. [CrossRef]
9. Boldea, I.; Tutelea, L.; Pitic, C. PM-Assisted Reluctance Synchronous Motor/Generator (PM-RSM) for Mild Hybrid Vehicles: Electromagnetic Design. *IEEE Trans. Ind. Appl.* **2004**, *40*, 492–498. [CrossRef]
10. Niazi, P.; Toliyat, H.; Goodarzi, A. Robust Maximum Torque per Ampere (MTPA) Control of PM-Assisted SynRM for Traction Applications. *IEEE Trans. Veh. Technol.* **2007**, *56*, 1538–1545. [CrossRef]
11. Trancho, E.; Ibarra, E.; Arias, A.; Kortabarria, I.; Jurgens, J.; Marengo, L.; Fricasse, A.; Gragger, J. PM-Assisted Synchronous Reluctance Machine Flux Weakening Control for EV and HEV Applications. *IEEE Trans. Ind. Electron.* **2018**, *65*, 2986–2995. [CrossRef]
12. Thounthong, P.; Sikkabut, S.; Poonnoy, N.; Mungporn, P.; Yodwong, B.; Kumam, P.; Bizon, N.; Nahid-Mobarakeh, B.; Pierfederici, S. Nonlinear Differential Flatness-Based Speed/Torque Control With State-Observers of Permanent Magnet Synchronous Motor Drives. *IEEE Trans. Ind. Appl.* **2018**, *54*, 2874–2884. [CrossRef]
13. Thounthong, P.; Pierfederici, S.; Davat, B. Analysis of Differential Flatness-Based Control for a Fuel Cell Hybrid Power Source. *IEEE Trans. Energy Convers.* **2010**, *25*, 909–920. [CrossRef]
14. Variani, M.; Tomsovic, K. Two-Level Control of Doubly Fed Induction Generator Using Flatness-Based Approach. *IEEE Trans. Power Syst.* **2016**, *31*, 518–525. [CrossRef]
15. Menhour, L.; d'Andrea-Novel, B.; Fliess, M.; Gruyer, D.; Mounier, H. An Efficient Model-Free Setting for Longitudinal and Lateral Vehicle Control: Validation through the Interconnected Pro-SiVIC/RTMaps Prototyping Platform. *IEEE Trans. Intell. Transp. Syst.* **2018**, *19*, 461–475. [CrossRef]
16. Fliess, M.; Join, C. Model-free control. *Int. J. Control* **2013**, *86*, 2228–2252. [CrossRef]
17. Kerdsup, B.; Takorabet, N.; Nahidmobarakeh, B. Design of Permanent Magnet-Assisted Synchronous Reluctance Motors with Maximum Efficiency-Power Factor and Torque per Cost. In Proceedings of the 2018 XIII International Conference on Electrical Machines (ICEM), Alexandroupoli, Greece, 3–6 September 2018.
18. Ding, T. Study and Optimization of Line-Start Permanent Magnet Motors. Ph.D. Dissertation, Université Henri Poincaré, Nancy, France, 2011.
19. Chen, X.; Wang, J.; Sen, B.; Lazari, P.; Sun, T. A High-Fidelity and Computationally Efficient Model for Interior Permanent-Magnet Machines Considering the Magnetic Saturation, Spatial Harmonics, and Iron Loss Effect. *IEEE Trans. Ind. Electron.* **2015**, *62*, 4044–4055. [CrossRef]
20. Sriprang, S.; Nahid-Mobarakeh, B.; Takorabet, N.; Pierfederici, S.; Kumam, P.; Bizon, N.; Taghavi, N.; Vahedi, A.; Mungporn, P.; Thounthong, P. Design and control of permanent magnet assisted synchronous reluctance motor with copper loss minimization using MTPA. *J. Electr. Eng.* **2020**, *71*, 11–19. [CrossRef]
21. Sriprang, S.; Nahid-Mobarakeh, B.; Takorabet, N.; Pierfederici, S.; Bizon, N.; Kuman, P.; Thounthong, P. Permanent Magnet Synchronous Motor Dynamic Modeling with State Observer-based Parameter Estimation for AC Servomotor Drive Application. *Appl. Sci. Eng. Prog.* **2019**, *12*. [CrossRef]
22. Veeser, F.; Braun, T.; Kiltz, L.; Reuter, J. Nonlinear Modelling, Flatness-Based Current Control, and Torque Ripple Compensation for Interior Permanent Magnet Synchronous Machines. *Energies* **2021**, *14*, 1590. [CrossRef]

sustainability

Review

State-of-the-Art Review on IoT Threats and Attacks: Taxonomy, Challenges and Solutions

Ritika Raj Krishna [1], Aanchal Priyadarshini [1], Amitkumar V. Jha [1], Bhargav Appasani [1], Avireni Srinivasulu [2] and Nicu Bizon [3,4,*]

1 School of Electronics Engineering, Kalinga Institute of Industrial Technology, Bhubaneswar 751024, India; ritikaraj11a@gmail.com (R.R.K.); aanchal233100@gmail.com (A.P.); amit.jhafet@kiit.ac.in (A.V.J.); bhargav.appasanifet@kiit.ac.in (B.A.)
2 Department of Electronics and Communication Engineering, K. R. Mangalam University, Gurugram 122103, India; avireni@ieee.org
3 Faculty of Electronics, Communication and Computers, University of Pitesti, 110040 Pitesti, Romania
4 Doctoral School, Polytechnic University of Bucharest, 313 Splaiul Independentei, 060042 Bucharest, Romania
* Correspondence: nicu.bizon@upit.ro

Abstract: The Internet of Things (IoT) plays a vital role in interconnecting physical and virtual objects that are embedded with sensors, software, and other technologies intending to connect and exchange data with devices and systems around the globe over the Internet. With a multitude of features to offer, IoT is a boon to mankind, but just as two sides of a coin, the technology, with its lack of securing information, may result in a big bane. It is estimated that by the year 2030, there will be nearly 25.44 billion IoT devices connected worldwide. Due to the unprecedented growth, IoT is endangered by numerous attacks, impairments, and misuses due to challenges such as resource limitations, heterogeneity, lack of standardization, architecture, etc. It is known that almost 98% of IoT traffic is not encrypted, exposing confidential and personal information on the network. To implement such a technology in the near future, a comprehensive implementation of security, privacy, authentication, and recovery is required. Therefore, in this paper, the comprehensive taxonomy of security and threats within the IoT paradigm is discussed. We also provide insightful findings, presumptions, and outcomes of the challenges to assist IoT developers to address risks and security flaws for better protection. A five-layer and a seven-layer IoT architecture are presented in addition to the existing three-layer architecture. The communication standards and the protocols, along with the threats and attacks corresponding to these three architectures, are discussed. In addition, the impact of different threats and attacks along with their detection, mitigation, and prevention are comprehensively presented. The state-of-the-art solutions to enhance security features in IoT devices are proposed based on Blockchain (BC) technology, Fog Computing (FC), Edge Computing (EC), and Machine Learning (ML), along with some open research problems.

Keywords: Internet of Things; security; threats; privacy; vulnerabilities; Blockchain

Citation: Krishna, R.R.; Priyadarshini, A.; Jha, A.V.; Appasani, B.; Srinivasulu, A.; Bizon, N. State-of-the-Art Review on IoT Threats and Attacks: Taxonomy, Challenges and Solutions. *Sustainability* **2021**, *13*, 9463. https://doi.org/10.3390/su13169463

Academic Editor: Zubair Baig

Received: 30 May 2021
Accepted: 18 August 2021
Published: 23 August 2021

1. Introduction

We live in a time when technology is an essential requirement for all humans, and the evidence is the increased dependence on technology in almost every aspect of our lives. Today's world is evolving with the rapidly growing Internet of Things (IoT)-based application [1]. The rise of the IoT has been a glorious phenomenon in recent years. The physical and virtual objects implanted with sensors, software, and other technologies are interlinked together in IoT [2]. It envisages communicating and sharing data with other devices and systems worldwide over the Internet. Further, IoT is like an array of network-enabled devices that exclude traditional computers such as laptops and servers.

IoT has sprawled everywhere, starting from the healthcare sector to the big industries. It is now implantable, wearable, and portable, resulting in a pervasive and interactive world [3]. It modifies the physical objects around us into smart objects, creating an information environment that increasingly changes human living standards. For instance, IoT devices track and collect essential measurements (such as blood pressure, blood sugar level, pulse rate, etc.) in real time, allowing emergency alerts to improve the odds of a patient's survival [4]. Moreover, autonomous and self-driving vehicles prevent drivers from deviating from paths or accidents while providing them assistance to reach their destinations. In addition, those definitions are expanded to provide automatic emergency alerts of the closest road and medical assistance in the event of an accident. IoT also covers many aspects of modern industries, including manufacturing, assembly, packing, logistics, smart cities, and aviation industries [5]. Some of the essential IoT-based application domains in health, commerce, communication, and entertainment are shown in Figure 1.

Figure 1. Important IoT application domains.

To implement IoT, the traditional technology had to undergo some major modifications. For example, to convert an isolated device into a transmitting device, there is a need to increase small computing devices' memory and processing capacity while dramatically reducing their size [6]. Further, the creation of various lightweight, secure protocols for communication between different IoT devices is equally important. The improvements to the conventional networks to help the operation of the IoT ecosystem have their own set of consequences. However, the unprecedented growth of interconnected devices has crippled the IoT ecosystem. Consequently, there exists enough scope for threats and attacks in IoT-based applications.

The Global Vice President at New Net Technologies (NNT), Dirk Schrader, stated that IoT-based computers have become the crown jewels of cybercriminals. He also said that less than 42% of businesses can detect insecure IoT devices. Hence, for researchers to develop well-grounded solutions to trace and avert these threats, they must first understand the threats and attacks to make the IoT environment safe, secure, and reliable. There are three significant aspects to consider while examining the IoT from a security perspective. To begin with, there are a massive number of smart devices, possibly billions. This suggests that the IoT would be the most complex man-made system ever in terms of the number of entities involved [7]. Second, they are essentially heterogeneous, with respect to the functionality, protocol stacks, radios, operating systems (some objects do not even have one), energy sources, identities, and so on [8]. Third, each smart object is owned by a company or a person, and it is managed by the same or a different company or individual. Millions of businesses and individuals are in control of a subset of the smart objects in their management domains. From the standpoint of protection, privacy, and trust, how this control is technically upheld is a critical issue.

The attack surface in the IoT domain has increased significantly, as have the possible threats to the protection of these entities in the domain [9]. For example, the security threats to the autonomous and self-driving industry may lead to disastrous consequences. Autonomous vehicles are vulnerable to sensor-based attacks. By manipulating the sensors (e.g., linear acceleration sensor, magnetic sensor, etc.), attackers may collect data, transfer malware to it, or trigger a malicious activity [10]. Furthermore, smartphones and embedded systems contribute to a digital ecosystem for global communication that simplifies lives by being sensitive, flexible, and responsive to human needs. However, on the other hand, security cannot be assured due to vulnerabilities in IoT. When a user's signal is disrupted or intercepted, their privacy may be jeopardized, and their information may be leaked.

The state-of-the-art survey on various aspects of IoT, including security, privacy, and robustness, has been presented in [11] by Chen et al. The authors focused on specific issues of IoT interface positioning and localization. The development of lightweight block cipher algorithms has been proposed to be used in devices for data encryption and decryption [12]. A desktop review and qualitative analysis have been performed by Gamudani et al. in [13] to compute performance analysis of attacks. Cryptographic approaches have been discussed in [14] as a method of ensuring long-term security approaches. Different layer architectures of IoT and security issues associated with them have been discussed with possible countermeasures using Blockchain (BC) in [15]. The survey on security aspects of IoT has been presented by Alaba et al. in [16], covering the scope of security countermeasures in some other allied paradigms, including Machine-to-Machine (M2M), Cyber-Physical System (CPS), and Wireless Sensor Networks (WSNs). In [17], Abomhara et al. discussed various applications of IoT and the security threats related to them, including vulnerabilities, intruders, and some other attacks. The threats concerning security and privacy in IoT architecture have been presented without counter measuring techniques by Kozlov et al. in [18].

The organization of the paper is as follows. The state-of-the-art motivation and contributions of this research are presented in Section 2. The background of the IoT as the foundation to the security threats and attacks is presented in Section 3. Section 4 deals with the IoT reference model and the protocol stack. The state-of-the-art review on the vulnerabilities with threats and attacks taxonomy in the IoT paradigm is presented in Section 5. Security goals and a roadmap in IoT are presented in Section 6. Section 7 deals with the state-of-the-art security solution for IoT framework using ubiquitous technologies, such as BC, FC, EC, and ML. Some of the open research problems are discussed in Section 8. The last section deals with the conclusion of the article with future scope for research.

2. State-of-the-Art, Motivation, and Contributions of This Research

2.1. Trends in Literature and Motivation

There is an ample amount of work in literature focusing on the IoT from various perspectives. Particularly, aspects such as applications, architecture, protocols, and standards are extensively covered in the literature. However, threats and attacks in IoT are comparatively less explored. The analysis from one of the world's largest databases, i.e., SCOPUS, can be used to understand the relevance of the particular aspects of IoT. If we search the number of articles in the SCOPUS database that focus on IoT architecture, IoT architecture and threats, and IoT architecture and attacks, then it can be corroborated from the search results that the threats and attacks analysis of IoT architecture is sparsely explored in the literature, which can be validated from the SCOPUS statistics seen in Figure 2. Further, there is rapid growth in the interest of the researchers towards threats and attacks analysis in the IoT architecture. This can be corroborated from the number of articles pertaining to the threats and attacks analysis in IoT architecture, which was four and zero, respectively, in 2010, and rapidly increased to 73 and 157, respectively, up to the third quarter of the year 2021 (approximately).

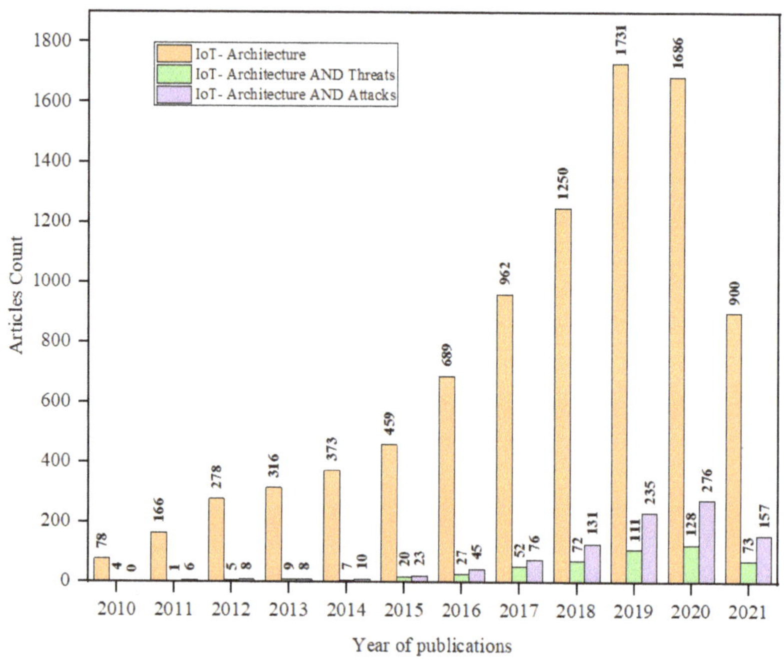

Figure 2. Literature statistics on IoT architecture, IoT architecture and threats, and IoT architecture and attacks.

A similar trend can be seen with respect to the protocols and standards in the IoT paradigm. The publication statistics obtained from the SCOPUS database for the articles on IoT protocols, IoT protocols and threats, and IoT protocols and attacks are shown in Figure 3. The plotted statistics reveal that the threats and attacks analysis in IoT protocols were sparsely explored in the past 10 years. Nevertheless, these aspects are gaining rapid momentum, which can be corroborated from the published articles on threats and attacks analysis in IoT protocols, which were six and five, respectively, in 2010, and have increased to 160 and 254, respectively, by the third quarter of 2021 (approximately). In a nutshell, the increasing interests of the researchers in the paradigm of IoT architecture and protocols, which are sparsely explored from threats and attacks point-of-view, is the motivating factor for the present work.

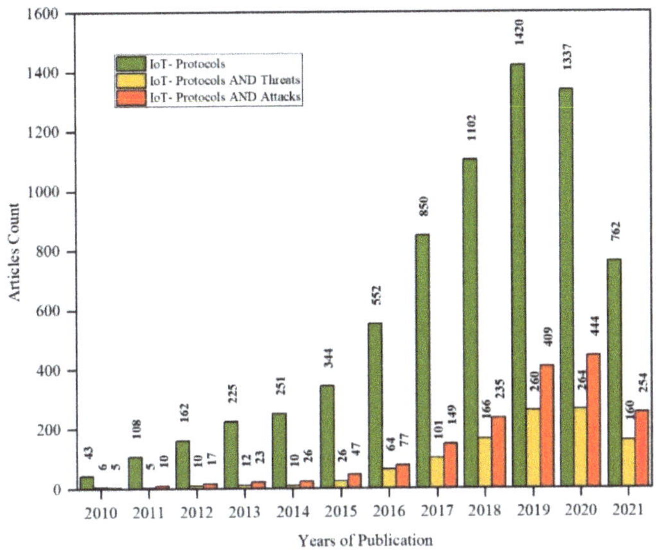

Figure 3. Literature statistics on IoT protocols, IoT protocols and threats, and IoT protocols and attacks.

The other motivating factor for the present work is the threats and attacks analysis and possible solutions using ubiquitous technologies, such as BC, Fog Computing (FC), Edge Computing (EC), and Machine Learning (ML). The threats and attacks analysis and possible solutions in architecture, protocols, and standards have gained significant momentum in the past few years, corroborating the upwards trend in the published articles as per the SCOPUS database statistics shown in Figure 4.

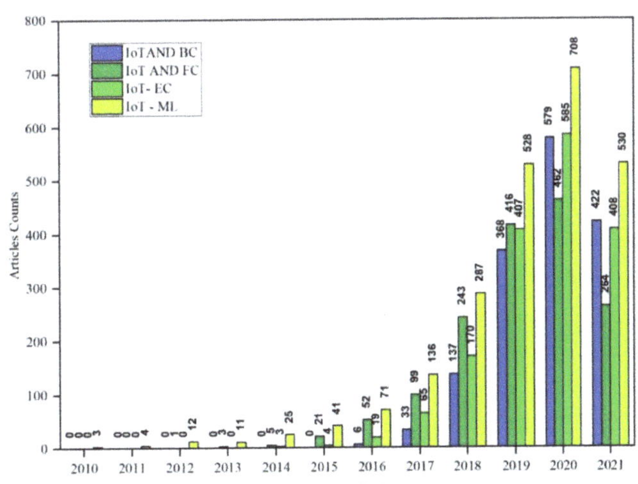

Figure 4. Literature statistics on IoT and BC, IoT and FC, IoT and EC, and IoT and ML from SCOPUS database.

IoT is far behind in realizing its true potential due to the lack of interoperability. The most comprehensive review on security standards and interoperability goals is presented by Lee et al. [19]. To comprehensively review the existing architectures, protocols, and standards is one of the promising means to address the interoperability issues in IoT and other challenges. If we look at the trend of the type of documents in the SCOPUS database, it can be seen that researchers are significantly contributing with Review articles (12.2%) being the third most in number behind Articles (50.8%) and Conference papers (29.1%). These statistics obtained from the SCOPUS database are shown in Figure 5. Conclusively, the present work is a review that comprehensively surveys the existing work and presents the possible solution in the context of threats and attacks pertaining to the architecture, protocols, and standards in the IoT paradigm.

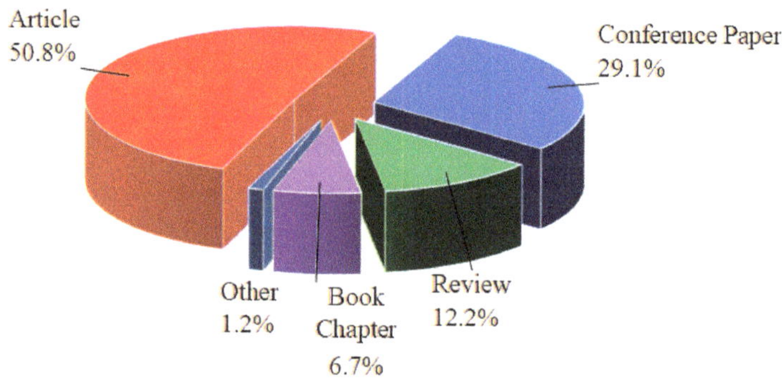

Figure 5. Literature trends from SCOPUS database.

2.2. Comparison with Existing Surveys

Several works have surveyed IoT, its architecture, its reference model, communication protocols, etc., from the perspective of security, threats, and vulnerabilities with possible countermeasure methodologies. In this section, some of the existing surveys are discussed and compared with the present work.

Many works in the literature cover the various aspects of the threats and attacks in IoT. Authors in [20–26] cover the taxonomy of the threats and attacks pertaining to the IoT. These works mainly focus on two broad categories: the architecture of IoT and protocols/standards in IoT. Despite covering the threats and attacks taxonomy, only a few of these works present the possible countermeasures. However, none of these works present countermeasures of threats and attacks based on ubiquitous technology such as BC, FC, EC, and ML. Such ubiquitous technologies in analyzing the threats and attacks have been surveyed in scattered ways by the authors in [27–37]. A comprehensive review of all these technologies to combat IoT threats and attacks is not available.

The comprehensive survey on security and attacks with possible countermeasures solutions has been presented by Abosata et al. in [20], where authors consider the application-specific IoT architecture belonging to industrial IoT. Mann et al. in [21] presented the classification of attacks pertaining to the IoT environment. For attack classification, authors have considered a three-layer architecture comprising devices, gateway, and cloud with respect to the possible attack type. The countermeasures have also been discussed. Ogonji has comprehensively presented a threat taxonomy for the IoT environment in [22], including two broad categories: security threat and privacy threat. The authors of this survey presented taxonomy and countermeasures for the three-layer domain-specific IoT architecture. The state-of-the-art survey on intrusion detection for mitigating the impact of threats and attacks on IoT systems has been presented by Zarpelão et al. [23]. The authors proposed four attributes in the survey: intrusion detection placement strategy, detection method, security threats, and validation. A similar extensive survey by Hajiheidari et al. discusses the state-of-the-art intrusion detection system for IoT environment with a detailed taxonomy of the attacks responsible for intrusion in IoT at various layers [24]. The seminal survey using the top-down approach on various aspects of security in the IoT environment from application-specific IoT architecture has been presented by Kouicem et al. [25]. The authors have also presented the detailed taxonomy of security solutions covering several application-specific IoT architectures, and they also proposed software-defined networking-based solutions to the security in IoT. Sun et al., in [26], focused on the physical layer of the IoT and presented a rich survey covering various security aspects of the protocols and standards, including the countermeasure methodologies. All these surveys are rich in content covering various aspects of threats and attacks in the context of IoT architectures or protocols. Despite presenting the possible countermeasures of threats and attacks, the countermeasures based on rapidly evolving technologies such as BC, FC, EC, ML, etc., have not been discussed to the authors' best knowledge. However, the concluding remarks of all these surveys identified some of the research gaps and provided a hint towards utilizing these technologies.

Elazhary [27] has presented an extensive survey on such computing technologies in the paradigm of IoT. Despite handling various aspects of IoT, particularly computation, processing, and analysis of voluminous IoT data, the security aspects of these data from the architecture point of view are not extensively covered in this survey. Taylor, P.J. et al., in [28], presents a seminal survey of using BC technology for providing the security countermeasures in IoT environment with some open challenges to incorporate other such technologies in the IoT for improving cybersecurity. BC as an infrastructure for IoT architecture with enhanced performance and security has been proposed by Memon et al. in [29]. In this survey, the authors have presented a comparative survey on cloud-based vs BC-based IoT architecture and identified some research gaps with some other similar technologies such as EC and FC. From the point of design objectives, a systematic survey on BC envisioning secure IoT infrastructure has been presented by Tran et al. in [30].

Fersi et al. developed a comprehensive survey in [31] about the scope of FC from the various aspects of the IoT, including enhanced data computing, network management, interoperability issues, security, etc. A similar review on FC from several perspectives, including threats and attacks countermeasures, has been presented by Atlam et al. [32]. Hamdan et al., in [33], comprehensively review the architecture of IoT based on EC. The survey is very rich from the architectural point of view; however, the threats and attacks analysis of such EC-based architecture is narrowly covered in this survey. Another pragmatic survey with EC-based architecture in IoT covering physical layer aspects is presented by Capra et al. in [34]. In this survey, the authors also cover the security aspects of hardware-based IoT architecture. Knowing the extraordinary effectiveness of the EC in IoT, the most seminal survey on the various simulator that can be used to validate the IoT model has been presented by Ashouri et al. in [35]. This survey is one of the best in its field, covering the EC-based simulation tools in IoT, which can even be exploited for modeling and analysis of threats and attacks in the IoT environment.

One of the most comprehensive surveys in the paradigm of ML to enable security and privacy in the IoT data ecosystem has been presented by Amiri et al. in [36]. This survey considers an ML-based approach for enhancing privacy in the IoT data ecosystem where a three-layer architecture comprising perception, network, and application layers of IoT has been considered. The authors also propose a similar approach of using BC with ML to enhance security on the IoT data ecosystem. The state-of-the-art review on the application of ML for intrusion detection in IoT environment has been presented by Adnan in [37]. The authors consider the three dominant attributes, namely, computational complexity, concept drift, and dimensionality, which are mitigated by integrating ML in IoT, envisioning the security of the IoT-based applications.

Some of the other seminal surveys in this context are summarized in Table 1.

Table 1. Some of the key literature surveys and research papers, and their scope.

Reference	Scope of Threats and Attacks Analysis		Technology Adopted for the Solution to Threats and Attacks				Research Gap and Open Challenges	Year
	Architecture	Protocols and Standards	BC	FC	EC	ML		
[38]	×	×	×	×	✓	×	✓	2021
[39]	×	×	×	×	×	×	×	2020
[40]	×	×	✓	×	×	×	✓	2019
[41]	×	✓	×	×	×	×	✓	2015
[42]	×	×	✓	×	×	×	✓	2019
[43]	✓	×	×	×	×	×	×	2017
[44]	✓	×	×	✓	✓	×	×	2017
[45]	×	✓	×	×	×	×	×	2019
[46]	×	×	×	×	✓	×	✓	2021
[47]	×	×	×	×	×	×	✓	2020
[48]	✓	×	✓	×	×	×	×	2019
[49]	✓	×	✓	×	×	×	✓	2020
[50]	×	×	×	×	×	×	✓	2019
[51]	✓	×	×	×	×	✓	✓	2020
[52]	✓	×	×	×	✓	×	✓	2017
[53]	✓	✓	×	×	×	×	✓	2019
[54]	×	✓	✓	×	×	×	✓	2020
[55]	✓	×	×	×	×	×	×	2021
[56]	×	✓	×	×	×	×	✓	2017
[57]	×	×	×	×	×	×	✓	2019
[58]	×	×	×	✓	✓	×	✓	2020
[59]	✓	×	×	✓	×	×	✓	2020
[60]	×	×	✓	×	×	×	✓	2019
[61]	×	✓	×	×	×	×	✓	2019
[62]	✓	×	×	×	×	✓	✓	2019
[63]	×	×	×	×	✓	×	✓	2018
[64]	×	×	×	✓	×	×	✓	2018
[65]	×	×	✓	×	×	✓	✓	2018
[66]	✓	✓	×	✓	✓	×	×	2019
[67]	✓	×	×	×	×	✓	×	2020
[68]	×	✓	✓	×	×	×	✓	2019
[69]	✓	×	✓	✓	✓	✓	✓	2019
[70]	✓	×	×	×	×	✓	✓	2020
[71]	✓	✓	×	×	×	×	×	2018
[72]	×	×	✓	×	×	×	✓	2020
[73]	✓	✓	×	×	×	×	✓	2020
[16]	✓	✓	×	×	×	×	✓	2017
[74]	×	×	×	×	×	×	✓	2019
[75]	×	✓	×	✓	×	×	✓	2019
This survey	✓	✓	✓	✓	✓	✓	✓	NA

These surveys are classified based on: (1) scope of threats and attacks analysis in IoT—architecture, protocols/standards, and general; and (2) possible technology adopted as a solution to the threats and attacks in IoT. The last entry of this table presents the scope of the present survey to highlight a clear comparative picture of the contributions of this survey.

2.3. Scope of the Present Survey and Contributions

As discussed in the previous sections, the threats and attacks analysis in IoT is scattered and none of the surveys so far, to the best knowledge of the authors, covers the threats and attacks taxonomy covering architecture, protocols, and standards of IoT with possible countermeasures using rapidly evolving ubiquitous technologies, such as BC, FC, EC, and

ML, simultaneously. The threats and attacks were discussed in general without focusing on architecture and protocols [38–40]. Security concerns were addressed in [38] using EC and in [40] using BC. The research gaps were identified, and some open research problems were proposed in [38,40]. The analysis on threats and attacks based on protocols and standards without addressing security solutions was shown in [41]. On the other hand, [42] neither covers the architecture nor protocols for analyzing the threats and attacks. However, security countermeasures were discussed using BC with open research problems in [42]. In [43], threats and attacks were analyzed based on architecture without any security countermeasures. In [44], threats and attacks were discussed based on architectures with possible security countermeasures using FC and EC, but it does not identify the research gaps. Similar observations can be made throughout the seminal existing surveys discussed in Table 1. A comparative analysis reveals that none of these surveys analyze the threats and attacks covering all aspects, i.e., architecture as well as protocols and standards. In addition, the security countermeasures have not been discussed in any one of the existing surveys using all four ubiquitous technologies, i.e., BC, FC, EC, and ML, simultaneously. An extensive survey on threats and attacks analysis in the context of IoT, its challenges, taxonomy, and possible technological solutions covering the most important aspects of the IoT, such as architecture, protocols, and standards, is presented in this work. The vital contributions of the paper are highlighted below:

- This survey envisages providing a deeper insight into the IoT from the perspective of threats and attacks.
- An information-rich survey on various aspects of IoT, including threats and attacks from the literature, is presented.
- This survey presents a five-layer IoT architecture and seven-layer IoT architecture, along with the existing three-layer architecture.
- A comprehensive survey on the communication standards and protocols corresponding to three-layer, five-layer, and seven-layer IoT architectures is presented.
- The multidimensional taxonomy of threats and attacks in IoT is proposed with impact assessment on its architecture.
- With respect to communication standards and protocols, the threats and attacks corresponding to each of the proposed architectures, i.e., three-layer, five-layer, and seven-layer IoT architectures, are comprehensively reviewed.
- The potential use of ubiquitous technologies, such as BC, FC, EC, and ML, are presented in the context of security enhancement in IoT.
- The research gap, challenges, and some open problems are presented which can be further explored in the IoT paradigm.

3. Elementary Overview of an IoT System

The IoT is an evolving notion as a vast network of interconnected devices and services that store, share, and process data to dynamically adapt to the environment. IoT offers an ocean of opportunities, and so, many organizations aim to have IoT services integrated into their business processes. Before discussing the security threats, vulnerabilities, attacks, etc., it is pertinent to have a keen understanding of the layout of IoT. The emerging IoT technology typically consists of three levels of hardware which are integrated using software [76]. IoT devices, controllers, and peripherals constitute the first level of IoT, gateways and networks are associated with the second level, whereas cloud servers and control devices are part of the third level of IoT. Such a typical IoT system is depicted in Figure 6, followed by a brief discussion of each level.

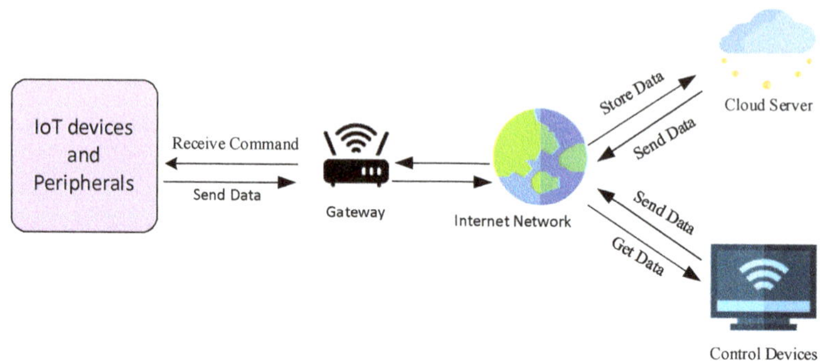

Figure 6. Elementary overview of an IoT system.

3.1. *IoT Devices, Controllers, and Peripherals*

The first level consists of IoT devices, controllers, and peripherals consisting of sensors, actuators, transducers, etc. Their basic function is to capture real-time data of the outer world and convert them into information for further analysis. These devices can be connected to or implanted in any device that needs to be tracked or mounted in the environment to control the device indirectly.

The IoT devices are embedded devices capable of transmitting information across a network to improve interactions with people and with other smart objects. These smart devices make up the bottom layer of the basic IoT architecture. One of the most important features of IoT devices is their ability to use multiple sensors for different applications. Sensors in IoT gadgets are generally coordinated through the sensor hubs. A sensor hub is a single point of connection that gathers and sends data from multiple sensors to the system processing unit. Gathering data is the foremost step [77]. A sensor hub uses various transport mechanisms such as Inter-Integrated Circuit (I2C) or Serial Peripheral Interface (SPI) to transfer data between the sensors and the applications. A communication channel between sensors and applications is established by these transmitting mechanisms that accumulate sensor data through IoT devices [78].

The vulnerabilities associated with some of the sensors in the IoT paradigm are described in Table 2.

Table 2. A few sensor types and their vulnerabilities.

Sensor Type	Sensor	Vulnerabilities
Motion Sensors	Accelerometer Gyroscope Linear Acceleration Sensor	Task Inference False Data Injection Malware Transmission
Environmental Sensors	Light Sensor Proximity Sensor Air Pressure Sensor Audio Sensor Temperature Sensor Soil Moisture sensor Noise Sensor	Eavesdropping Task Inference Smudge Attack False Data Injection Transferring Malware DoS Information Leakage
Position Sensors	GPS Magnetic Sensor	Location Inference Eavesdropping False Data Injection

Sensors are vulnerable to numerous security attacks and threats which might be internal or external depending upon their features [79]. To name a few, information tampering, Man-In-The-Middle Attack (MITM), Distributed Denial of Service (DDoS), jamming, etc., are some of the notable threats to the IoT sensors.

3.2. *Gateways and Networks*

A gateway for IoT is a system or software program that connects the cloud to controller development boards, actuators, and smart devices [80]. It builds a bridge between the cloud and IoT devices. It systematically connects the field to the cloud. An IoT gateway, either a software application or a hardware appliance, is responsible for transmitting data between the cloud and IoT devices. It serves as a network router, connecting IoT devices to the cloud. It is capable of handling both inbound and outbound traffic. Inbound traffic is used for system management tasks, including upgrading device firmware, while outbound traffic is used to transfer IoT data to the cloud. The IoT gateway provides services to safely accumulate, operate, and filter data for analysis. It aids in the secure and safe transport of confederated data produced by the systems and the devices from the edge to the cloud. Ethernet, Wi-Fi, or a 4G/3G modem are used to link the IoT gateway to the cloud [81]. For data exchange and command transfer, a two-way communication channel is developed with the cloud. In an IoT environment, sensors and devices must logically communicate with other devices through the gateway or redirect the necessary data to the cloud. Some of the key functionalities of the IoT gateway are enumerated below.

- Facilitating contact with non-Internet linked or legacy devices.
- Data pre-processing, cleansing, filtering, and optimization.
- Data caching, buffering, and streaming.
- Aggregation of data.
- M2M (Machine-to-Machine) communications.
- Networking features and live data hosting.
- Data visualization and analytics.
- Security feature in data exchange.

Glancing over the number of functions and responsibilities of the IoT gateway, one can easily quote that it is essential to have a secure gateway network to carry out all the enlisted functions safely and efficiently. The gateway is prone to several different kinds of attacks which can be classified into five categories [82]:

- Physical Attack: Unauthorized access to gateway hardware or any unaccredited geographical movement.
- Software Attack: Trojan, Worms, virus, jamming, denial of services.
- Network Attack: Node capture, node subversion, node malfunctioning, message corruption, routing attacks, false node.
- Cryptanalysis Attack: Known-plaintext, Man-In-The-Middle-Attack (MITM), cipher-text only, chosen plaintext.
- Side Channel Attack: Micro probing, reverse engineering.

The state-of-the-art discussion on these attacks is comprehensively discussed later in this article.

3.3. *Cloud Servers and Control Device*

Smart devices of the IoT are being deployed at a rapid rate. However, the amount of data they produce makes it difficult to store and process in the local platforms. The unstructured IoT data can be easily stored in a public cloud infrastructure [83]. The scalability provided by cloud computing offers a solution to this problem. Cloud computing provides flexible computing and storage tools that can be used to assist in data management. As a result, this technology can be used to analyze data generated by sensors and IoT devices. Many of the major cloud providers use object storage technology to offer low-cost, scalable storage systems. Cloud computing allows businesses to store and analyze data easily and in real time, enabling them to get the most out of their data. According to a survey conducted by Information Week [84], 65% of respondents said that "the opportunity to satisfy business demands easily" was one of the most significant factors for a company to migrate to the cloud. Since they have high-speed networks with no data ingress fees, the public cloud is an excellent place to store the vast quantities of IoT data generated by businesses. However, the public cloud has plenty to do. Big data analysis applications that consume and process vast amounts of unstructured content have been added to the product offerings of cloud service providers. This enables companies that can potentially process data more efficiently than a private data center to build highly scalable IoT applications. Depending on the device's networking features, devices can connect to the cloud in a variety of ways. Some of these are cellular, satellite, Wi-Fi, Low Power Wide Area Networks (LPWAN) such as NB-IoT, and direct access to the Internet through Ethernet.

While the cloud has acquired universal popularity, and most IoT applications use cloud services for data storage and retrieval. However, questions about whether cloud technologies are genuinely safe and reliable are continuing to be debated. Nevertheless, cloud risks should also be addressed. The cloud is a public platform used by many people, and there could be malicious users on the cloud who pose a risk to IoT data. The cloud is vulnerable to several attacks such as SQL injection, DDoS, weak authentication, malicious applications, back doors, exploits, etc. [85]. An extensive survey on these aspects is discussed later in this survey.

4. IoT Reference Model and Protocol Stack

4.1. Three-Layer Reference Model

The mitigation of security threats and attacks in IoT can be achieved by understanding the IoT reference model and protocol stack in-depth. There is no widely agreed-upon framework for the IoT [86]. However, different architectures have been suggested by different researchers [87]. The most basic architecture being followed widely is the three-layer reference model consisting of perception layer, network layer, and the application layer, which is illustrated in Figure 7a. The functionality of each of these layers is briefly summarized below.

- Perception Layer: The perception layer is also often known as the physical layer. The layer deals with the various sensors affixed to the IoT devices. Sensor nodes, RFID Sensors, and other sensory technologies are provided by this layer [88]. The sensors in this layer gather data and transfer it to the network layer. Physical quantities such as temperature, humidity, light intensity, sound, etc., are measured by the sensors, which are pre-processed before they send the information to the network layer. The perception layer is primarily responsible for the data collection and its transmission to the network layer. Devices linked in short-range networks can collaborate with the help of the perception layer.
- Network Layer: The network layer is made up of network components that enable communication to take place. It facilitates the data exchanged between the IoT devices. The network layer serves as a connection between the perception layer and the application layer. It is in charge of IoT networking, which involves connecting and translating IoT devices over a network. The network layer's job is to route and relay the data obtained by the perception layer over the network. The data are sent over the Internet to other computers or IoT hubs. Wi-Fi, Bluetooth, 3G/LTE, Zigbee, Lora, and other network technologies are examples of commonly used network technologies [89].
- Application Layer: The application layer is the topmost layer of the IoT architecture, and it is responsible for accomplishing the final purpose of community service. The application layer collects data from the network layer and uses them to accomplish the ultimate objective of delivering the IoT infrastructure's intended service. The application layer is liable for offering types of assistance and decides a bunch of conventions for message passing at the application level. The application layer serves as a bridge between applications and end clients, allowing them to communicate. It defines the allocation of resources and computation in data production, processing, screening, and feature selection. The application layer is a client-driven layer that performs various tasks for the clients and offers customized assistance as per a client's pertinent requirements [90]. This IoT layer brings together the industries to create high-level intelligent application solutions such as disaster monitoring, health monitoring,

translation, fortune, medical, environmental monitoring, and global management for all intelligent applications.

4.2. Five-Layer Reference Model

The architecture of IoT has been further improved by decomposing the responsibilities and functionalities of the existing three-layer architecture, resulting in a five-layer architecture [91]. A five-layer architecture consisting of a perception layer, network layer, service layer, operation layer, and application layer is proposed, which is different from that proposed by [92]. The pictorial representation of the five-layer reference model is as shown in Figure 7b. It is worthy to note that the application layer is segregated into three layers, namely, service layer, operation layer, and application layer. The functionalities of the service, operation, and application layers are briefly summarized below, whereas the perception layer and network layer hold the same responsibilities.

- Service layer: This layer envisages facilitating the use of heterogeneous IoT devices, tools, testbeds, platforms, etc., for a wide range of IoT applications. The processing of the data from the network layers is also its responsibility. Generally, the data at this layer are voluminous, for which processing, computing, and analyzing are some key challenges to be handled by this layer.
- Operation layer: This is an important layer, especially from the business point of view in IoT. The supervision of services offered by IoT, creating business models, visualization of the data, decision-making, etc., are some of the key responsibilities of this layer. Ensuring QoS across all layers is one of the vital responsibilities associated with this layer. This layer is also responsible for real-time monitoring, control, and evaluation of various application-specific parameters in an IoT environment.
- Application layer: This layer is primarily responsible for providing service to the end-users related to particular applications. There exists a wide range of applications envisaged using IoT, viz., smart city, smart home, smart agriculture, industry 4.0, healthcare, environmental monitoring, etc. This is the layer through which end users usually interact and pay for the service provided to them.

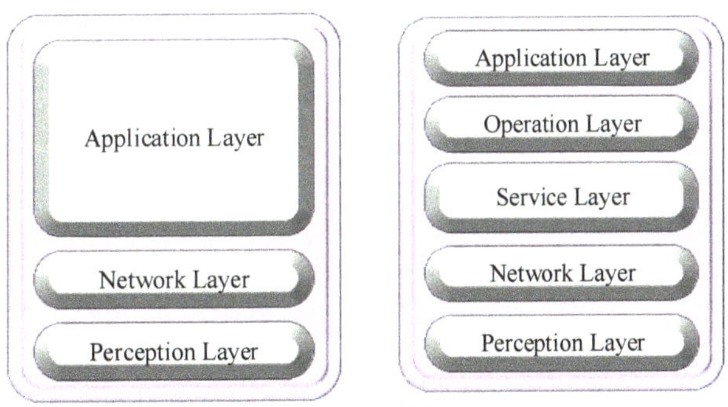

(**a**) Three-Layer Architecture. (**b**) Five-Layer Architecture.

Figure 7. Three-layer vs proposed five-layer architecture of IoT.

4.3. Seven-Layer Reference Model

Even though the architectures of IoT are either application-specific or domain-specific, we propose a more generic IoT architecture that comprises seven layers. The seven-layer generic IoT reference model comprises a perception layer, abstraction layer, network layer, transport layer, computing layer, operation layer, and application layer. The representation of the seven-layer reference model is as shown in Figure 8. Further, the functionality of each of the layers is briefly described below.

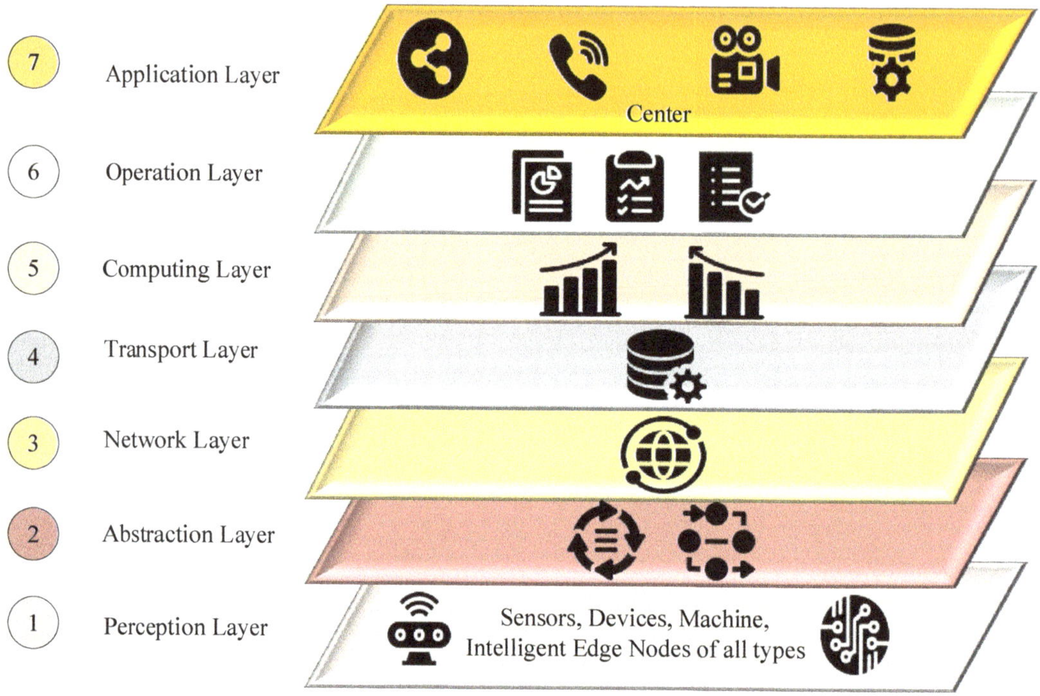

Figure 8. The proposed seven-layer architecture of IoT.

- Perception layer: This is the first level which consists of different IoT sensing and actuating devices such as sensors, actuators, RFID tags, controllers, etc. Being the first layer, the devices at this layer must adhere to the security protocols and standards to ensure they mitigate security threats to other layers originating from the perception layer.
- Abstraction layer: The IoT is based on a large and diverse set of items, each with specialized functionalities accessible through its dialect. Thus, this layer envisages harmonizing the potentials of other devices by providing a common language, protocol, and standard-based solutions.
- Network layer: This layer is responsible for providing various networking-related solutions to IoT devices. Routing, forwarding, security, etc., are some of the key responsibilities of this layer.
- Transport layer: This layer is responsible for transmitting the data from one service to other services within the application. The security at the transport layer is one of the key responsibilities in IoT in addition to the QoS.
- Computing layer: Voluminous data are generated and shared in IoT-based applications. The computing, processing, and analysis of such voluminous data is very cumbersome in general. Thus, this layer is associated to deal with such challenges in IoT. The integration of several burgeoning technologies such as cloud computing, big data, FC, EC, deep learning, machine learning, etc., is seen as promising at this layer for improving performance and security in IoT-based applications.
- Operation layer: This is an important layer, especially from the business point of view in IoT. The supervision of services offered by IoT, creating business models, visualization of the data, decision-making, etc., are some of the key responsibilities of this layer. Ensuring QoS in all layers is one of the vital responsibilities associated

with this layer. This layer is also responsible for real-time monitoring, control, and evaluation of various application-specific parameters in an IoT environment.

- Application layer: This layer is primarily responsible for providing service to the end-users related to particular applications. There exists a wide range of applications envisaged using IoT, viz., smart city, smart home, smart agriculture, industry 4.0, healthcare, environmental monitoring, etc. This is the layer through which end users usually interact and pay for the service provided to them.

4.4. IoT Protocols and Standards

In the Internet of Things, the communication protocol is a bunch of rules set down for exchanging information between electronic gadgets. Since IoT devices are more resource-limited/dependent than traditional network devices, the protocol stack in an IoT network must be different from the traditional OSI model. IoT protocols are supposed to be small and compact. The IoT protocol stack can be considered as an augmented version of the layered TCP/IP protocol stack [93]. In recent times, many standardization efforts have been seen to reduce the efforts of all stakeholders of the burgeoning IoT, such as service providers, developers, manufacturers, programmers, operators, etc. To this extent, although there are numerous players, some of the prominent organizations involved are EPC global, the European Telecommunications Standards Institute (ETSI), Internet Engineering Task Force (IETF), World Wide Web Consortium (W3C), and Institute of Electrical and Electronics Engineers (IEEE). The protocols can be broadly grouped into four categories: application protocol, service discovery protocol, connectivity and networking protocol, and other dominant protocols [94]. Some of the widely explored protocols under these categories are summarized in Table 3, whereas a detailed discussion can be found in the seminal work carried out in [41,57], of which we briefly describe some of the key protocols in the following subsections.

Table 3. Protocols at various layers of IoT architecture with key functionality.

<table>
<tr><th>Broad Category</th><th colspan="2">Dominant Protocols</th><th>Functionality</th></tr>
<tr><td>Application protocol</td><td colspan="2">CoAP, DDS, AMQP, MQTT, MQTT-SN, XMPP, HTTP REST</td><td>Services to end-users for various applications</td></tr>
<tr><td>Service discovery protocol</td><td colspan="2">mDNS, DNS-SD</td><td>Domain name resolution, client pairing for service discovery</td></tr>
<tr><td rowspan="4">Connectivity and networking protocol</td><td>Routing protocol</td><td>RPL</td><td>Routing in low power lossy networks</td></tr>
<tr><td>Network layer protocol</td><td>6LoWPAN, IPv4, IPv6</td><td>To provide networking for effective communication in IoT over the existing IPv4 and IPv6 infrastructure</td></tr>
<tr><td>Data link layer protocol</td><td>IEEE 802.15.4</td><td>To provide channel access, coordination, scheduling, and resource management tasks.</td></tr>
<tr><td>Connectivity protocol</td><td>LTE-A, EPC global, IEEE 802.15.4, Z-Wave</td><td>To interconnect IoT devices at the perception layer for effective communication</td></tr>
<tr><td>Other dominant protocols</td><td colspan="2">IEEE 1888.3, IPSec, IEEE 1905.1</td><td>To provide interoperability, security in an IoT environment</td></tr>
</table>

- CoAP: CoAP stands for Constrained Framework Protocol. COAP is a transfer protocol similar to HTTP, but it is designed to help devices with limited resources communicate [95]. This protocol is used to communicate between low-resource IoT devices and high-resource Internet-connected devices. CoAP is a binary protocol that communicates using UDP. The semantics of CoAP are designed to be very similar to those of HTTP [96]. It has less data overhead because it is a binary protocol, and because it uses UDP, it has more flexibility in communication models and can minimize latency. One of the advantages of using HTTP semantics on top of CoAP UDP rather than HTTP TCP is that a computer can easily communicate with the cloud and other devices on the local network using the same protocol language. One of the benefits of using HTTP semantics on top of CoAP UDP rather than HTTP TCP is that a machine can communicate with the cloud and other local network devices using the same protocol language [97].
- mDNS: Multicast DNS (mDNS) is having responsibility similar to the Domain Name Space (DNS) protocol in TCP/IP. This is responsible for mapping IP addresses and names among IoT devices. Since mDNS can be used without extra configuration or memory locally, it is quite flexible with speedy response [98].
- RPL: It is an abbreviation of routing protocols for low power and noisy network (RPL). It was established to assist the creation of a robust topology across lossy lines to provide minimal routing needs [99]. The point-to-multipoint, multipoint-to-point, and point-to-point traffic models are all supported by this routing protocol [100].
- 6LoWPAN: 6LoWPAN is an abbreviated form of IPv6 over Low power Wireless Personal Area. 6LoWPAN is a low-power wireless mesh network with individual IPv6 addresses for each node [101]. This enables the node to link to the Internet directly using open standards. Data are sent as packets in the form of a wireless sensor network. The protocol is used for transporting IPv6 packet data over IEEE 802.15.4 and other networks. It offers end-to-end IPv6 access, allowing it to provide direct connectivity to a wide range of networks, including the Internet. The 6LoWPAN protocol includes a layer that aids in the adaptation of resource-constrained devices to the IP environment [102]. This allows Internet access to sensor devices. Under the low power wide area network, LoRa (Long Range) and SigFox are some of the new emerging technologies.
- IEEE 802.15.4: The IEEE 802.15.4 protocol specifies a Medium Access Control (MAC) sublayer and a physical layer (PHY) for low-rate wireless personal area networks (LR-WPAN). Some of the notable characteristics are low data rate, low cost, low power consumption, and high throughput [103]. It also provides excellent security features and can support many smart IoT devices over the networks. However, the QoS feature is not guaranteed by this protocol.
- LTE-A: It stands for Long Term Evolution- Advanced (LTE-A) and it is based on cellular communication technology. Due to the utilization of sprawling existing infrastructure, it is a cost-effective and most affordable solution for IoT. Its performance is better than some other cellular-based technology in the IoT paradigm [104].
- IEEE 1905.1: The security protocols which are best for traditional Internet-based communication seem to be inappropriate for providing security in the IoT environment. Since resource constraints are among of the prominent challenges in IoT, security protocols must be built in a way that is not resource hungry. IEEE 1905.1 is designed to solve interoperability issues in IoT. Particularly, it envisages integrating heterogenous technologies with the digital home network. Interestingly, with IEEE 1905.1 as an interoperable protocol, IEEE 802.3, IEEE 802.11, IEEE 1901, and MoCA can coexist together in an IoT environment [105].
- UDP: UDP is a connectionless protocol; here, the sender sends data without waiting for the receiver to establish a link. They are connectionless datagrams that allow for the transmission of smaller packets and cycles with less overhead and a faster wake-up time [106].

- EXI: Efficient XML Interchange is abbreviated as EXI. This is an XML representation in a small package. To support XML applications on resource-constrained devices, EXI is described as a technique that uses less bandwidth and improves encoding/decoding efficiency. EXI compression aids in the reduction of document content by creating small tags internally based on the current XML schema, processing level, and context. It assures the tags are optimized for data representation. The document is in binary format, with all of the document's data tags encoded using event codes. Event codes are binary tags that keep their value only in the EXI stream where they are allocated.

5. IoT Vulnerabilities, Security Threats, and Attacks

With the unprecedented growth in IoT devices with rapidly evolving technologies, the new generation IoT-based applications are at risk. Nevertheless, there is an increasing consciousness that the new age of cell phones, computers, and other gadgets might be powerless against malware and assault. Thus, the vulnerabilities, security, and attacks must be comprehensively analyzed to make envisioned IoT a reality.

5.1. Vulnerability

Vulnerabilities are the defects in a framework's design or usefulness that permits the attacker to execute orders, access unapproved information, and launch distributed denial-of-service (DDoS) attack [107]. Attackers can utilize IoT gadgets with existing issues to infiltrate the networks. DNS rebinding attacks, which allow for the processing and ex-filtration of data from internal networks to new side-channel attacks, such as infrared laser inducted attacks against smart devices in homes and workplaces, are among the risks. In IoT systems, vulnerabilities can be found in several places [108].

Hardware and software systems are two central components of IoT frameworks, vulnerable to design flaws. Regardless of whether bugs are identified due to compatibility and interoperability of the equipment or efforts to remedy them, hardware flaws are very difficult to detect and even more difficult to repair. Computer bugs may exist in operating systems, programming software, and control software. Human elements and programming complexity are two factors that contribute to software configuration defects. Human flaws are normally the source of technical vulnerabilities [109]. Miscommunication between the developer and clients, lack of resources, skills, and experience, and a failure to manage and monitor the system can result from a poor understanding of the specifications introducing vulnerabilities in the IoT framework. Thus, vulnerability poses indispensable threats and attacks in the IoT environment. What follows next is the taxonomy of threats and attacks in IoT.

5.2. Taxonomy of Threats and Attacks in IoT

A threat is an activity that exploits a system's security flaws and has a negative effect on it. Humans and the environment are the two main sources of security threats [110,111]. As an example, seismic tremors, typhoons, floods, and fires are all natural hazards that can cause serious damage to computer systems. Few shields can be used to protect against traumatic events since these naturally occurring events cannot be prevented. Backup and contingency planning, for example, are the best ways to protect stable infrastructures from common threats. Human threats are those that humans create, such as malicious threats that are either internal (someone has allowed access) or external (individuals or organizations operating outside the network) in nature and seek to damage or disrupt a system. Following are the different types of human threats:

- Unstructured threats: These are made up mainly of novice people who use the readily available hacking software.
- Structured threats: People aware of system vulnerabilities and can comprehend, build, and exploit code and scripts are known as structured risks.
- Advanced Persistent Threats (APT): A coordinated assault is an example of advanced persistent threats. APT is a sophisticated network attack that seeks to steal data

from high-value information in industries such as manufacturing, banking, and national defense [112].

A taxonomy of threats posing a big concern from a security perspective in the IoT environment is shown in Figure 9.

Compared to the threat that can be intentional or unintentional, the attack is always intentional and malicious to cause damage. Several security attacks persist in the IoT framework, which can be analyzed with respect to the proposed IoT reference model. A taxonomy of attacks in IoT has been presented in Figure 10. These threats and attacks pose severe challenges to the IoT environment from a security perspective. The security concern due to various threats and attacks are categorically described in the following subsections.

5.3. Security Concern Due to Threats and Attacks at Different Layers

5.3.1. Security Concern at Perception Layer

Since current sensor management systems and protection schemes are insufficient to protect the sensors, an attacker may use them in various ways. In general, sensor-based threats refer to passive and active malicious actions that are attempted by the manipulation of sensors for their malicious purposes. Different kinds of threats and attacks which cause serious security challenges at the perception layer are eavesdropping, battery drainages, hardware failure, malicious data injection, Sybil threat, disclosure of critical information, device compromise, node cloning, node capture, side-channel attack (SCA), tag cloning, Radio Frequency (RF) jamming, node injection, exhaustion, node outage, etc. Some of these security threats and attacks are briefly discussed below. Further, a detailed discussion on these threats and attacks is comprehensively covered in [49,57,65,67].

- Eavesdropping: Attackers can sniff the traffic generated by IoT data flow to gather users' critical information by setting similar IoT devices.
- Malicious Data Injection: False sensor data injection is a form of attack in which the sensor data used in IoT applications are forged or modified for malicious purposes. False sensor data may be injected into devices by physical access or clandestine use of different networking mediums such as Bluetooth, Wi-Fi, GPS, etc. For instance, a spoof attack in a vehicle equipped with a GPS system. To change the location of the car, the attacker sends a forged GPS signal to the system. This conceals the vehicle's true location, allowing the attacker to attack the targeted vehicle with any physical attack [113].
- Sybil Attack: The malicious nodes in this can have multiple identities of a genuine node by either impersonating it or with a fake identity through duplication. One such malicious node may have several identities simultaneously or at different instances.
- Disclosure of Critical Information: Sensors used in IoT gadgets can disclose sensitive information such as passwords, secret keys, credit card credentials, and so on. These details may be used to violate user privacy or to build a database for future attacks. One such example of this attack is eavesdropping. It is a kind of attack where a pernicious application records a discussion subtly by misusing sound sensors and extracts data from the discussion. An attacker can save the recorded discussion on a gadget or tune in to the discussion continuously. Soundcomber is one of the current instances of eavesdropping over the receiver of a cell phone. In this model, a pernicious application secretively records when a discussion is initiated from the gadget. Since the recording is carried out behind the scenes, a client is completely unaware of the chronicle [114].

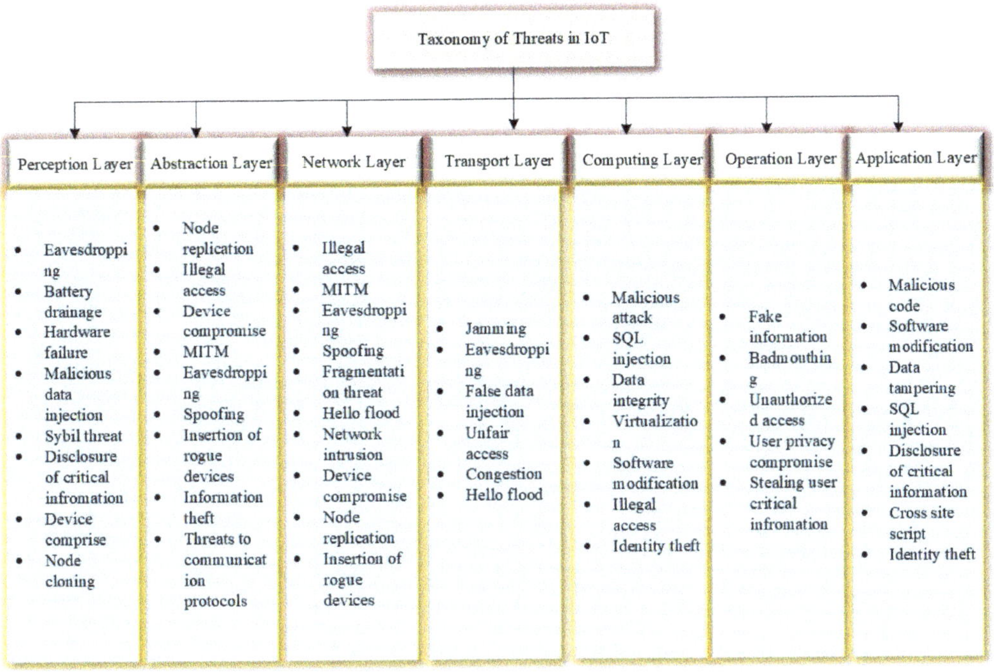

Figure 9. Taxonomy of threats in IoT.

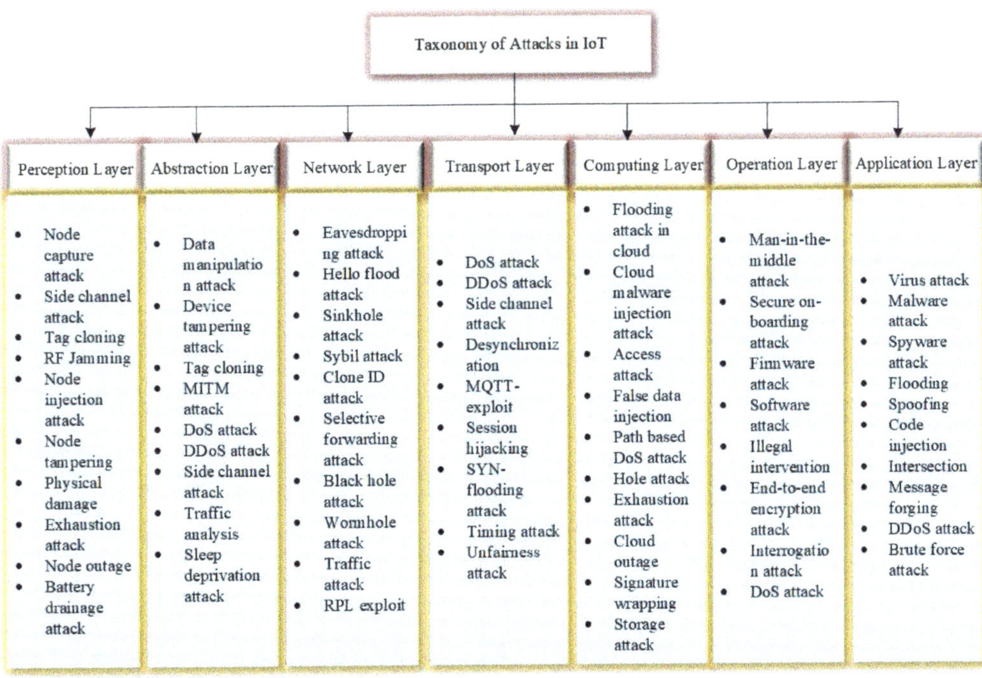

Figure 10. Taxonomy of attacks in IoT.

- Side-Channel Attacks: The assailant gathers information and performs the reverse engineering process to collect the encryption credentials of an IoT device while the encryption process is under way. This information cannot be collected from plaintext or ciphertext during the encryption process, but from the encryption devices. Side-channel attacks the use of certain or all data to acquire the key the device uses. Some instances of such attacks include timing attacks, power or failure analysis, and electromagnetic attacks. The opponent uses data leaks and collects block cipher keys. In the event of the attacks, an intrusion prevention system such as Boolean masking can be directed.
- Malicious Data Injection: Attackers take advantage of flaws in communication protocols to insert data into the network [115]. The intruder will tamper with the information required to control the device if the protocol does not verify the integrity of the data. The injection attack may result in code execution or system control from afar.
- Node cloning: In most cases, IoT devices such as sensor nodes and CCTV cameras are developed without hardware defects, given the lack of standardization of the IoT device design. Therefore, for unauthorized purposes, these devices can be easily forged and replicated. This is also known as the cloning of nodes. It can take place in either of the two phases, i.e., production and during operations. An internal attacker can replace an original device with an unauthorized, pre-programmed object in the former case. A node can be captured and cloned during the operational phase. Capturing nodes could further remove security parameters and substitute firmware replacement attacks.
- Exhaustion attack: Jamming or DoS attacks that have been mentioned before could lead to attacks of exhaustion. In particular, energy consumption can affect the battery-operated devices if an assailant attacks the network continuously. Repeated retransmission attempts could cause collisions with IoT MAC protocols leading to high-energy depletion. Exhaustion is a dot attack and is connected with deactivation assaults, reducing the size of the network and removing nodes permanently from the network.

5.3.2. Security Concern at Abstraction Layer

Different kinds of threats and attacks which cause serious security challenges at the abstraction layer are node replication, illegal access, device compromise, MITM, eavesdropping, spoofing, insertion of rogue devices, information theft, a threat to the communication protocols, data manipulation, device tampering, tag cloning, DoS, DDoS, SCA, traffic analysis, and sleep deprivation. Some of these security threats and attacks are briefly discussed below. Further, a detailed discussion on these threats and attacks is comprehensively covered in [49,57,65,67].

- Illegal access: IoT equipment often operates without any physical protection in an untrusted environment, such as traffic light sensors, environmentally friendly sensors, agricultural sensors, smart city sensors, and much more. Problems such as illegal access and malicious change of data may arise during the processing of sensitive data.
- MITM: Man-in-the-Middle (MITM) is a system that listens in on traffic between a smart device and a gateway [116]. All traffic will be routed via the attacker's computer using the ARP poisoning technique. This attack can be avoided if the attacker is unable to see the traffic in progress. As a result, encryption is required in the protocol [117].
- Spoofing: To initiate a spoofing attack, an attacker can mimic a node. Due to its manner of assault, a spoofing attack is one of the high-risk attacks. A transmission could be recorded using a compatible portable reader. Because the attacker is impersonating the node, the retransmission may appear to be coming from a legitimate node. This threat could exist in all three layers of the IoT. Spoofing attacks that impersonate nodes are classified as authentication attacks, and they also breach the privacy principle.
- Threat to communication protocols: The fact that most current wireless communication protocols adhere to the OSI layered protocol architecture and the physical layer encryption is not reinforced with additional security methods in the upper layers of

the communication presents additional issues in IoT/CPS security design. Cellular technologies such as UMTS, GSM, and LTE, on the other hand, have their own set of security challenges. Because radio baseband stacks are implemented openly, mobile networks are vulnerable to hacking and cyber-attacks. Furthermore, aggressive attackers can use "IMSI Catching" to compromise GSM and UMTS networks.

- Tag cloning: An opponent can readily clone RFID tags by gaining direct access to a device or via reverse engineering to obtain the essential information. A tag cloning attack was described in the literature, in which an RFID scanner was unable to discriminate between legitimate and compromised tags.
- Denial-of-Service (DoS): It is a type of attack in which a device or application is maliciously denied normal operation. DoS attacks can be active attacks where an application or task is strongly denied or where passive attacks can stop another ongoing task on the device by attacking one application [118].
- DDoS: Any IoT device, network, or software program could be shut down by a distributed denial-of-service (DoS) attack, rendering the service inaccessible to its consumers. These attacks can take many different shapes. One method of attack is to generate a large amount of network traffic and send a massive request to the victim. The main goal of this attack is to make the target consumers' devices, software, network services, and resources unavailable. Furthermore, the attacker may be able to obtain sensitive information from users. DDoS attacks are more harmful than DoS attacks, which use many attacking platforms to infiltrate one or more systems
- Traffic analysis: For attackers, a network's traffic pattern may be as useful as the substance of data packets. Analyzing traffic patterns can provide valuable information about the networking topology. In WSNs, the sink nodes closer to the base station generate more transmissions than the other nodes because they relay more packets than the nodes further away. Similarly, clustering is a key scaling strategy in WSNs, and cluster heads are busier than the rest of the network's nodes. For adversaries, detection of the base station, nearby nodes, or cluster heads may be very beneficial since a denial-of-service attack or packet eavesdropping against these nodes might be very useful
- Sleep deprivation: The denial of a sleep attack on a battery-powered device will result in energy depletion. Collision attacks or repetitive handshaking, i.e., repeatedly shaking hands, can be used to carry out this attack. Request to Send (RTS) and Clear to Send (CTS) manipulate flow control signals, stopping the node from entering the stage of sleep.

5.3.3. Security Concern at Network Layer

Gateways and networking systems assist in the routing and networking of data packets to their intended destinations. If the gateway communicates using wireless protocols, the attacker will use wireless attacks to link to the gateway or internal network. As a result, the attacker will be able to carry out further attacks, such as ARP poisoning, MITM, packet injection, and sniffing. Different kinds of threats and attacks which cause serious security challenges at the network layer are illegal access, MITM, eavesdropping, spoofing, fragmentation, hello flood, network intrusion, device compromise, node replication, insertion of rogue devices, sinkhole attack, Sybil attack, clone ID attack, selective forwarding attack, blackhole attack, wormhole attack, traffic attack, and RPL exploits. Some of these security threats and attacks are briefly discussed below. Further, a detailed discussion on these threats and attacks is presented in [49,57,65,67].

- Hello flood: Message flooding is amongst the biggest network layer threats. By sending multiple route establishment requests to a network or node. The nodes in the network interpret a hello message as coming from within and mark it as a communication route.
- Sinkhole: By using this approach, an attacker compromises a network's central node and overrides it in a bid to render it unavailable. An attack that uses sinkholes is

more likely to cause a major incident than a tempering attack, which involves a few affected nodes. As a result of sinkhole attacks, the whole infrastructure base could be controlled.

- Blackhole: If the malicious node experiences a Blackhole attack, it will drop all packets encountered and the entire network will be affected. Since it absorbs all routing data, it is considered a high-impact attack. By flooding malicious routing information, an intruder tries to hijack the most efficient route to a destination. Upon transmission through the malicious route, the source node continues to send packets, and the attacker drops all packets, preventing any traffic from being forwarded to the destination.
- Traffic Analysis: The attacker analyses the traffic and saves a copy for later use in this attack. As a result, the interface can be managed using the traffic that was previously communicating with the gateway. The traffic or data that have been checked are reused in a different context [119].
- Wormhole: This network attack would intercept traffic in one location and redirect it to another. As a result, it causes network congestion and efficiency problems.
- Selective forwarding: An attacker launches an SF attack by entering a network and dropping packets. Some packets are dropped casually, while others are selectively forwarded. Consequently, packet dropping can be difficult to figure out in IoT networks due to their lossy nature. As a consequence, the entire network may suffer bandwidth degradation and delay.
- RPL exploit: The IoT is made up of devices with limited resources, such as battery power, memory, and computational power. RPL is a new network layer routing protocol developed for these types of networks (routing protocol for low power and lossy networks). RPL is a lightweight routing protocol that does not contain all of the features of typical routing protocols. RPL was developed specifically for data sinks (multi-point to point communications) and has lately been adopted by the IoT. In such attacks, spiteful nodes can seek to redirect paths when data are transferred. Sinkhole attacks are a kind of routing attack in which an opponent advertises and hire a node to drive traffic [120]. Wormhole attacks can also pose a serious threat to IoT systems if associated with other attacks such as sinkhole attacks [121]. A wormhole is an out-of-band link that allows easy packet transfer between two nodes. An attacker will try to circumvent the basic security protocols in an IoT application by creating a wormhole between a compromised node and a computer on the Internet.

5.3.4. Security Concern at Transport Layer

Different kinds of threats and attacks which cause serious security challenges at the transport layer are jamming, eavesdropping, false data injection, unfair access, congestion, hello flood, DoS, DDoS, SCA, desynchronization, MQTT exploit, session hijacking, SYQ-flooding, timing attack, etc. Some of these security threats and attacks are briefly discussed below. Further, a detailed discussion on these threats and attacks is comprehensively covered in [49,57,65,67].

- Desynchronization: De-synchronizing the transmissions between two nodes allows an attacker to break actual links between them. Trying to send fabricated messages to both sides of communication, such as false flag types of messages, is an example of this type of attack. By forcing them to lose their synchronization, they will lose their ability to communicate.
- Session hijacking: In session hijacking, an attacker steals the session ID and pretends to be the legitimate user to take over a user's online session. The attacker can spoof the user's session ID and do anything the authorized user can do on the network once the attacker obtains it.

5.3.5. Security Concern at Computing Layer

This part of the IoT infrastructure supports data storage and computer remote control. If cloud servers are not properly configured, they can then lead to the server and smart devices being exploited. Different kinds of threats and attacks which cause serious security challenges at the computing layer are malicious attack, SQL injection, data integrity, virtualization, software modification, illegal access, identity theft, flooding attack in cloud, cloud malware injection, access attack, false data injection, path-based DoS, hole attack, exhaustion attack, cloud outage, signature wrapping, storage attack, etc. Some of these security threats and attacks are briefly discussed below. Further, a detailed discussion on these threats and attacks is comprehensively covered in [49,57,65,67].

- Malicious Attack: As workers in the company download untrustworthy malicious software programs from the Internet, there is a good chance that the machine will be hacked. The malware would spread across the internal network, putting the whole company under its influence. The attacker would use the compromised machine to hack the IoT system connected to the network. As a result, it may result in economic loss and abasement of the company's reputation.
- SQL injection: SQL injection is a web security flaw that permits an attacker to meddle with a web application's database queries. It permits an attacker to access the information that they would not usually be able to reclaim. This may incorporate information belonging to different clients or whatever other information the application can access. An attacker may alter or erase these data, resulting in the application's content being permanently altered. In certain circumstances, an attacker can improvise a SQL injection attack to alter the basic server or other back-end foundation or carry out a distributed denial-of-service (DDoS) attack [122].
- Illegal Access: It is one of the major challenges faced by companies providing cloud services. Most enterprise proprietors are unfamiliar with cloud-based technology, which opens them to a variety of data breaks that can affect their tasks. Since cloud computing is built to be simple to use and share, it is difficult for businesses to ensure that data are only available to legal parties. On the off chance that IoT gadgets do not properly configure, the entire network will be damaged. Additionally, companies using cloud-based computing lack complete control over their networks, which requires configuring and protecting their cloud deployments on security controls provided by their cloud service providers (CSP).
- Storage Attack: It can be very difficult to detect and deal with cryptojacking. The main problem here is that hackers will slow down the activity of the device as they use the cloud storage resources, but it will continue to operate. This means it may seem that nothing is malicious and that the machines are probably just struggling with their processing capacity. Many teams in IT experience the symptoms of cryptojacking as an upgrade fault or as a sluggish Internet link, so the real issue is much longer to be resolved.
- Access Attack: Advanced persistent threat is another term for an access attack. An unauthorized individual or adversary gains access to the IoT network in this form of attack. The intruder will remain undetected in the network for an extended period. Rather than causing network harm, the ultimate goal of such a type of attack is to steal valuable information. IoT applications receive and transmit valuable data regularly, making them particularly vulnerable to such attacks.
- Software modification: An IoT device can be compromised by modifying its software or firmware by using physical or remote access to take unauthorized actions. By patching or substituting code, or by making code extensions, the vulnerability can be exploited further.

5.3.6. Security Concern at Operation Layer

Different kinds of threats and attacks which cause serious security challenges at the operation layer are fake information, badmouthing, unauthorized access, users' privacy compromise, stealing users' critical information, MITM, secure on-boarding, firmware attack, software attack, illegal intervention, end-to-end encryption attack, interrogation attack, DoS, etc. Some of these security threats and attacks are briefly discussed below. Further, a detailed discussion on these threats and attacks is comprehensively covered in [49,57,65,67].

- Illegal Intervention: Cloud services are typically provided, monitored, and managed through APIs and software user interfaces. Although, cloud service providers are engaged diligently to improve APIs and interfaces, this boom has additionally extended safety dangers related to them. Cloud specialist organizations utilize a particular structure to give APIs to developers, making their frameworks more endangered against an attacker. In 2018, the social media platform Facebook suffered a security breach that affected around 50 million users due to a flaw [123]. API flaws, particularly when linked to user interfaces, may provide the attacker a direct path to steal employee or client credentials.
- **Unauthorized Access:** Access control is an approval system that permits authentic clients to acquire information. Multi-client access and simultaneous altering of design systems ought to be vigorous against multi-client access. When numerous clients can alter the designs of different segments of the IoT frameworks, simultaneous execution of setup changes and simultaneous altering of arrangement records effectively leads to temperamental framework status. In IoT applications, access control is important because if access is compromised, the entire IoT framework becomes susceptible to attacks.

5.3.7. Security Concern at Application Layer

The application layer manages the services offered to the clients. This layer serves applications such as telehealth, industrial automation, smart metering, and so on. This layer has its own set of security concerns that are unique to each program. Different kinds of threats and attacks which cause serious security challenges at the application layer are malicious code, software modification, data tampering, SQL injection, disclosure of critical information, cross-site script, identity theft, virus attack, malware attack, spyware attack, flooding, spoofing, code injection, intersection, message forging, DDoS attack, brute force attack, etc. Some of these security threats and attacks are briefly discussed below. Further, a detailed discussion on these threats and attacks is comprehensively covered in [49,57,65,67].

- Malicious code: Malicious codes or targeted malware can easily exploit the vulnerabilities of IoT devices through the Internet, which allows hackers to compromise those devices. Further attacks can be launched on other endpoints/networking apps via the infected devices.
- Software Modification: Minor changes can lead to more complicated problems. Unexpected environment changes along with minor framework alterations and system changes may have unanticipated consequences. As the arrangement of framework develops, these results can spread to more concerning issues. If the programming mechanism is not secured, the attacker will be able to reprogram IoT devices remotely. This could result in the IoT network being hacked.
- Data tampering: During an attack of this type, the information on the end device is misrepresented by an attacker. Invaders retrieve data format and type, then insert tamper detection measures and recreate the original data. Due to this, there is considerable doubt about the precision of data collected over the network.
- Cross-site script: XSS (cross-site script) is a technique attackers use to insert malicious code into a website that is otherwise trusted. If an XSS attack is successful, the IoT system will be under the complete influence of the attacker.
- Identity Thefts: IoT systems deal with plenty of personal and sensitive information. Clients will hesitate to enlist their personal information on IoT applications if these applications are helpless against information burglary. Some of the protocols and

methodologies used to protect IoT applications from information burglary include data isolation, data encryption, privacy management, user and network authentication, etc.

- Virus attack: The objective of these attacks is to breach the confidentiality of the system. The risk of these attacks is significantly higher for smartphones, sinks, or gateways in IoT networks. Hence, IoT applications must seriously consider mitigating viruses and malware.
- Spyware attack: Installed on IoT devices without consent, spyware is an installation program that collects information. Using this type of attack, attackers are looking to gather sensitive information about users by monitoring their behavior. Signature, behavior, and specification-based techniques are some common approaches to spyware detection.
- Code Injection: Attackers usually use the simplest or easiest way to break into a device or network. If the device is endangered to spiteful scripts and misdirection as a result of inadequate code tests, it will be the first point of entry for an attacker.
- Intersection: System integrity is a critical feature of the IoT framework. When a system's integrity is compromised, there is a high risk of safety and security threats. High activity stress or irregular process conditions, network or device failures, multiple warnings, executing previously unexecuted error path code or system recovery code, or wrongly executed commands do not cause the system to crash. This necessitates extensive research.
- Brute force attack: A brute force attack involves systematically trying and guessing every possible passphrase or password combination to gain access to the system. Crypto-analysts are ultimately able to identify the correct one which allows them access to the system.

To summarize, the different threats and attacks are reported in Table 4, along with their scope in IoT architecture and protocols, their impact, and references focusing on different detection, prevention, and mitigation strategies. With reference to this table, the following abbreviations are used: PL—Perception Layer, AbsL—Abstraction Layer, NL—Network Layer, TL—Transport Layer, CL—Computing Layer, OL—Operation Layer, AL—Application Layer, AP—Application Protocols, SDP—Service Discovery Protocols, RP—Routing Protocols, NLP—Network Layer Protocols, DLLP—Data Link Layer Protocols, CP—Connectivity Protocols, ODP—Other Dominant Protocols.

Table 4. The scope and panoramic view of threats and attacks with detection, prevention, or mitigation strategies in IoT architecture.

Threats/Attacks	Scope on Different Layers in IoT Architecture	Scope on Different Protocols and Standards of IoT	Impact	References Focusing Detection, Prevention, or Mitigation Strategies
Eavesdropping	PL, AbsL, NL, TL	AP, SDP, RP, NLP, DLLP, CP, ODP	Affect user's privacy and confidentiality	[124,125]
Battery drainage	PL	CP	Drain the batters of IoT devices at a much faster rate	[126]
Hardware failure	PL	NLP, CP	Affect the service due to failure causing unreliability	[127,128]
Malicious data injection	PL,	AP, SDP, ODP	Can harm applications services	[129]
Sybil threat	PL, NL	SDP, CP	Enhances packet drop probability	[130,131]
Disclosure of critical information	PL, AL	AP, SDP,	Affect user's privacy	[132]
Node cloning	PL	SDP, DLLP, NP, CP	Can copy the functions, data, etc., of a particular node or even capture a node	[133,134]

Table 4. *Cont.*

Threats/Attacks	Scope on Different Layers in IoT Architecture	Scope on Different Protocols and Standards of IoT	Impact	References Focusing Detection, Prevention, or Mitigation Strategies
Side-channel attack	PL, AbsL	SDP, RP	Indirect attack on node leaking sensitive information	[135]
RF jamming	PL, TL	CP	Cause interreference, and DoS	[136,137]
Physical damage	PL	CP, DLLP	Affect service of a node	[127]
Exhaustion attack	PL, CL	SDP, NLP, DLLP, CP	Affect network lifetime	[138]
Node outage	PL	SDP, CP	Causing unreliability	[126]
Node replication	AbsL	ODP, SDP, NLP	Injecting huge traffic flow	[139]
Illegal access	AbsL, NL, CL, OL	AP, SDP	Can steal user's confidential·data	[126]
Device compromise	AbsL, NL	AP, SDP, RP, NLP, DLLP, CP, ODP	Affect credibility of device	[140]
MITM	AbsL, NL, OL	AP, SDP, RP, NLP, DLLP, CP, ODP	Affect network resources and authenticity	[141]
Spoofing	AbsL, NL, AL	AP, SDP, RP, NLP, DLLP, CP	Affect trust and confidentiality	[142]
Threats to communication protocols	AbsL	CP, NLP, DLLP	Affect connectivity	[143]
Tag cloning	AbsL	SDP, RP	Affect authenticity	[144]
DoS	AbsL, TL, OL	AP, SDP, RP, NLP	Affect service availability resulting in huge losses	[145]
DDoS	AbsL, TL, AL	AP, SDP, RP, NLP	Affect reliability, and availability	[146]
Traffic analysis	AbsL, NL	NLP	Affect user's privacy and confidentiality	[124]
Sleep deprivation attack	AbsL	SDP, DLLP	Affect the network lifetime	[147]
Fragmentation threat	NL	NLP, RP	Affect data integrity	[148]
Hello flood	NL	NLP, RP	Creates unnecessary traffic in the system	[149]
Network intrusion	NL	NLP, RP, CP	Affect the network resources	[79]
Insertion of rogue devices	NL	NLP, RP, ODP, CP	Affect network security and data integrity	[150]
Sinkhole	NL	NLP, RP	Result in network failure	[151]
Clone ID attack	NL	NLP, RP	Results in other network attacks	[152]
Selective forwarding attack	NL	NLP, RP, DLLP	Affect data integrity	[153]
Blackhole attack	NL	NLP, RP	Affect entire network	[154]
Wormhole attack	NL	SDP, RP, NLP, DLLP, CP, ODP	Affect entire network	[155,156]
RPL exploit	NL	RP	Affect routing of packets	[157,158]
False data injection	TL, CL	DLLP, CP	Affect the legitimate information	[129]
Unfair access	TL	AP, SDP, RP, NLP, DLLP, CP, ODP	Affect the performance	[145]
Congestion	TL	AP, ODP	Can cause more packet drop and latency	[126]
Desynchronization	TL	SDP, RP, NLP, DLLP	Affect data integrity	[159]
MQTT-exploit	TL	AP	Affect transmission of packets	[160]
Session hijacking	TL	SDP, RP, NLP, DLLP, CP	Exploitation and tampering with the legitimate session	[161]

Table 4. *Cont.*

Threats/Attacks	Scope on Different Layers in IoT Architecture	Scope on Different Protocols and Standards of IoT	Impact	References Focusing Detection, Prevention, or Mitigation Strategies
SYN-flooding	TL	SDP, RP, NLP	Affect node resources such as energy and memory	[162]
Timing attack	TL	SDP, RP, CP	Leads to SCA	[163]
SQL injection	CL, AL	AP, SDR, ODP	Affect SQL database	[164]
Data integrity	CL	AP	Affect credibility of data	[141]
Virtualization	CL	AP, SDP, ODP	Affect data protection	[165]
Software modification	CL	AP, SDP, ODP	Affect entire application resources	[166]
Identity theft	CL	AP, SDP	Affect user's privacy, and data confidentiality	[167]
Access attack	CL	AP, SDP, RP, NLP, DLLP, CP, ODP	Can steal valuable data from the network	[126]
Cloud outage	CL	AP, SDP, RP, NLP, DLLP, CP, ODP	Affect cloud-related services	[166]
Signature wrapping	CL	AP, SDP	Affect signature algorithm resulting in eavesdropping attack	[168]
Storage attack	CL	AP, SDP, NLP	Affect the data storage	[166]
Path-based DoS attack	CL	RP, NLP	Affect application layer similar to DoS	[169]
Badmouthing	OL	AP, ODP	Affect the credibility	[126]
Unauthorized access	OL	AP, SDP, CP	Can result in stealing of critical information	[167]
User privacy compromise	OL	AP	Affect the privacy of users	[167]
Secure on-boarding attack	OL	AP, SDP, NLP, DLLP, CP	Can cause eavesdropping during on-boarding of new devices	[170]
Firmware attack	OL	AP, SDP, RP, NLP, DLLP, CP	Affect low-level control software of IoT	[171]
Software attack	OL	AP, CP	Affect software of IoT	[171]
End-to-end encryption attack	OL	AP	Affect privacy and integrity of the end-users	[172]
Interrogation attack	OL	AP, SDP, RP, NLP, DLLP, CP, ODP	Affect the channel resources	[173]
Malicious code	AL	AP	Can cause illegitimate access to the IoT resources	[174]
Virus attack	AL	AP	Affect high-end IoT devices	[175,176]
Malware attack	AL	AP, SDP	Affect high-end IoT devices causing user's security concern	[175,176]
Spyware attack	AL	AP, SDP	Indirect harm to users	[177]
Intersection	AL	AP, SDP	Affect privacy	[178]
Message forging	AL	AP	Can steal critical information	[179]
Brute force attack	AL	AP, SDP, ODP	Affect the user's privacy and can steal critical login information	[174]

6. Security Goals and Roadmap in IoT

There are certain security objectives that IoT must essentially meet to provide undisputed services. For smooth functioning, IoT applications require secure connections with proper authentication mechanisms and data confidentiality. To ensure information security, one needs to implement the CIA triad—data Confidentiality, Integrity, and Availability.

Threats and violations in any of these areas can result in substantial damage to the system, compromise its integrity, and disrupt its activity. To be efficient in implementing effective IoT security, the following primary security objectives must be considered. These security objectives can be achieved with effective methodologies for detection, prevention, and mitigation of threats and attacks pertaining to the IoT ecosystem, described in the next section.

- Confidentiality: Confidentiality is an important security feature in the Internet of Things, but it is not always required, for example, in cases where data are exchanged with the public. In the vast majority of situations and cases, sensitive data must not be disclosed or read by unauthorized persons [180]. Sensitive information about patient data, company information, and possibly military information, as well as security accreditations, should all be kept private from unauthorized users. Confidentiality should be granted such that the information gathered or distributed is safe and only accessible to approved users. Data collected by a computer or a sensor should not be sent to other devices unless they are properly encrypted. To prevent malicious actors from accessing the collected data, only encrypted messages should be sent to neighboring devices. A data encryption system transforms each bit of data into ciphertext, followed by a two-step verification process in which two devices/components permit access only if the authentication test is passed by both the devices, and a biometric verification in which the user is uniquely identifiable and biometric authentication in which the person can be identified by his or her fingerprints.
- Integrity: Integrity should be offered to ensure data validity. Data integrity is critical since data recipients must be able to verify whether data obtained from other devices are authentic. In most cases, integrity is a necessary security property for IoT users to receive reliable services [181]. Different IoT systems have different levels of trustworthiness. As an example, because of data sensitivities, a patient observation framework would have high trustworthiness testing against arbitrary mistakes. It is integrated into the network to protect cybercrimes data in the communication process so that data manipulation cannot be carried out without the danger detected by the device. Two error detection methods are used to ensure data integrity in the inspection and cyclic redundancy search. For continuous data sync for backup purposes, a version control system is used.
- Authentication and authorization: Authenticity is related to credibility, and it means that each system in the network should be able to recognize and authenticate other devices. Since the IoT is made up of so many devices, it is critical to be able to recognize them; otherwise, malicious devices might use spoofing to target IoT networks. Due to the design of IoT settings, the possible communication between the device and device (M2M) is exacerbated by the problem of authentication in IoT connectivity. Different authentication criteria in different systems require different solutions. Some solutions, such as bank card or bank device authentication, require a high level of reliability. However, others will need to be foreign, such as e-Passport, while others will need to be local. Only approved entities (any authenticated entity) can conduct such network operations using the authorization property [182].
- Availability: The primary aim of every IoT protection system is to make data available to users promptly. The consumer should be able to obtain data from the resources right away, not only in usual circumstances but also in emergencies. Firewalls are installed in the network to protect against attacks on services such as denial-of-service attacks, which prevent data from reaching the end-user [183].
- Accountability: Accountability provides redundancy and responsibility for some activities, tasks, and the preparation of the execution of network security policies while designing security strategies to be used in a safe network [184]. Accountability cannot prevent attacks on its own, but it does help ensure that other security measures are functioning properly. Integrity and confidentiality, for example, can be rendered worthless if they are not subjected to transparency. Often, in a disapproved event, an

entity's behavior can also be traced through an accountability system, which can help determine the inside story of what occurred and who was ultimately responsible.

7. Scope of Security Enhancements in IoT with Burgeoning Technologies

Now, we review the state-of-the-art methodology to enhance security and privacy in an IoT environment using a few of the ubiquitous technologies such as BC, FC, EC, and ML. Despite some other technologies such as cloud computing, Big Data, embedded system, digital twin, etc., the trend in the literature unanimously shows that BC, FC, EC, and ML have huge potential to answer the security concern in the IoT ecosystem. Further, these technologies are indispensable for the IoT ecosystem, which motivates researchers to address the security concern based on these ubiquitous technologies.

7.1. BC for IoT

BC technology is a network of peer-to-peer nodes that stores transactional records, known as blocks; these blocks consisting of numerous public databases are known as the "chain". The fundamental principle of BC is based on a distributed ledger. IoT devices collect real-time data from sensors, and BC ensures the security of data by deploying a decentralized, distributed, and shared ledger [185]. Any transaction in this ledger is signed with the owner's digital signature, which verifies the transaction and protects it from tampering. As a result, the data in the digital ledger are extremely stable. The BC entries are both chronological and time-stamped. In the ledger, each entry is linked to the previous entry by applying cryptographic hash keys. Individual transactions are stored in a Merkle tree, and the tree's root hash is stored in the BC. Individual transactions are represented by T1, T2, T3, and Tn in the diagram. The cryptographically hashed transactions are stored on the leaf node represented as H1, H2, H3, and so on. The hashes of the child nodes are combined to create a new root hash. The BC stores the final root hash (i.e., Ha and Hb). It can be confirmed whether the transactions associated with the root hash are secure or not, by just verifying the root node. If a single transaction is modified, all hash values on that side of the tree will be affected. The miners verify all the transactions and then a key is produced that allows the most recent transaction to be included in the ledger. This procedure renders the most recent transaction available to all network nodes. It is very difficult and time-consuming for the attackers to hack the blocks as each block is secured using cryptographic hash keys [186]. The miners are only mining to gain their bonuses and have no personal stake in the transactions. The identity of the transaction's owners is unknown to the miners. Furthermore, several miners are working on the same collection of transactions, and they are in fierce competition to link the transactions to the BC. These characteristics enable the BC to serve as a safe, distributed, tamper-proof, and open data system for IoT data. The entire process of a transaction from its inception to its commitment to the distributed chain is elucidated in Figure 11.

In academia and industry, various platforms and frameworks are being built to support the development and maintenance of BC. Ethereum, Hyperledger Cloth, Ripple, and other platforms are examples of this kind [187]. Nevertheless, the simplified general architecture of the BC is as shown in Figure 12.

The following are the key characteristics of the BC that can be exploited to enhance security and privacy in IoT.

- To create blocks, a consensus algorithm is used; involved individuals (typically miners) verify the transactions' coherence and validity.
- A monetary competition exists for block certification and the computation of a new branch, which is based on algorithms such as Proof of Work (PoW) or Proof of Stake (PoS).
- There is no third party to rely on; each individual produces his or her own keys.
- All ledger elements, such as blocks, and transactions are stored in the database.

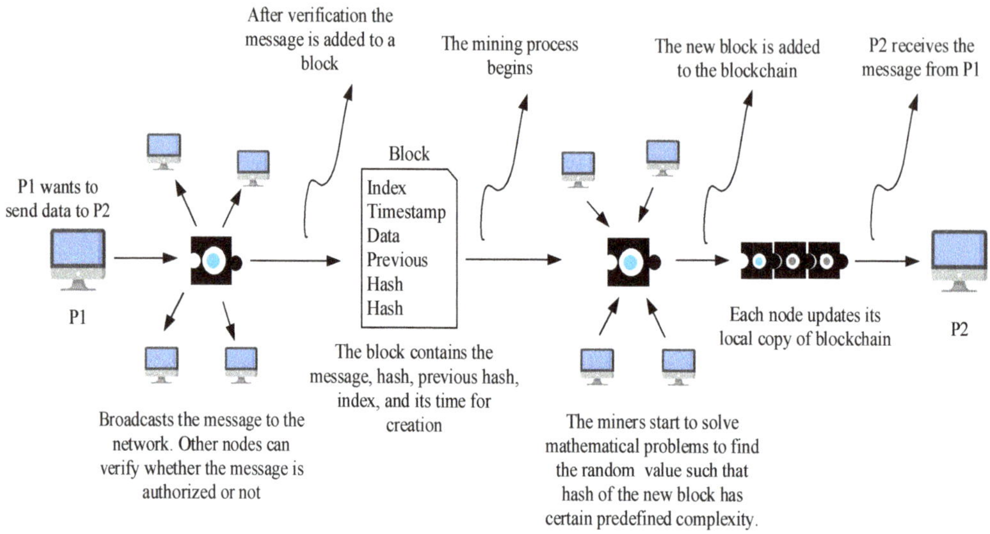

Figure 11. Basics of BC for enhancing security and privacy in IoT.

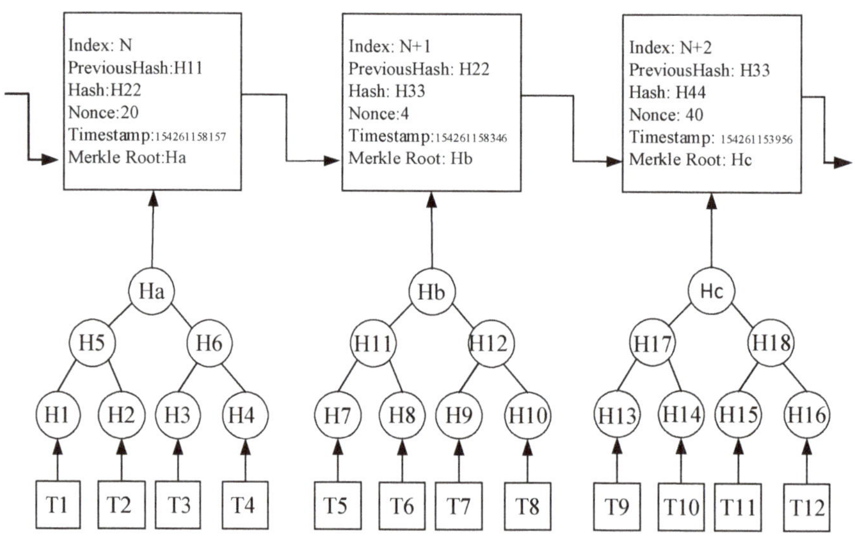

Figure 12. The architecture of BC.

The use of BC in IoT applications has several benefits. The followings are a summary of the main advantages of using BC in IoT applications.

- BC can be used to store the data from IoT devices: The IoT technologies incorporate a wide range of devices that are all interconnected. This arrangement is additionally associated with the cloud to empower IoT applications to be accessed from anywhere. BC is a promising method for storing and protecting such an enormous amount of data. BC is an apt solution for storing and transmitting data regardless of the layer in an IoT application.
- Terminating the centralized cloud server system: BC boosts the security of IoT frameworks by removing the centralized cloud server and establishing a peer-to-peer net-

work framework. Data pirates are mostly interested in centralized cloud servers. BC enables the distribution of data across all the nodes of the network and encrypts them.

- Forestalling illegal access: Several IoT applications necessitate a lot of contact between different nodes on a regular basis. Since BC communication is based on public and private keys, data can only be accessed by the intended party or node. If an unintended person accesses the data, the content will be nonsensical because it is protected with keys. As a result, the BC data system attempts to address a variety of security problems that IoT applications face.
- A solution for resource-constrained devices: Because of the limited resources, IoT devices are unable to store large ledgers. There have been different works toward this path to work with the assistance of BC. One of the potential solutions for IoT devices to use BC is proxy-based architecture. By setting up the proxy servers, the data can be stored in an encrypted format and the encrypted resources can be downloaded via proxy servers.
- Forestalling spoofing attack: Spoofing is a type of attack where a foreign node enters the IoT ecosystem and tries to emulate the existing nodes to be seen as a member of the original framework. This foreign node can monitor or inject malicious data into the network. The BC technology appears to be a potential solution for preventing such attacks. Each genuine client or gadget is enlisted on BC, and gadgets can undoubtedly recognize and validate each other.
- Forestalling data loss: IoT devices acquire the danger of losing information. There is a possibility that the data are lost by the sender and the recipient due to natural environmental causes. The utilization of BC can forestall such losses as it is impossible to eliminate a block once it is included in the chain.

7.2. FC for IoT

The Internet infrastructure is being challenged by an unprecedented amount of data generated by IoT. The integration of IoT and the cloud has led to the development of numerous new possibilities on how to process, store, manage, and secure data. These benefits do not fully address all of the problems associated with the IoT. Cloud computing and FC complement each other rather than replace each other [188].

Computing in the fog enables processing, storage, and intelligence control to come within the proximity of the data devices. It uses two frameworks, namely Fog-Device Framework and Fog Cloud Framework [189]. With the Fog-Devices framework, different services can be delivered to a user without involving any cloud servers. Whereas the simple decisions in the Fog-Cloud-Device framework occur at the fog layer, the complex ones occur at the cloud level [190]. The architecture of the Fog-Cloud-Device framework is shown in Figure 13.

The convenience and flexibility of this structure make it possible to offer cloud computing at the network edge. The result is a reduction in distance and improved efficiency while decreasing the amount of data required to be transported into the cloud for processing, analysis, and storage. Comparing the FC with cloud-only models, data traffic between the cloud and network edge is reduced by 90%, and response times for users are cut by 20% [191]. This flexible structure extends cloud computing services to the edge of the network. Thus, it reduces the distance across the network, improves efficiency, and decreases the amount of data needed to transport to the cloud for processing, analysis, and storage.

Using fog technology, data are collected at nodes referred to as fog nodes, and the nodes can process 40 percent [192]. It reduces the latency of IoT devices by offloading traffic from the core network. According to its time sensitivity, data are directed to the cloud, fog, or aggregation nodes. By providing cryptographic computations to IoT applications, fog nodes help secure communication [193].

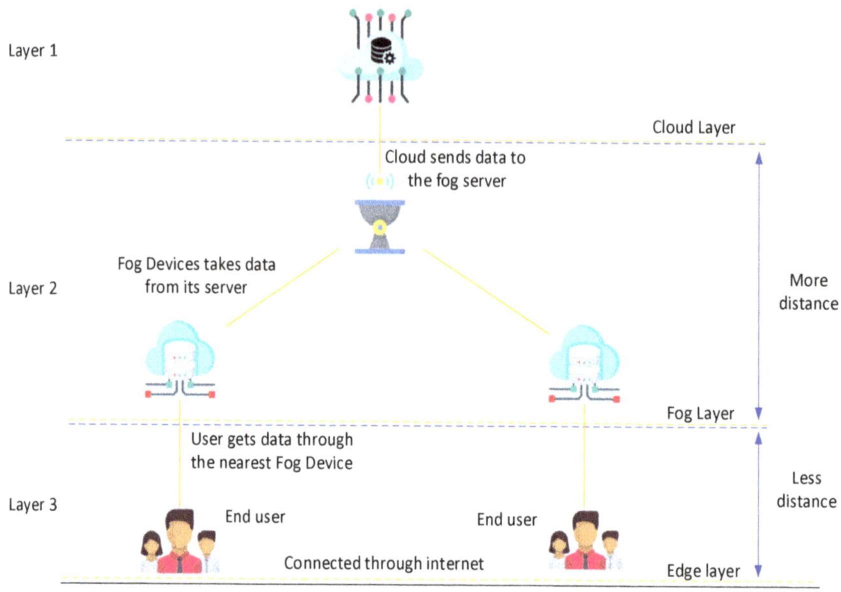

Figure 13. An elementary overview of FC.

FC can provide some solutions to counteract certain security threats and attacks as discussed in the earlier section. More details are provided below to demonstrate how FC can counteract these threats.

- Incident response services: Some critical applications cannot be stopped completely to resolve malware issues. When the system is running, fog nodes can help with such resolutions. It is possible to program fog nodes to provide incident response services in real time. As soon as the fog nodes detect suspicious data or requests, they can generate a warning flag for the end-user or the IoT system. Using FC, malware can be detected and problems resolved in transit. Some of the real-time services include identity recognition, intrusion detection, access management, etc.
- Resource-constraint issues: IoT devices are typically resource-constrained, which makes them an ideal target for attackers. By damaging edge devices, attackers try to exploit weak points and worm their way in. Fog nodes can support edge devices so those devices will not be attacked. For protection, fog nodes can provide more sophisticated security functions, as well.
- Eavesdropping: Rather than routing the information throughout the network, usage of fog nodes enables communication with only the end-user and fog nodes. Because the network traffic is reduced, there is less opportunity for adversaries to eavesdrop.
- Data transit attacks: Data management and storage are much more efficient when using secure fog nodes instead of IoT devices. Fog nodes provide a greater level of protection for data than end-user devices for storing data.
- Man-in-the-middle attack: A fog serves as a layer of security between the cloud and the end-user. A fog layer stands in between all threats or attacks on IoT systems, and in this layer, unusual activity can be identified and mitigated before it reaches the system.

7.3. EC for IoT

Both FC and EC share similar responsibilities, such as reducing latency, reducing the volume of the data sent to the cloud, enhancing computational efficacy, incorporating heterogeneity, etc., with a common objective to bring intelligence and computing possibly as close as to the data source. However, they are not the same. They differ in the way they operate and handle the data. For example, usually, FC takes place on the devices to which sensors are connected, such as switches, routers, gateways, access points, etc. On the other hand, EC takes place at the sensors themselves or devices which are at a one-hop distance from the sensor. Thus, the FC nodes are at more distance than the EC nodes.

Contrary to the EC, the data are transmitted from sensors to the FC nodes for processing and then sent back to the edge nodes for appropriate actions. Nevertheless, EC and FC are widely used by many companies as an extension of cloud computing. The main difference between cloud, fog, and edge stems from the location where intelligence and power computation are conducted. In the cloud, more data are processed, and users are comparatively located at a greater distance, requiring a much higher level of data processing [194]. EC uses a small edge server to overcome the problems associated with cloud computing, placed between the user and the cloud.

Figure 14 shows the EC architecture's device components, which include edge devices, fog nodes, and cloud data centers [195]. The processing power and analytical capability are provided at the edge itself in an EC framework. An application comprises devices that communicate among themselves and collaborate to calculate data [196]. The IoT application can then minimize the amount of data sent to the outside, whether to cloud or fog nodes, and this will improve the application's security.EC reduces communication costs, as all the data do not have to be moved to the cloud.

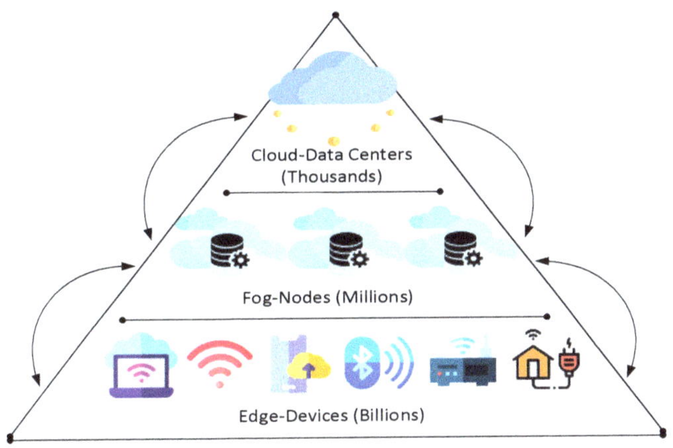

Figure 14. An elementary architecture of EC.

Looking at the threats and attacks causing serious security concerns to the IoT system, the following are possible solutions that can be achieved by incorporating EC with IoT.

- Data Compliance Issues: Data movement outside of borders is prohibited by many countries due to their restrictive regulatory acts, such as the GDPR (General Data Protection Regulation) of the European Union. Data sovereignty laws can be followed through EC, which keeps the data inside organizations [197]
- Data Breaches: Data are stored and processed entirely within local networks or devices in EC. In this case, no data are transferred from the source to the processor. Therefore, there is no risk of data theft or data breaches since the data are not in transit [198].
- Bandwidth Issues: Most of the data generated by IoT applications are raw and relatively of low value. As well as having a high bandwidth cost, the process of moving all the data to the cloud is also very hard in terms of security. The use of EC can enable data processing to be conducted at the edge nodes rather than sending the data to a cloud service [199].

- Safety issues: Physical safety can be compromised even if there is just a slight delay in responses. In the case of sensors that send all of their data and wait for the cloud to act, it may be too late to prevent injuries or deaths. Therefore, to achieve faster responses, devices can be deployed with EC to examine the abnormalities, process the data, and send them to the data center.

7.4. ML for IoT

In recent years, the field of ML has been of major interest. For their development, many domains use ML, and it is also used for IoT security. ML seems to be an excellent way of protecting IoT devices against cyber assaults by offering an approach other than traditional methods to defend against attacks. ML refers to intelligent approaches that use example data or previous experience through learning to optimize performance criteria. Different ML algorithms have been developed to provide some non-traditional solutions to these challenges.

The basic requirement in IoT is the securing of all network-connected systems and devices. The role of ML is to use, train, and prevent data loss in IoT equipment to detect anomalies or to detect any unwanted activity in IoT systems. Consequently, ML provides a promising platform to overcome the problems in securing IoT devices.

Looking at the threats and attacks causing serious security concerns to the IoT system, the following are possible solutions that can be achieved by incorporating ML with IoT.

- DoS Attack: DoS attacks on IoT or IoT devices are a major concern. A multilayer perceptron (MLP) protocol to secure networks against DoS attacks serves as an approach for preventing such attacks [200]. Pavani, K. et al. proposed to create an MLP to improve the safety of wireless networks through particle swarm optimization and a backpropagation algorithm [201]. ML technologies help increase the accuracy of deductions and secure IoT devices vulnerable to DoS attacks.
- Eavesdropping: Attackers can sweep messages while data are being transmitted. ML techniques such as Q learning-based offloading strategy [202] or Bayesian non-parametric techniques [203] can be used to protect against such attacks. ML techniques such as Q-learn and Dyna-Q can be used to protect devices from eavesdropping, as well. Experimental evaluation and strengthening education of those schemes are presented in [204].
- Digital Fingerprinting: Digital fingerprinting is a promising solution for safe IoT systems and for the end-user to have enough confidence in applications. Digital fingerprints are widely used for smartphones, payments, car and home doors, etc. Digital fingerprinting is a dominant bio-metric identification method thanks to its low cost, reliability, acceptability, and high level of safety [205]. Aside from the advantages of digital fingerprinting, the efficiency of using this technology in IoT is varied, including fingerprint classification, improved image, and functional matching.

So far, from the discussion, it can be inferred that there is a huge potential for security enhancement in IoT using burgeoning technologies such as BC, FC, EC, and ML. The scope of possible security enhancements in IoT through the integration of these ubiquitous burgeoning technologies sprawling the appropriate layers is summarized in Table 5. Some of the research papers in literature focusing on security solutions in different capacities covering various aspects of IoT based on BC, FC, EC, and ML are shown in Figure 15. In this figure, three applications of IoT are considered, namely, healthcare, smart devices, and smart grid, for which some of the papers are presented from the literature which covers the security solutions in different capacities based on BC, FC, EC, and ML.

Table 5. Scope of security enhancement in IoT using burgeoning technologies.

Burgeoning Technology	Scope of Security Enhancement in IoT
BC	Due to its key operational characteristics such as decentralized behavior, encryption-based communication, distributed functionality, inbuilt cryptography, authenticated access, etc., it offers security solutions against several threats and attacks across multiple layers of the IoT such as malicious data injection, disclosure of critical information, device compromise, node cloning, tag cloning, exhaustion attack, illegal access, information theft, spoofing, data manipulation, false data injection, unfair access, session hijacking, unfairness attack, fake information, unauthorized access, stealing users critical information, illegal interventions, software modification, message forging, brute force attack, etc.
FC	With abundant capabilities in processing, storing, managing the voluminous data, it offers security solutions against various threats and attacks such as eavesdropping, hardware failure, disclosure of critical information, device compromise, node capture attack, node tampering, battery drainages attack, node replication, illegal access, MITM, information theft, data manipulation, DoS, DDoS, false data injection, session hijacking, malicious attack, data integrity, virtualization, illegal access, cloud malware injection, illegal intervention, etc.
EC	The real-time services such as identity recognition, intrusion detection, access management, etc., enable EC to enhance security against several threats and attacks such as eavesdropping, battery drainage, hardware failure, node capture, DoS, DDoS, jamming, malicious attack, SQL injection, data integrity, virtualization, illegal access, flooding attack in the cloud, access attack, signature wrapping, etc.
ML	With enormous success in the paradigm of speech recognition, fraud detection, computer vision, spam detection, computer networks, etc., it is envisaged to solve several threats and attacks persisting to IoT. Some of these include device compromise, Sybil threat, node cloning, node capture, RF jamming, battery drainage attack, node replication, MITM, information theft, threats to communication protocols, DoS, DDoS, SCA, hello flood, congestion, MQTT-exploit, hole attack, firmware attack, illegal intervention, SQL injection, cross-site script, intersection, etc.

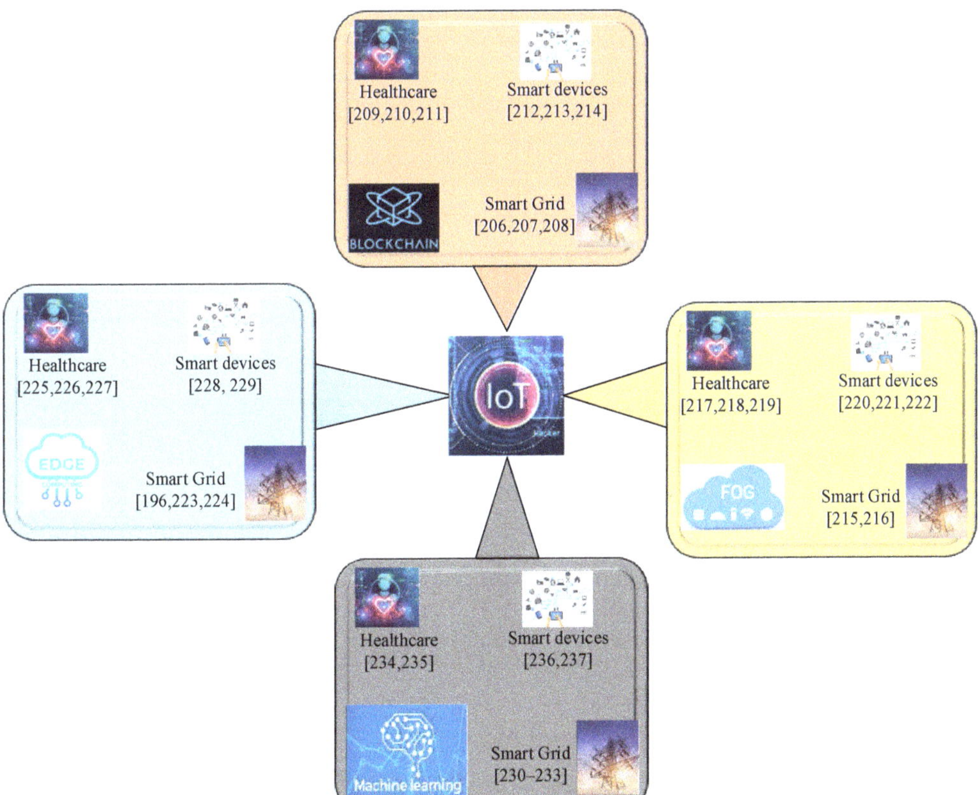

Figure 15. Some of the application domains of IoT and related work focusing scope for the security enhancement using burgeoning technologies [196,206–237].

8. Open Research Problems

Despite a successful journey so far, the IoT has many technological challenges and research issues that are yet to be explored. Some of the prominent research challenges are enumerated below.

- There is no generic validated architecture of IoT so far, i.e., most of the architectures are either domain-specific or application-specific. Thus, the security enhancement methodology may not fit the most generic architecture.
- The detailed protocols stack and its interoperability are still being explored. Due to immaturity, the exhaustive security aspects for protocols and the standards are far behind the actual realization.
- The amount of abstraction in security, the formal language to be utilized for policy encoding, and contextual IoT features to be considered for optimal usage of software-defined networking-based security mechanisms in a secure IoT framework is still an open problem.
- Given the inherent trade-off between flexibility, performance, and cost, the granularity of protection methods poses an open challenge in the provision of network slices specialized for IoT applications.
- The implementation methods and utilization of software and hardware are critical factors in BC to enhance security in IoT using BC. Being public in nature, the transactions of IoT data are still a problem of security concern that can be revealed to the public in general [238].
- Since FC is an extension to cloud computing, some of the serious concerns of cloud computing such as security and privacy are inherent to the FC, which are being extensively explored in the literature.
- EC poses serious security and privacy concerns since most of the computations are generally performed at the edge devices. However, most of the edge devices are resource-constrained in an IoT system, which may not be able to compute, analyze, and process the data securely.
- There are enormous ML algorithms. The selection of suitable algorithms is of vital importance, because choosing the incorrect algorithm will result in "garbage" output and a loss of effort, effectiveness, and accuracy. Similarly, selecting the incorrect data set will result in "garbage" input and inaccurate results. Thus, the correct data sets and appropriate algorithms are critically important, which can be explored for securing IoT environment using machine learning.
- The systematic review on vulnerabilities of BC, FC, EC, and ML and their mapping with impact on IoT are some of the research problems which can be further explored.
- The optimum resource sharing in FC is another research area that can be explored to avoid the burden on the cloud for processing the voluminous data during heavy traffic conditions.
- Further, the scope of other technologies such as artificial intelligence, big data, etc., must be also analyzed, which were shown to have great potential in IoT-based applications [239].

9. Conclusions

The introduction of smart computing devices using IoT has made day-to-day lives more convenient. Data analytics, automation, and smart devices have all benefited from the introduction of IoT into human life. Nevertheless, the unprecedented growth in IoT has also been crippled with many vulnerabilities and challenges. Further, the IoT's heterogeneous design expands the attack surface and adds new challenges to an already vulnerable IoT network. The successful compromise of the system's security may have fatal consequences for users. The overall security of the device must be considered to ensure that critical vulnerabilities are mitigated. Policies and protocols must be enforced as much as possible to deter threats and attacks. In this paper, we have presented a most comprehensive survey on IoT from the perspective of security threats and attacks. Further, modern threats and attacks on the emerging IoT infrastructure, security flaws, and countermeasures are discussed in this paper. In addition, a roadmap of using ubiquitous technologies, viz., BC, FC, EC, and ML, for enhancing security in IoT are comprehensively discussed in this paper.

However, due to IoT devices' heterogeneous existence and limitations, any resolution would be ineffective and obsolete. Consequently, due to the evolving nature of technology, it is estimated that more countermeasures and vulnerabilities will be revealed in the near future. As future work, the authors are working on ML and IoT integration to enhance IoT-based applications' security under dynamically varying conditions.

Author Contributions: Conceptualization, A.V.J. and B.A.; methodology, A.V.J.; software, R.R.K.; validation, A.P., R.R.K. and A.V.J.; formal analysis, A.V.J. and A.S.; investigation, R.R.K.; resources, R.R.K.; data curation, A.P. and A.S.; writing—original draft preparation, R.R.K.; writing—review and editing, A.V.J. and N.B.; visualization, N.B.; supervision, B.A. and N.B.; project administration, A.V.J., B.A. and N.B.; funding acquisition, N.B. All authors have read and agreed to the published version of the manuscript.

Funding: There is no funding available for this research.

Institutional Review Board Statement: Not applicable.

Informed Consent Statement: Not applicable.

Conflicts of Interest: The authors declare no conflict of interest.

References

1. Jha, A.V.; Appasani, B.; Ghazali, A.N. Performance Evaluation of Network Layer Routing Protocols on Wireless Sensor Networks. In Proceedings of the 2019 International Conference on Communication and Electronics Systems (ICCES), Coimbatore, India, 17–19 July 2019; pp. 1862–1865. [CrossRef]
2. Tiwary, A.; Mahato, M.; Chidar, A.; Chandrol, M.K.; Shrivastava, M.; Tripathi, M. Internet of Things (IoT): Research, architectures and applications. *Int. J. Future Revolut. Comput. Sci. Commun. Eng.* **2018**, *4*, 23–27.
3. González-Zamar, M.D.; Abad-Segura, E.; Vázquez-Cano, E.; López-Meneses, E. IoT Technology Applications-Based Smart Cities: Research Analysis. *Electronics* **2020**, *9*, 1246. [CrossRef]
4. Internet of Things in Healthcare: Applications, Benefits, and Challenges. Available online: https://www.peerbits.com/blog/internet-of-things-healthcare-applications-benefits-and-challenges.html (accessed on 12 April 2021).
5. Cvar, N.; Trilar, J.; Kos, A.; Volk, M.; Stojmenova Duh, E. The Use of IoT Technology in Smart Cities and Smart Villages: Similarities, Differences, and Future Prospects. *Sensors* **2020**, *20*, 3897. [CrossRef] [PubMed]
6. Ryan, P.J.; Watson, R.B. Research Challenges for the Internet of Things: What Role Can OR Play? *Systems* **2017**, *5*, 24. [CrossRef]
7. Jha, A.V.; Appasani, B.; Ghazali, A.N.; Pattanayak, P.; Gurjar, D.S.; Kabalci, E.; Mohanta, D.K. Smart grid cyber-physical systems: Communication technologies, standards and challenges. *Wirel. Netw.* **2021**, *27*, 2595–2613. [CrossRef]
8. Jha, A.V.; Mishra, S.K.; Appasani, B.; Ghazali, A.N. Communication Networks for Metropolitan E-Health Applications. *IEEE Potentials* **2021**, *40*, 34–42. [CrossRef]
9. Lin, H.; Bergmann, N.W. IoT Privacy and Security Challenges for Smart Home Environments. *Information* **2016**, *7*, 44. [CrossRef]
10. Rajendran, G.; Nivash, R.S.R.; Parthy, P.P.; Balamurugan, S. Modern security threats in the Internet of Things (IoT): Attacks and Countermeasures. In Proceedings of the International Carnahan Conference on Security Technology (ICCST), Chennai, India, 1–3 October 2019; pp. 1–6. [CrossRef]
11. Chen, L.; Thombre, S.; Järvinen, K.; Lohan, E.S.; Alén-Savikko, A.; Leppäkoski, H.; Bhuiyan, M.Z.H.; Bu-Pasha, S.; Ferrara, G.N.; Honkala, S.; et al. Robustness, security and privacy in location-based services for future IoT: A survey. *IEEE Access* **2017**, *5*, 8956–8977. [CrossRef]
12. Shin, H.; Lee, H.K.; Cha, H.Y.; Heo, S.W.; Kim, H. IoT security issues and light weight block cipher. In Proceedings of the International Conference on Artificial Intelligence in Information and Communication (ICAIIC), Okinawa, Japan, 11–13 February 2019; pp. 381–384.
13. Gamundani, A.M. An impact review on internet of things attacks. In Proceedings of the International Conference on Emerging Trends in Networks and Computer Communications (ETNCC), Windhoek, Namibia, 17–20 May 2015; pp. 114–118. [CrossRef]
14. Kumar, N.; Madhuri, J.; Channe Gowda, M. Review on security and privacy concerns in Internet of Things. In Proceedings of the International Conference on IoT and Application (ICIOT), Nagapattinam, India, 19–20 May 2017; pp. 1–5. [CrossRef]
15. Khan, M.A.; Salah, K. IoT security: Review, blockchain solutions, and open challenges. *Future Gener. Comput. Syst.* **2018**, *82*, 395–411. [CrossRef]
16. Alaba, F.A.; Othman, M.; Hashem, I.A.T.; Alotaibi, F. Internet of Things security: A survey. *J. Netw. Comput. Appl.* **2017**, *88*, 10–28. [CrossRef]

17. Abomhara, M.; Køien, G.M. Cyber security and the internet of things: Vulnerabilities, threats, intruders and attacks. *J. Cyber Secur. Mobil.* **2015**, *4*, 65–88. [CrossRef]
18. Kozlov, D.; Veijalainen, J.; Ali, Y. Security and privacy threats in IoT architectures. *BODYNETS* **2012**, 256–262. [CrossRef]
19. Lee, E.; Seo, Y.D.; Oh, S.R.; Kim, Y.G. A Survey on Standards for Interoperability and Security in the Internet of Things. *IEEE Commun. Surv. Tutor.* **2021**, *23*, 1020–1047. [CrossRef]
20. Abosata, N.; Al-Rubaye, S.; Inalhan, G.; Emmanouilidis, C. Internet of Things for System Integrity: A Comprehensive Survey on Security, Attacks and Countermeasures for Industrial Applications. *Sensors* **2021**, *21*, 3654. [CrossRef]
21. Mann, P.; Tyagi, N.; Gautam, S.; Rana, A. Classification of Various Types of Attacks in IoT Environment. In Proceedings of the 12th International Conference on Computational Intelligence and Communication Networks (CICN), Bhimtal, India, 3 November 2020; pp. 346–350. [CrossRef]
22. Ogonji, M.M.; Okeyo, G.; Wafula, J.M. A survey on privacy and security of Internet of Things. *Comput. Sci. Rev.* **2020**, *38*, 10031. [CrossRef]
23. Zarpelão, B.B.; Miani, R.S.; Kawakani, C.T.; de Alvarenga, S.C. A survey of intrusion detection in Internet of Things. *J. Netw. Comput. Appl.* **2017**, *84*, 25–37. [CrossRef]
24. Hajiheidari, S.; Wakil, K.; Badri, M.; Navimipour, N.J. Intrusion Detection Systems in the Internet of Things: A Comprehensive Investigation. *Comput. Netw.* **2019**, *160*, 165–191. [CrossRef]
25. Kouicem, D.E.; Bouabdallah, A.; Lakhlef, H. Internet of things security: A top-down survey. *Comput. Netw.* **2018**, *141*, 199–221. [CrossRef]
26. Sun, L.; Du, Q. A Review of Physical Layer Security Techniques for Internet of Things: Challenges and Solutions. *Entropy* **2018**, *20*, 730. [CrossRef]
27. Elazhary, H. Internet of Things (IoT), mobile cloud, cloudlet, mobile IoT, IoT cloud, fog, mobile edge, and edge emerging computing paradigms: Disambiguation and research directions. *J. Netw. Comput. Appl.* **2019**, *128*, 105–140. [CrossRef]
28. Taylor, P.J.; Dargahi, T.; Dehghantanha, A.; Parizi, R.M.; Choo, K.K.R. A systematic literature review of blockchain cyber security. *Digit. Commun. Netw.* **2020**, *6*, 147–156. [CrossRef]
29. Memon, R.A.; Li, J.P.; Ahmed, J.; Nazeer, M.I.; Ismail, M.; Ali, K. Cloud-based vs. blockchain-based IoT: A comparative survey and way forward. *Front. Inform. Technol. Electron. Eng.* **2020**, *21*, 563–586. [CrossRef]
30. Tran, N.K.; Babar, M.A.; Boan, J. Integrating blockchain and Internet of Things systems: A systematic review on objectives and designs. *J. Netw. Comput. Appl.* **2020**, *173*, 102844. [CrossRef]
31. Fersi, G. Fog computing and Internet of Things in one building block: A survey and an overview of interacting technologies. *Cluster Comput.* **2021**, 1–31. [CrossRef]
32. Atlam, H.F.; Walters, R.J.; Wills, G.B. Fog Computing and the Internet of Things: A Review. *Big Data Cogn. Comput.* **2018**, *2*, 10. [CrossRef]
33. Hamdan, S.; Ayyash, M.; Almajali, S. Edge-Computing Architectures for Internet of Things Applications: A Survey. *Sensors* **2020**, *20*, 6441. [CrossRef]
34. Capra, M.; Peloso, R.; Masera, G.; Ruo Roch, M.; Martina, M. Edge Computing: A Survey on the Hardware Requirements in the Internet of Things World. *Future Internet* **2019**, *11*, 100. [CrossRef]
35. Ashouri, M.; Lorig, F.; Davidsson, P.; Spalazzese, R. Edge Computing Simulators for IoT System Design: An Analysis of Qualities and Metrics. *Future Internet* **2019**, *11*, 235. [CrossRef]
36. Amiri-Zarandi, M.; Dara, R.A.; Fraser, E. A survey of machine learning-based solutions to protect privacy in the Internet of Things. *Comput. Secur.* **2020**, *96*, 101921. [CrossRef]
37. Adnan, A.; Muhammed, A.; Abd Ghani, A.A.; Abdullah, A.; Hakim, F. An Intrusion Detection System for the Internet of Things Based on Machine Learning: Review and Challenges. *Symmetry* **2021**, *13*, 1011. [CrossRef]
38. Alwarafy, A.; Al-Thelaya, K.A.; Abdallah, M.; Schneider, J.; Hamdi, M. A Survey on Security and Privacy Issues in Edge-Computing-Assisted Internet of Things. *IEEE Internet Things J.* **2020**, *8*, 4004–4022. [CrossRef]
39. Qiu, J.; Tian, Z.; Du, C.; Zuo, Q.; Su, S.; Fang, B. A survey on access control in the age of internet of things. *IEEE Internet Things J.* **2020**, *7*, 4682–4696. [CrossRef]
40. Parmar, M.S.; Shah, P.P. Uplifting Blockchain Technology for Data Provenance in Supply Chain. *Int. J. Adv. Sci. Technol.* **2020**, *29*, 5922–5938. [CrossRef]
41. Granjal, J.; Monteiro, E.; Silva, J.S. Security for the internet of things: A survey of existing protocols and open research issues. *IEEE Commun. Surv. Tutor.* **2015**, *17*, 1294–1312. [CrossRef]
42. Ferrag, M.A.; Derdour, M.; Mukherjee, M.; Derhab, A.; Maglaras, L.; Janicke, H. Blockchain technologies for the internet of things: Research issues and challenges. *IEEE Internet Things J.* **2018**, *6*, 2188–2204. [CrossRef]
43. Yang, Y.; Wu, L.; Yin, G.; Li, L.; Zhao, H. A survey on security and privacy issues in Internet-of-Things. *IEEE Internet Things J.* **2017**, *4*, 1250–1258. [CrossRef]
44. Lin, J.; Yu, W.; Zhang, N.; Yang, X.; Zhang, H.; Zhao, W. A survey on internet of things: Architecture, enabling technologies, security and privacy, and applications. *IEEE Internet Things J.* **2017**, *4*, 1125–1142. [CrossRef]
45. Malik, M.; Dutta, M.; Granjal, J. A survey of key bootstrapping protocols based on public key cryptography in the Internet of Things. *IEEE Access* **2019**, *7*, 27443–27464. [CrossRef]

46. Alshehri, F.; Muhammad, G. A comprehensive survey of the Internet of Things (IoT) and AI-based smart healthcare. *IEEE Access* **2021**, *9*, 3660–3678. [CrossRef]
47. Stoyanova, M.; Nikoloudakis, Y.; Panagiotakis, S.; Pallis, E.; Markakis, E.K. A survey on the internet of things (IoT) forensics: Challenges, approaches, and open issues. *IEEE Commun. Surv. Tutor.* **2020**, *22*, 1191–1221. [CrossRef]
48. Viriyasitavat, W.; Da Xu, L.; Bi, Z.; Hoonsopon, D. Blockchain technology for applications in internet of things—mapping from system design perspective. *IEEE Internet Things J.* **2019**, *6*, 8155–8168. [CrossRef]
49. Butun, I.; Österberg, P.; Song, H. Security of the Internet of Things: Vulnerabilities, attacks, and countermeasures. *IEEE Commun. Surv. Tutor.* **2019**, *22*, 616–644. [CrossRef]
50. Cha, S.C.; Hsu, T.Y.; Xiang, Y.; Yeh, K.H. Privacy enhancing technologies in the Internet of Things: Perspectives and challenges. *IEEE Internet Things J.* **2018**, *6*, 2159–2187. [CrossRef]
51. Al-Garadi, M.A.; Mohamed, A.; Al-Ali, A.K.; Du, X.; Ali, I.; Guizani, M. A survey of machine and deep learning methods for internet of things (IoT) security. *IEEE Commun. Surv. Tutor.* **2020**, *22*, 1646–1685. [CrossRef]
52. Mosenia, A.; Jha, N.K. A comprehensive study of security of internet-of-things. *IEEE Trans. Emerg. Top. Comput.* **2016**, *5*, 586–602. [CrossRef]
53. Neshenko, N.; Bou-Harb, E.; Crichigno, J.; Kaddoum, G.; Ghani, N. Demystifying IoT security: An exhaustive survey on IoT vulnerabilities and a first empirical look on Internet-scale IoT exploitations. *IEEE Commun. Surv. Tutor.* **2019**, *21*, 2702–2733. [CrossRef]
54. Nguyen, D.C.; Pathirana, P.N.; Ding, M.; Seneviratne, A. Integration of blockchain and cloud of things: Architecture, applications and challenges. *IEEE Commun. Surv. Tutor.* **2020**, *22*, 2521–2549. [CrossRef]
55. Serror, M.; Hack, S.; Henze, M.; Schuba, M.; Wehrle, K. Challenges and opportunities in securing the industrial internet of things. *IEEE Trans. Ind. Inform.* **2020**, *17*, 2985–2996. [CrossRef]
56. Ngu, A.H.; Gutierrez, M.; Metsis, V.; Nepal, S.; Sheng, Q.Z. IoT middleware: A survey on issues and enabling technologies. *IEEE Internet Things J.* **2016**, *4*, 1–20. [CrossRef]
57. Makhdoom, I.; Abolhasan, M.; Lipman, J.; Liu, R.P.; Ni, W. Anatomy of threats to the internet of things. *IEEE Commun. Surv. Tutor.* **2018**, *21*, 1636–1675. [CrossRef]
58. Qiu, T.; Chi, J.; Zhou, X.; Ning, Z.; Atiquzzaman, M.; Wu, D.O. Edge computing in industrial internet of things: Architecture, advances and challenges. *IEEE Commun. Surv. Tutor.* **2020**, *22*, 2462–2488. [CrossRef]
59. Hamad, S.A.; Sheng, Q.Z.; Zhang, W.E.; Nepal, S. Realizing an internet of secure things: A survey on issues and enabling technologies. *IEEE Commun. Surv. Tutor.* **2020**, *22*, 1372–1391. [CrossRef]
60. Ali, M.S.; Vecchio, M.; Pincheira, M.; Dolui, K.; Antonelli, F.; Rehmani, M.H. Applications of blockchains in the Internet of Things: A comprehensive survey. *IEEE Commun. Surv. Tutor.* **2018**, *21*, 1676–1717. [CrossRef]
61. Meneghello, F.; Calore, M.; Zucchetto, D.; Polese, M.; Zanella, A. IoT: Internet of threats? A survey of practical security vulnerabilities in real IoT devices. *IEEE Internet Things J.* **2019**, *6*, 8182–8201. [CrossRef]
62. Chaabouni, N.; Mosbah, M.; Zemmari, A.; Sauvignac, C.; Faruki, P. Network intrusion detection for IoT security based on learning techniques. *IEEE Commun. Surv. Tutor.* **2019**, *21*, 2671–2701. [CrossRef]
63. Yu, W.; Liang, F.; He, X.; Hatcher, W.G.; Lu, C.; Lin, J.; Yang, X. A survey on the edge computing for the Internet of Things. *IEEE Access* **2017**, *6*, 6900–6919. [CrossRef]
64. Ni, J.; Zhang, K.; Lin, X.; Shen, X. Securing fog computing for internet of things applications: Challenges and solutions. *IEEE Commun. Surv. Tutor.* **2017**, *20*, 601–628. [CrossRef]
65. Restuccia, F.; D'Oro, S.; Melodia, T. Securing the internet of things in the age of machine learning and software-defined networking. *IEEE Internet Things J.* **2018**, *5*, 4829–4842. [CrossRef]
66. Omoniwa, B.; Hussain, R.; Javed, M.A.; Bouk, S.H.; Malik, S.A. Fog/edge computing-based IoT (FECIoT): Architecture, applications, and research issues. *IEEE Internet Things J.* **2018**, *6*, 4118–4149. [CrossRef]
67. Khanam, S.; Ahmedy, I.B.; Idris, M.Y.I.; Jaward, M.H.; Sabri, A.Q.B.M. A survey of security challenges, attacks taxonomy and advanced countermeasures in the internet of things. *IEEE Access* **2020**, *8*, 219709–219743. [CrossRef]
68. Alotaibi, B. Utilizing blockchain to overcome cyber security concerns in the internet of things: A review. *IEEE Sens. J.* **2019**, *19*, 10953–10971. [CrossRef]
69. Hassija, V.; Chamola, V.; Saxena, V.; Jain, D.; Goyal, P.; Sikdar, B. A survey on IoT security: Application areas, security threats, and solution architectures. *IEEE Access* **2019**, *7*, 82721–82743. [CrossRef]
70. Hussain, F.; Hussain, R.; Hassan, S.A.; Hossain, E. Machine learning in IoT security: Current solutions and future challenges. *IEEE Commun. Surv. Tutor.* **2020**, *22*, 1686–1721. [CrossRef]
71. Benkhelifa, E.; Welsh, T.; Hamouda, W. A Critical Review of Practices and Challenges in Intrusion Detection Systems for IoT: Toward Universal and Resilient Systems. *IEEE Commun. Surv. Tutor.* **2018**, *20*, 3496–3509. [CrossRef]
72. Sengupta, J.; Ruj, S.; Bit, S.D. A Comprehensive Survey on Attacks, Security Issues and Blockchain Solutions for IoT and IIoT. *J. Netw. Comput. Appl.* **2020**, *149*, 102481. [CrossRef]
73. Yugha, R.; Chithra, S. A survey on technologies and security protocols: Reference for future generation IoT. *J. Netw. Comput. Appl.* **2020**, *169*, 102763. [CrossRef]
74. Bhoyar, P.; Sahare, P.; Dhok, S.B.; Deshmukh, R.B. Communication technologies and security challenges for internet of things: A comprehensive review. *AEU-Int. J. Electron. Commun.* **2019**, *99*, 81–99. [CrossRef]

75. Bellavista, P.; Berrocal, J.; Corradi, A.; Das, S.K.; Foschini, L.; Zanni, A. A survey on fog computing for the Internet of Things. *Pervasive Mob. Comput.* **2019**, *52*, 71–99. [CrossRef]
76. Peña-López, I. ITU Internet Report 2005: The Internet of Things. Available online: https://www.comminit.com/global/content/itu-internet-reports-2005-internet-things (accessed on 12 April 2021).
77. Sikder, A.K.; Petracca, G.; Aksu, H.; Jaeger, T.; Uluagac, A.S. A survey on sensor-based threats to internet-of-things (IoT) devices and applications. *arXiv* **2018**, arXiv:1802.02041v1.
78. Hongsong, C.; Zhongchuan, F.; Dongyan, Z. Security and trust research in m2m system. In Proceedings of the 2011 IEEE International Conference on Vehicular Electronics and Safety, Beijing, China, 10–12 July 2011; pp. 286–290. [CrossRef]
79. Kumar, S.A.; Vealey, T.; Srivastava, H. Security in internet of things: Challenges, solutions and future directions. In Proceedings of the 49th Hawaii International Conference on System Sciences (HICSS), Koloa, HI, USA, 5–8 January 2016; pp. 5772–5781. [CrossRef]
80. Lee, S.K.; Bae, M.; Kim, H. Future of IoT networks: A survey. *Appl. Sci.* **2017**, *7*, 1072. [CrossRef]
81. Chen, H.; Jia, X.; Li, H. A Brief Introduction to IoT Gateway. In Proceedings of the IET International Conference on Communication Technology and Application (ICCTA), Beijing, China, 14–16 October 2011; pp. 1–4. [CrossRef]
82. Tan, H.; Tsudik, G.; Jha, S. MTRA: Multi-Tier randomized remote attestation in IoT networks. *Comput. Secur.* **2019**, *81*, 78–93. [CrossRef]
83. Internet of Things Challenges in Storage and Data. Available online: https://www.computerweekly.com/news/252450705/Internet-of-things-challenges-in-storage-and-data (accessed on 25 April 2021).
84. 12 Benefits of Cloud Computing. Available online: https://www.salesforce.com/in/products/platform/best-practices/benefits-of-cloud-computing/ (accessed on 25 April 2021).
85. Li, X.; Wang, Q.; Lan, X.; Chen, X.; Zhang, N.; Chen, D. Enhancing Cloud-Based IoT Security Through Trustworthy Cloud Service: An Integration of Security and Reputation Approach. *IEEE Access* **2019**, *7*, 9368–9383. [CrossRef]
86. Kepçeoğlu, B.; Murzaeva, A.; Demirci, S. Performing energy consuming attacks on IoT devices. In Proceedings of the 27th Telecommunications Forum (TELFOR), Belgrade, Serbia, 26–27 November 2019; pp. 1–4. [CrossRef]
87. Bilal, M. A Review of Internet of Things Architecture, Technologies and Analysis Smartphone-based Attacks Against 3D printers. *arXiv* **2017**, arXiv:1708.04560. Available online: https://arxiv.org/abs/1708.04560 (accessed on 25 April 2021).
88. Dodig, I.; Cafuta, D.; Kramberger, T.; Cesar, I. A Novel Software Architecture Solution with a Focus on Long-Term IoT Device Security Support. *Appl. Sci.* **2021**, *11*, 4955. [CrossRef]
89. Capella, J.V.; Campelo, J.C.; Bonastre, A.; Ors, R. A Reference Model for Monitoring IoT WSN-Based Applications. *Sensors* **2016**, *16*, 1816. [CrossRef]
90. Sadiku, M.N.; Tembely, M.; Musa, S.M. Home area networks: A primer. *Int. J.* **2017**, *7*, 208. [CrossRef]
91. Atzori, L.; Iera, A.; Morabito, G. The Internet of Things: A survey. *Comput. Netw.* **2010**, *54*, 2787–2805. [CrossRef]
92. Swamy, S.N.; Kota, S.R. An Empirical Study on System Level Aspects of Internet of Things (IoT). *IEEE Access* **2020**, *8*, 188082–188134. [CrossRef]
93. Lombardi, M.; Pascale, F.; Santaniello, D. Internet of Things: A General Overview between Architectures, Protocols and Applications. *Information* **2021**, *12*, 87. [CrossRef]
94. Al-Fuqaha, A.; Guizani, M.; Mohammadi, M.; Aledhari, M.; Ayyash, M. Internet of Things: A Survey on Enabling Technologies, Protocols, and Applications. *IEEE Commun. Surv. Tutor.* **2015**, *17*, 2347–2376. [CrossRef]
95. Shelby, Z.; Hartke, K.; Bormann, C.; Frank, B. Constrained Application Protocol (CoAP), Draft-Ietf-Corecoap-18, Work in Progress. sl: IETF. 2013. Available online: http://tools.ietf.org/html/draft-ietf-corecoap-18 (accessed on 1 July 2021).
96. IoT Standards and Protocols Guide—Protocols of the Internet of Things. Available online: https://www.avsystem.com/blog/iot-protocols-and-standards/ (accessed on 25 April 2021).
97. Bormann, C.; Castellani, A.P.; Shelby, Z. Coap: An application protocol for billions of tiny internet nodes. *IEEE Internet Comput.* **2012**, *16*, 62–67. [CrossRef]
98. Cheshire, S.; Krochmal, M. Multicast DNS. *RFC* **2013**, *6762*. Available online: https://www.rfc-editor.org/info/rfc6762 (accessed on 1 July 2021). [CrossRef]
99. Vasseur, J.; Agarwal, N.; Hui, J.; Shelby, Z.; Bertrand, P.; Chauvenet, C. RPL: The IP routing protocol designed for low power and lossy networks. *IPSO Alliance* **2011**, 1–20. Available online: http://www.cse.chalmers.se/edu/year/2019/course/DAT300/PAPERS/rpl.pdf (accessed on 1 July 2021).
100. Winter, T.; Thubert, P.; Brandt, A.; Hui, J.W.; Kelsey, R.; Levis, P.; Alexander, R.K. RPL: IPv6 routing protocol for low-power and lossy networks. *RFC* **2012**, *6550*, 1–157. Available online: https://datatracker.ietf.org/doc/html/rfc6550 (accessed on 1 July 2021).
101. Yang, Z.; Yue, Y.; Yang, Y.; Peng, Y.; Wang, Z.; Liu, W. Study and Application on the Architecture and Key Technologies for IOT. In Proceedings of the International Conference on Multimedia Technology, Hangzhou, China, 26–28 July 2011; pp. 747–751. [CrossRef]
102. Palattella, M.R.; Accettura, N.; Vilajosana, X.; Watteyne, T.; Grieco, L.A.; Boggia, G.; Dohler, M. Standardized protocol stack for the internet of (important) things. *IEEE Commun. Surv. Tutor.* **2012**, *15*, 1389–1406. [CrossRef]

103. IEEE 802 Working Group. *IEEE Standard for Local and Metropolitan Area Networks—Part 15.4: Low-Rate Wireless Personal Area Networks (LR-WPANs)*; IEEE Std 802.15.4-2011 (Revision of IEEE Std 802.15.4-2006); IEEE: Manhattan, NY, USA, 2011; pp. 1–314. [CrossRef]
104. Hasan, M.; Hossain, E.; Niyato, D. Random access for machine-to-machine communication in LTE-advanced networks: Issues and approaches. *IEEE Commun. Mag.* **2013**, *51*, 86–93. [CrossRef]
105. IEEE 802 Working Group. *IEEE Standard for a Convergent Digital Home Network for Heterogeneous Technologies*; IEEE Std 1905.1-2013; IEEE: Manhattan, NY, USA, 2013; pp. 1–93. [CrossRef]
106. User Datagram Protocol(UDP). Available online: https://www.geeksforgeeks.org/user-datagram-protocol-udp/ (accessed on 25 April 2021).
107. Pipkin, D.L. *Halting the Hacker: A Practical Guide to Computer Security*, 2nd ed.; Prentice Hall Professional: Hoboken, NJ, USA, 2003.
108. Bertino, E.; Martino, L.D.; Paci, F.; Squicciarini, A.C. Web services threats, vulnerabilities, and countermeasures. In *Security for Web Services and Service-Oriented Architectures*; Springer: Heidelberg, Germany, 2019; pp. 25–44. [CrossRef]
109. Kizza, J.M. *Guide to Computer Network Security*, 1st ed.; Springer: Heidelberg, Germany, 2009; pp. 387–411. [CrossRef]
110. Dahbur, K.; Mohammad, B.; Tarakji, A.B. A survey of risks, threats and vulnerabilities in cloud computing. In Proceedings of the 2011 International Conference on Intelligent Semantic Web-Services and Applications, New York, NY, USA, 18–20 April 2011; pp. 1–6. [CrossRef]
111. Rainer, R.K.; Cegielski, C.G. Ethics, privacy, and information security. In *Introduction to Information Systems: Supporting and Transforming Business*; John Wiley & Sons: Hoboken, NJ, USA, 2010; Volume 3, pp. 70–121.
112. Tankard, C. Advanced persistent threats and how to monitor and deter them. *Netw. Secur.* **2011**, *2011*, 16–19. [CrossRef]
113. Coffed, J. *The Threat of Gps Jamming: The Risk to An Information Utility*; EXELIS: Herndon, VA, USA, 2014; pp. 6–10.
114. Tippenhauer, N.O.; Pöpper, C.; Rasmussen, K.B.; Capkun, S. On the requirements for successful GPS spoofing attacks. In Proceedings of the 18th ACM Conference on COMPUTER and Communications Security, Chicago, IL, USA, 17–21 October 2011; pp. 75–86. [CrossRef]
115. Uluagac, A.S.; Subramanian, V.; Beyah, R. Sensory channel threats to cyber physical systems: A wake-up call. In Proceedings of the IEEE Conference on Communications and Network Security, San Francisco, CA, USA, 29–31 October 2014; pp. 301–309. [CrossRef]
116. Ge, M.; Hong, J.B.; Guttmann, W.; Kim, D.S. A framework for automating security analysis of the internet of things. *J. Netw. Comput. Appl.* **2017**, *83*, 12–27. [CrossRef]
117. Anthi, E.; Ahmad, S.; Rana, O.; Theodorakopoulos, G.; Burnap, P. Eclipse. IoT: A secure and adaptive hub for the Internet of Things. *Comput. Secur.* **2018**, *78*, 477–490. [CrossRef]
118. Sanchez Alcon, J.A.; López, L.; Martínez, J.F.; Rubio Cifuentes, G. Trust and privacy solutions based on holistic service requirements. *Sensors* **2016**, *16*, 16. Available online: https://www.mdpi.com/1424-8220/16/1/16 (accessed on 25 April 2021). [CrossRef] [PubMed]
119. Mauro, C.; Pallavi, K.; Rabbani, M.M.; Ranise, S. Attestation-enabled secure and scalable routing protocol for IoT networks. *Ad Hoc Netw.* **2020**, *98*, 102054. [CrossRef]
120. Prabadevi, B.; Jeyanthi, N. Distributed Denial of service attacks and its effects on Cloud environment-a survey. In Proceedings of the International Symposium on Networks, Computers and Communications, Hammamet, Tunisia, 17–19 June 2014; pp. 1–4. [CrossRef]
121. Varga, P.; Plosz, S.; Soos, G.; Hegedus, C. Security threats and issues in automation IoT. In Proceedings of the IEEE 13th International Workshop on Factory Communication Systems (WFCS), Trondheim, Norway, 31 May–2 June 2017; pp. 1–6. [CrossRef]
122. Qian, L.; Zhu, Z.; Hu, J.; Liu, S. Research of SQL injection attack and prevention technology. In Proceedings of the International Conference on Estimation, Detection and Information Fusion (ICEDIF), Harbin, China, 10–11 January 2015; pp. 303–306. [CrossRef]
123. Everything You Need to Know About Facebook's Data Breach Affecting 50M Users. Available online: http://https//techcrunch.com/2018/09/28/everything-you-need-to-know-about-facebooks-data-breach-affecting-50m-users/ (accessed on 26 April 2021).
124. Zou, Y.; Wang, G. Intercept Behavior Analysis of Industrial Wireless Sensor Networks in the Presence of Eavesdropping Attack. *IEEE Trans. Ind. Inform.* **2016**, *12*, 780–787. [CrossRef]
125. Chan, H.; Perrig, A.; Song, D.X. Random key predistribution schemes for sensor networks. In Proceedings of the IEEE Symposium Security Privacy, Berkeley, CA, USA, 11–14 May 2003; pp. 197–213. [CrossRef]
126. Abomhara, M.; Køien, G.M. Security and privacy in the Internet of Things: Current status and open issues. In Proceedings of the IEEE International Conference Privacy Security Mobile System (PRISMS), Aalborg, Denmark, 11–14 May 2014; pp. 1–8. [CrossRef]
127. Ashraf, Q.M.; Habaebi, M.H. Autonomic schemes for threat mitigation in Internet of Things. *J. Netw. Comput. Appl.* **2015**, *49*, 112–127. [CrossRef]
128. Znaidi, W.; Minier, M.; Babau, J.P. *An Ontology for Attacks in Wireless Sensor Networks*; RR-6704; INRIA: Rocquencourt, France, 2008.
129. Ye, F.; Luo, H.; Lu, S.; Zhang, L. Statistical en-route filtering of injected false data in sensor networks. *IEEE J. Sel. Areas Commun.* **2005**, *23*, 839–850.

130. Newsome, J.; Shi, E.; Song, D.; Perrig, A. The sybil attack in sensor networks: Analysis & defenses. In Proceedings of the ACM Third International Symposium on Information Processing in Sensor Networks, Berkeley, CA, USA, 27 April 2004; pp. 259–268.
131. Sarigiannidis, P.G.; Karapistoli, E.D.; Economides, A.A. Detecting sybil attacks in wireless sensor networks using UWB ranging-based information. *Expert Syst. Appl.* **2015**, *42*, 7560–7572. [CrossRef]
132. Savola, R.M.; Abie, H.; Sihvonen, M. Towards metrics-driven adaptive security management in e-health IoT applications. In Proceedings of the 7th International Conference Body Area Network, Brussels, Belgium, 24–26 February 2012; pp. 276–281.
133. Choi, H.; Zhu, S.; Porta, T.F.L. SET: Detecting node clones in sensor networks. In Proceedings of the IEEE 3rd Int. Confernce Security Privacy Commun. Netw. Workshops (SecureComm), Nice, France, 17–21 September 2007; pp. 341–350. [CrossRef]
134. Xing, K.; Liu, F.; Cheng, X.; Du, D.H.C. Real-time detection of clone attacks in wireless sensor networks. In Proceedings of the 28th International Conference on Distributed Computing Systems (ICDCS), Beijing, China, 17–20 June 2008; pp. 3–10. [CrossRef]
135. Standaert, F.X. *Introduction to side-channel attacks. Secure Integrated Circuits and Systems*; Springer: Boston, MA, USA, 2010; pp. 27–42. ISBN 978-0-387-71829-3.
136. Wood, A.D.; Stankovic, J.A.; Son, S.H. JAM: A jammed-area mapping service for sensor networks. In Proceedings of the 24th IEEE Real-Time Systems Symposium, Cancun, Mexico, 5 December 2003; pp. 286–297. [CrossRef]
137. Hussein, A.A.; Leow, C.Y.; Rahman, T.A. Robust multiple frequency multiple power localization schemes in the presence of multiple jamming attacks. *PLoS ONE* **2017**, *12*, e0177326. [CrossRef]
138. Shabana, K.; Fida, N.; Khan, F.; Jan, S.R.; Rehman, M.U. Security issues and attacks in wireless sensor networks. *Int. J. Adv. Res. Comput. Sci. Electron. Eng.* **2016**, *5*, 81.
139. Ho, J.-W.; Wright, M.; Das, S.K. Fast detection of replica node attacks in mobile sensor networks using sequential analysis. In Proceedings of the IEEE INFOCOM, Rio de Janeiro, Brazil, 19–25 April 2009; pp. 1773–1781. [CrossRef]
140. Wurm, J.; Hoang, K.; Arias, O.; Sadeghi, A.-R.; Jin, Y. Security analysis on consumer and industrial IoT devices. In Proceedings of the 21st IEEE Asia South Pacific Design Automation Conference (ASP-DAC), Macao, China, 25–28 January 2016; pp. 519–524. [CrossRef]
141. Puthal, D.; Nepal, S.; Ranjan, R.; Chen, J. Threats to networking cloud and edge datacenters in the Internet of Things. *IEEE Cloud Comput.* **2016**, *3*, 64–71. [CrossRef]
142. Koh, J.Y.; Nevat, I.; Leong, D.; Wong, W.C. Geo-spatial location spoofing detection for Internet of Thing. *IEEE Internet Things J.* **2016**, *3*, 971–978. [CrossRef]
143. Lough, D.L. *A Taxonomy of Computer Attacks with Applications to Wireless Networks*; Virginia Polytechnic Institute and State University: Blacksburg, VA, USA, 2001.
144. Bu, K.; Xu, M.; Liu, X.; Luo, J.; Zhang, S.; Weng, M. Deterministic detection of cloning attacks for anonymous RFID systems. *IEEE Trans. Ind. Informat.* **2015**, *11*, 1255–1266. [CrossRef]
145. Wood, A.D.; Stankovic, J.A. Denial of service in sensor networks. *Computer* **2002**, *35*, 54–62. [CrossRef]
146. Mirai "Internet of Things" Malware From Krebs DDoS Attack Goes Open Source. Available online: https://nakedsecurity.sophos.com/2016/10/05/mirai-internet-of-things-malware (accessed on 30 June 2021).
147. Liu, Y.; Li, Y.; Man, H. MAC layer anomaly detection in ad hoc networks. In Proceeding of the 6th Annual IEEE SMC Information Assurance Workshop (IAW), West Point, NY, USA, 15–17 June 2005; pp. 402–409. [CrossRef]
148. Riaz, R.; Kim, K.-H.; Ahmed, H.F. Security analysis survey and framework design for IP connected LoWPANs. In Proceedings of the IEEE International Symposium Autonomous Decentralized Systems (ISADS), Athens, Greece, 23–25 March 2009; pp. 1–6.
149. Hamid, M.A.; Rashid, M.; Hong, C.S. Routing security in sensor network: Hello flood attack and defense. In Proceedings of the IEEE ICNEWS, Phoenix Park, Korea, 20–22 February 2006; pp. 2–4.
150. Murphy, J. Enhanced Security Controls for IBM Watson IoT Platform, Armonk. Available online: https://developer.ibm.com/iotplatform/2016/09/23/enhanced-securitycontrols-for-ibm-watson-iot-platform/ (accessed on 30 June 2021).
151. Teng, L.; Zhang, Y. SERA: A secure routing algorithm against sinkhole attacks for mobile wireless sensor networks. In Proceedings of the IEEE 2nd International Conference on Computer Modeling Simulation (ICCMS), Sanya, China, 22–24 January 2010; pp. 79–82.
152. Sathish, R.; Scholar, P.G. Dynamic detection of clone attack in wireless sensor networks. In Proceedings of the International Conference on Communication Systems Network Technologies, Gwalior, India, 6–8 April 2013; pp. 501–505.
153. Karlof, C.; Wagner, D. Secure routing in wireless sensor networks: Attacks and countermeasures. *Ad Hoc Netw.* **2003**, *1*, 293–315. [CrossRef]
154. Karakehayov, Z. Using reward to detect team black-hole attacks in wireless sensor networks. In Proceedings of the Workshop on Real World Wireless Sensor Network, Stockholm, Sweden, 20–21 June 2005; pp. 20–21.
155. Wang, W.; Bhargava, B.K. Visualization of wormholes in sensor networks. In Proceedings of the 3rd ACM Workshop Wireless Security, Philadelphia, PA, USA, 1 October 2001; pp. 51–60. [CrossRef]
156. Kaissi, R.Z.E.; Kayssi, A.; Chehab, A.; Dawy, Z. DAWWSEN: A Defense Mechanism Against Wormhole Attacks in Wireless Sensor Networks. Ph.D. Thesis, American University of Beirut, Beirut, Lebanon, 2005.
157. Perrey, H.; Landsmann, M.; Ugus, O.; Wählisch, M.; Schmidt, T.C. TRAIL: Topology Authentication in RPL. In Proceedings of the ACM International Conference on Embedded Wireless System and Network (EWSN), Graz, Austria, 15–17 February 2016; pp. 59–64.
158. Dvir, A.; Holczer, T.; Buttyán, L. Vera-version number and rank authentication in RPL. In Proceedings of the IEEE 8th International Conference on Mobile Ad Hoc Sensor Systems (MASS), Valencia, Spain, 17–22 October 2011; pp. 709–714. [CrossRef]

159. Accettura, N.; Piro, G. Optimal and secure protocols in the IETF 6TiSCH communication stack. In Proceedings of the IEEE 23rd International Symposium on Industrial Electronics (ISIE), Istambul, Turkey, 1–4 June 2014; pp. 1469–1474. [CrossRef]
160. Singh, M.; Rajan, M.; Shivraj, V.; Balamuralidhar, P. Secure MQTT for Internet of Things (IoT). In Proceedings of the IEEE 5th International Conference on Communication Systems and Network Technologies (CSNT), Gwalior, India, 4–6 April 2015; pp. 746–751. [CrossRef]
161. Song, S.; Choi, H.-K.; Kim, J.-Y. A secure and lightweight approach for routing optimization in mobile IPv6. *EURASIP J. Wirel. Commun. Netw.* **2009**, *2009*, 1–10. [CrossRef]
162. Neisse, R.; Steri, G.; Baldini, G. Enforcement of security policy rules for the Internet of Things. In Proceedings of the IEEE 10th International Confernce on Wireless and Mobile Computing Networking and Communications (WiMob), Larnaca, Cyprus, 8–10 October 2014; pp. 165–172. [CrossRef]
163. Xbox 360 Timing Attack. Available online: http://beta.ivc.no/wiki/index.php/Xbox_360_Timing_Attack (accessed on 30 June 2021).
164. Zhang, Q.; Wang, X. SQL injections through back-end of RFID system. In Proceedings of the International Symposium on Computer Network and Multimedia Technology, Wuhan, China, 18–20 January 2009; pp. 1–4.
165. Farris, I.; Taleb, T.; Khettab, Y.; Song, J. A survey on emerging SDN and NFV security mechanisms for IoT systems. *IEEE Commun. Surv. Tutor.* **2019**, *21*, 812–837. [CrossRef]
166. Singh, J.; Pasquier, T.; Bacon, J.; Ko, H.; Eyers, D. Twenty security considerations for cloud-supported Internet of Things. *IEEE Internet Things J.* **2016**, *3*, 269–284. [CrossRef]
167. Bose, T.; Bandyopadhyay, S.; Ukil, A.; Bhattacharyya, A.; Pal, A. Why not keep your personal data secure yet private in IoT: Our lightweight approach. In Proceedings of the IEEE 10th International Conference on Intelligent Sensors, Sensor Networks and Information Processing (ISSNIP), Singapore, 7–9 April 2015; pp. 1–6.
168. Kumar, J.; Rajendran, B.; Bindhumadhava, B.S.; Babu, N.S.C. XML wrapping attack mitigation using positional token. In Proceedings of the International Confernce Public Key Infrastructure and its Applications (PKIA), Bangalore, India, 14–15 November 2017; pp. 36–42.
169. Deng, J.; Han, R.; Mishra, S. Defending against path-based dos attacks in wireless sensor networks. In Proceedings of the 3rd ACM Workshop Security Ad Hoc Sensor Network, Alexandria, VA, USA, 14–15 November 2005; pp. 89–96.
170. Gupta, H.; Oorschot, P.C.V. Onboarding and Software Update Architecture for IoT Devices. In Proceedings of the 17th International Conference on Privacy, Security and Trust (PST), Fredericton, NB, Canada, 26–28 August 2019; pp. 1–11. [CrossRef]
171. Skorobogatov, S. Fault attacks on secure chips: From glitch to flash. In *Design and Security of Cryptographic Algorithms and Devices (ECRYPT II)*; University of Cambridge: Cambridge, UK, 2011; pp. 1–64.
172. Stanciu, A.; Balan, T.-C.; Gerigan, C.; Zamfir, S. Securing the IoT gateway based on the hardware implementation of a multi pattern search algorithm. In Proceedings of the International Conference on Optimization of Electrical and Electronic Equipment (OPTIM) & 2017 Intl Aegean Conference on Electrical Machines and Power Electronics (ACEMP), Brasov, Romania, 25–27 May 2017; pp. 1001–1006.
173. MohammadI, S.; Jadidoleslamy, H. A comparison of link layer attacks on wireless sensor networks. *Int. J. Appl. Graph Theory Wirel. Ad Hoc Netw. Sens. Netw.* **2011**, *3*, 35–56.
174. Swamy, S.N.; Jadhav, D.; Kulkarni, N. Security threats in the application layer in IoT applications. In Proceedings of the International Confernce IoT Social, Mobile, Analytics Cloud (I-SMAC), Palladam, India, 10–11 February 2017; pp. 477–480.
175. Sharmeen, S.; Huda, S.; Abawajy, J.H.; Ismail, W.N.; Hassan, M.M. Malware Threats and Detection for Industrial Mobile-IoT Networks. *IEEE Access* **2018**, *6*, 15941–15957. [CrossRef]
176. Ham, H.-S.; Kim, H.-H.; Kim, M.-S.; Choi, M.-J. Linear SVM-based Android malware detection for reliable IoT services. *J. Appl. Math.* **2014**, *2014*, 594501. [CrossRef]
177. Kaur, P.; Sharma, S. Spyware detection in Android using hybridization of description analysis permission mapping and interface analysis. *Procedia Comput. Sci.* **2015**, *46*, 794–803. [CrossRef]
178. Wolinsky, D.I.; Syta, E.; Ford, B. Hang with your buddies to resist intersection attacks. In Proceedings of the ACM SIGSAC Conference on Computer and Communications Security (CCS), Berlin, Germany, 4–8 November 2013; pp. 1153–1166.
179. Grover, J.; Laxmi, V.; Gaur, M.S. Attack models and infrastructure supported detection mechanisms for position forging attacks in vehicular ad hoc networks. *CSI Trans. ICT* **2013**, *1*, 261–279. [CrossRef]
180. Sicari, S.; Rizzardi, A.; Grieco, L.A.; Coen-Porisini, A. Security, privacy and trust in Internet of Things: The road ahead. *Comput. Netw.* **2015**, *76*, 146–164. [CrossRef]
181. Cherian, M.; Chatterjee, M. Survey of security threats in iot and emerging countermeasures. In Proceedings of the International Symposium on Security in Computing and Communication, Bangalore, India, 19–22 September 2018; pp. 591–604.
182. Sepulveda, J.; Willgerodt, F.; Pehl, M. SEPUFSoC: Using PUFs for memory integrity and authentication in multi-processors system-on-chip. In Proceedings of the GLSVLSI '18: Proceedings of the 2018 on Great Lakes Symposium on VLSI, Chicago, IL, USA, 23–25 May 2018; pp. 39–44. [CrossRef]
183. Bîrleanu, F.G.; Bizon, N. Reconfigurable computing in hardware security–a brief review and application. *J. Electr. Eng. Electron. Control Comput. Sci.* **2016**, *2*, 1–12.
184. Katsikogiannis, G.; Kallergis, D.; Garofalaki, Z.; Mitropoulos, S.; Douligeris, C. A policy-aware Service Oriented Architecture for secure machine-to-machine communications. *Ad Hoc Netw.* **2018**, *80*, 70–80. [CrossRef]
185. Laplante, P.A. Blockchain and the Internet of Things in the industrial sector. *IEEE Comput. Soc.* **2018**, *20*, 15–18.

186. Orman, H. Blockchain: The emperors new PKI? *IEEE Internet Comput.* **2018**, *22*, 23–28. [CrossRef]
187. Henry, R.; Herzberg, A.; Kate, A. Blockchain access privacy: Challenges and directions. *IEEE Secur. Priv.* **2018**, *16*, 38–45. [CrossRef]
188. Fog Computing: Focusing on Mobile Users at the Edge. Available online: https://arxiv.org/abs/1502.01815 (accessed on 1 July 2021).
189. Dastjerdi, A.V.; Buyya, R. Fog Computing: Helping the Internet of Things Realize Its Potential. *Computer* **2016**, *49*, 112–116. [CrossRef]
190. Sehgal, V.K.; Patrick, A.; Soni, A.; Rajput, L. Smart human security framework using Internet of Things, cloud and fog computing. Intelligent Distributed Computing. *Springer* **2015**, *321*, 251–263. [CrossRef]
191. Feasibility of Fog Computing. Available online: https://arxiv.org/abs/1701.05451 (accessed on 31 June 2021).
192. IoT Agenda. IoT and Big Data Analytics. Available online: https://internetofthingsagenda.techtarget.com/ (accessed on 30 June 2021).
193. Alwaris, A.; Alhothaily, A.; Hu, C.; Cheng, X. Fog Computing for the Internet of Things: Security and Privacy Issues. *IEEE Comput.* **2017**, *21*, 34–42. [CrossRef]
194. Mollah, M.B.; Azad, M.A.K.; Vasilakos, A. Secure data sharing and searching at the edge of cloud-assisted Internet of Things. *IEEE Cloud Comput.* **2017**, *4*, 34–42. [CrossRef]
195. Alrowaily, M.; Lu, Z. Secure edge computing in IoT systems: Review and case studies. In Proceedings of the IEEE/ACM Symposium on Edge Computing (SEC), Seattle, WA, USA, 25–27 October 2018; pp. 440–444. [CrossRef]
196. Li, Y.; Wang, S. An energy-aware edge server placement algorithm in mobile edge computing. In Proceedings of the IEEE International Confernce Edge Comput. (EDGE), San Francisco, CA, USA, 2–7 July 2018; pp. 66–73. [CrossRef]
197. 6 Significant Issues That Edge Computing in IoT Solves. Available online: https://internetofthingsagenda.techtarget.com/feature/6-significant-issues-that-edge-computing-in-IoT-solves (accessed on 30 June 2021).
198. Premsankar, G.; Di Francesco, M.; Taleb, T. Edge computing for the Internet of Things: A case study. *IEEE Internet Things J.* **2018**, *5*, 1275–1284. [CrossRef]
199. Abbas, N.; Zhang, Y.; Taherkordi, A.; Skeie, T. Mobile edge computing: A survey. *IEEE Internet Things J.* **2018**, *5*, 450–465. [CrossRef]
200. Pavani, K.; Damodaram, A. Intrusion detection using MLP for MANETs. In Proceedings of the Third International Conference on Computational Intelligence and Information Technology (CIIT 2013), Mumbai, India, 18–19 October 2013; pp. 440–444. [CrossRef]
201. Kulkarni, R.V.; Venayagamoorthy, G.K. Neural network based secure media access control protocol for wireless sensor networks. In Proceedings of the 2009 International Joint Conference on Neural Networks, Atlanta, GA, USA, 14–19 June 2009; pp. 1680–1687. [CrossRef]
202. Xiao, L.; Xie, C.; Chen, T.; Dai, H.; Poor, H.V. A mobile offloading game against smart attacks. *IEEE Access* **2016**, *4*, 2281–2291. [CrossRef]
203. Xiao, L.; Yan, Q.; Lou, W.; Chen, G.; Hou, Y.T. Proximity-based security techniques for mobile users in wireless networks. *IEEE Trans. Inf. Forensics Secur.* **2013**, *8*, 2089–2100. [CrossRef]
204. Xiao, L.; Li, Y.; Han, G.; Liu, G.; Zhuang, W. PHY-layer spoofing detection with reinforcement learning in wireless networks. *IEEE Trans. Veh. Technol.* **2016**, *65*, 10037–10047. [CrossRef]
205. Spirina, K. Biometric Authentication: The Future of IoT Security Solutions. Available online: https://www.IoTevolutionworld.com/IoT/articles/438690-biometricauthenticationfuture-IoT-security-solutions.html (accessed on 9 February 2019).
206. Blanco-Novoa, Ó.; Fernández-Caramés, T.; Fraga-Lamas, P.; Castedo, L. An electricity price-aware open-source smart socket for the Internet of energy. *Sensors* **2017**, *17*, 643. [CrossRef]
207. Zhang, Y.; Wen, J. An IoT electric business model based on the protocol of bitcoin. In Proceedings of the 18th International Conference on Intelligence in Next Generation Networks, Paris, France, 17–19 February 2015; pp. 184–191. [CrossRef]
208. Lundqvist, T.; Blanche, A.; Andersson, H.R.H. Thing-to-thing electricity micro payments using blockchain technology. In Proceedings of the Global Internet of Things Summit (GIoTS), Geneva, Switzerland, 6–9 June 2017; pp. 1–6. [CrossRef]
209. Salahuddin, M.A.; Al-Fuqaha, A.; Guizani, M.; Shuaib, K.; Sallabi, F. Softwarization of Internet of Things infrastructure for secure and smart healthcare. *arXiv* **2018**, arXiv:1805.11011. Available online: https://arxiv.org/abs/1805.11011 (accessed on 1 July 2021).
210. Bocek, T.; Rodrigues, B.B.; Strasser, T.; Stiller, B. Blockchains everywhere-a use-case of blockchains in the pharma supply-chain. In Proceedings of the IFIP/IEEE Symposium on Integrated Network and Service Management (IM), Lisbon, Portugal, 8–12 May 2017; pp. 772–777. [CrossRef]
211. Shae, Z.; Tsai, J.J.P. On the Design of a Blockchain Platform for Clinical Trial and Precision Medicine. In Proceedings of the IEEE 37th International Conference on Distributed Computing Systems (ICDCS), Atlanta, GA, USA, 5–8 June 2017; pp. 1972–1980. [CrossRef]
212. Lei, A.; Cruickshank, H.; Cao, Y.; Asuquo, P.; Ogah, C.P.A.; Sun, Z. Blockchain-Based Dynamic Key Management for Heterogeneous Intelligent Transportation Systems. *IEEE Internet Things J.* **2017**, *4*, 1832–1843. [CrossRef]
213. Huh, S.; Cho, S.; Kim, S. Managing IoT devices using blockchain platform. In Proceedings of the 19th International Conference on Advanced Communication Technology (ICACT), PyeongChang, Koreea, 19–22 February 2017; pp. 464–467. [CrossRef]
214. Samaniego, M.; Deters, R. Internet of Smart Things-IoST: Using Blockchain and CLIPS to Make Things Autonomous. In Proceedings of the IEEE International Conference on Cognitive Computing (ICCC), Honulul, HI, USA, 25–30 June 2017; pp. 9–16. [CrossRef]

215. Faruque, M.A.A.; Vatanparvar, K. Energy Management-as-a-Service Over Fog Computing Platform. *IEEE Internet Things J.* **2016**, *3*, 161–169. [CrossRef]
216. SGao, S.; Peng, Z.; Xiao, B.; Xiao, Q.; Song, Y. SCoP: Smartphone energy saving by merging push services in Fog computing. In Proceedings of the IEEE/ACM 25th International Symposium on Quality of Service (IWQoS), Vilanova i la Geltrú, Spain, 14–16 June 2017; pp. 1–10. [CrossRef]
217. Dubey, H.; Monteiro, A.; Constant, N.; Abtahi, M.; Borthakur, D.; Mahler, L. Fog computing in medical Internet-of-Things: Architecture implementation and applications. In *Handbook of Large-Scale Distributed Computing in Smart Healthcare*, 1st ed.; Khan, S.U., Zomaya, A.Y., Abbas, A., Eds.; Springer International Publishing: Berlin/Heidelberg, Germany, 2017; pp. 281–321, ISBN 978-3-319-58280-1.
218. Rahmani, A.M.; Gia, T.N.; Negash, B.; Anzanpour, A.; Azimi, I.; Jiang, M. Exploiting smart e-health gateways at the edge of healthcare Internet-of-Things: A fog computing approach. *Future Gener. Comput. Syst.* **2018**, *78*, 641–658. [CrossRef]
219. Gia, T.N.; Jiang, M.; Rahmani, A.M.; Westerlund, T.; Liljeberg, P.; Tenhunen, H. Fog computing in healthcare Internet of Things: A case study on ecg feature extraction. In Proceedings of the 2015 IEEE International Conference on Computer and Information Technology, Liverpool, UK, 26–28 October 2015; pp. 356–363. [CrossRef]
220. Shi, W.; Cao, J.; Zhang, Q.; Li, Y.; Xu, L. Edge Computing: Vision and Challenges. *IEEE Internet Things J.* **2016**, *3*, 637–646. [CrossRef]
221. Ni, J.; Zhang, A.; Lin, X.; Shen, X.S. Security, Privacy, and Fairness in Fog-Based Vehicular Crowdsensing. *IEEE Commun. Mag.* **2017**, *55*, 146–152. [CrossRef]
222. Markakis, E.K.; Karras, K.; Zotos, N.; Sideris, A.; Moysiadis, T.; Corsaro, A.; Pallis, E. EXEGESIS: Extreme Edge Resource Harvesting for a Virtualized Fog Environment. *IEEE Commun. Mag.* **2017**, *55*, 173–179. [CrossRef]
223. Huang, Y.; Lu, Y.; Wang, F.; Fan, X.; Liu, J.; Leung, V.C. An Edge Computing Framework for Real-Time Monitoring in Smart Grid. In Proceedings of the 2018 IEEE International Conference on Industrial Internet (ICII), Seattle, WA, USA, 21–23 October 2018; pp. 99–108. [CrossRef]
224. Oyekanlu, E.; Nelatury, C.; Fatade, A.O.; Alaba, O.; Abass, O. Edge computing for industrial IoT and the smart grid: Channel capacity for M2M communication over the power line. In Proceedings of the IEEE 3rd International Conference on Electro-Technology for National Development (NIGERCON), Owerri, Nigeria, 7–10 November 2017; pp. 1–11. [CrossRef]
225. Muhammed, T.; Mehmood, R.; Albeshri, A.; Katib, I. UbeHealth: A Personalized Ubiquitous Cloud and Edge-Enabled Networked Healthcare System for Smart Cities. *IEEE Access* **2018**, *6*, 32258–32285. [CrossRef]
226. Barik, R.K.; Dubey, H.; Mankodiya, K. SOA-FOG: Secure service-oriented edge computing architecture for smart health big data analytics. In Proceedings of the IEEE Global Conference on Signal and Information Processing (GlobalSIP), Montreal, UK, 14–16 November 2017; pp. 477–481. [CrossRef]
227. Singh, D.; Tripathi, G.; Alberti, A.M.; Jara, A. Semantic edge computing and IoT architecture for military health services in battlefield. In Proceedings of the 14th IEEE Annual Consumer Communications & Networking Conference (CCNC), Las Vegas, NV, USA, 8–11 January 2017; pp. 185–190. [CrossRef]
228. Aman, M.N.; Sikdar, B.; Chua, K.C.; Ali, A. Low Power Data Integrity in IoT Systems. *IEEE Internet Things J.* **2018**, *5*, 3102–3113. [CrossRef]
229. Gope, P.; Sikdar, B. Lightweight and Privacy-Preserving Two-Factor Authentication Scheme for IoT Devices. *IEEE Internet Things J.* **2019**, *6*, 580–589. [CrossRef]
230. Ahmed, S.; Lee, Y.; Hyun, S.; Koo, I. Feature Selection–Based Detection of Covert Cyber Deception Assaults in Smart Grid Communications Networks Using Machine Learning. *IEEE Access* **2018**, *6*, 27518–27529. [CrossRef]
231. Ozay, M.; Esnaola, I.; Vural, F.T.Y.; Kulkarni, S.R.; Poor, H.V. Machine Learning Methods for Attack Detection in the Smart Grid. *IEEE Trans. Neural Netw. Learn. Syst.* **2016**, *27*, 1773–1786. [CrossRef] [PubMed]
232. Zhang, D.; Han, X.; Deng, C. Review on the research and practice of deep learning and reinforcement learning in smart grids. *CSEE J. Power Energy Syst.* **2018**, *4*, 362–370. [CrossRef]
233. Hossain, E.; Khan, I.; Un-Noor, F.; Sikander, S.S.; Sunny, M.S.H. Application of Big Data and Machine Learning in Smart Grid and Associated Security Concerns: A Review. *IEEE Access* **2019**, *7*, 13960–13988. [CrossRef]
234. Mercer, C. How Machine Learning Will Change Society. Available online: https://www.techworld.com/picture-gallery/tech-innovation/5-ways-machine-learning-will-change-society-3666674 (accessed on 1 July 2021).
235. Chen, M.; Hao, Y.; Hwang, K.; Wang, L.; Wang, L. Disease prediction by machine learning over big data from healthcare communities. *IEEE Access* **2017**, *5*, 8869–8879. [CrossRef]
236. Vito, S.D.; Francia, G.D.; Esposito, E.; Ferlito, S.; Formisano, F.; Massera, E. Adaptive machine learning strategies for network calibration of IoT smart air quality monitoring devices. *Pattern Recognit. Lett.* **2020**, *136*, 264–271. [CrossRef]
237. Punithavathi, P.; Geetha, S.; Karuppiah, M.; Islam, S.K.F.; Hassan, M.M.; Choo, K.K.R. A lightweight machine learning-based authentication framework for smart IoT devices. *Inf. Sci.* **2019**, *484*, 255–268. [CrossRef]
238. Bigini, G.; Freschi, V.; Lattanzi, E. A Review on Blockchain for the Internet of Medical Things: Definitions, Challenges, Applications, and Vision. *Future Internet* **2020**, *12*, 208. [CrossRef]
239. Sepasgozar, S.; Karimi, R.; Farahzadi, L.; Moezzi, F.; Shirowzhan, S.M.; Ebrahimzadeh, S.; Hui, F.; Aye, L. A Systematic Content Review of Artificial Intelligence and the Internet of Things Applications in Smart Home. *Appl. Sci.* **2020**, *10*, 3074. [CrossRef]

sustainability

MDPI

Article

Performance Investigation of a Solar Photovoltaic/Diesel Generator Based Hybrid System with Cycle Charging Strategy Using BBO Algorithm

Anurag Chauhan [1,*], Subho Upadhyay [2], Mohd. Tauseef Khan [1], S. M. Suhail Hussain [3] and Taha Selim Ustun [3,*]

1 Department of Electrical Engineering, Rajkiya Engineeirng College, Banda 210201, Uttar Pradesh, India; mtkhan@recbanda.ac.in
2 Department of Electrical Engineering, Dayalbagh Educational Institute, Agra 282005, Uttar Pradesh, India; subhome2k10@dei.ac.in
3 Fukushima Renewable Energy Institute (FREA), National Institute of Advanced Industrial Science and Technology (AIST), Koriyama, Fukushima 963-0298, Japan; suhail@ieee.org
* Correspondence: anurag@recbanda.ac.in (A.C.); selim.ustun@aist.go.jp (T.S.U.)

Abstract: In the current scenario, sustainable power generation received greater attention due to the concerns of global warming and climate change. In the present paper, a Solar Photovoltaic/Diesel Generator/ Battery-based hybrid system has been considered to meet the electrical energy demand of a remote location of India. The cost of the energy of hybrid system is minimized using a Biogeography-based Optimization (BBO) algorithm under the constraints of power reliability, carbon emission and renewable energy fraction. Load following and cycle charging strategies have been considered in order to investigate the performance analysis of the proposed hybrid system. Further, different component combinations of specifications available on the market are presented for detail analysis. The minimum cost of energy of the proposed hybrid system is obtained as 0.225 $/kWh.

Keywords: renewable energy; solar; diesel generator; battery; BBO

Citation: Chauhan, A.; Upadhyay, S.; Khan, M..T.; Hussain, S.M.S.; Ustun, T.S. Performance Investigation of a Solar Photovoltaic/Diesel Generator Based Hybrid System with Cycle Charging Strategy Using BBO Algorithm. *Sustainability* **2021**, *13*, 8048. https://doi.org/10.3390/su13148048

Academic Editors: Nicu Bizon, Mamadou Baïlo Camara and Bhargav Appasani

Received: 7 June 2021
Accepted: 14 July 2021
Published: 19 July 2021

1. Introduction

In the current scenario, renewable energy has been recognized as the most effective tool in addressing climate change and global warming [1,2]. The installation cost of solar and wind energy is decreasing day by day and becoming competitive with fossil fuels [3,4]. Additionally, the use of renewable energy technologies offers low carbon emission in the environment with reserves of fossil fuels. Further, integration of two or more renewable energy sources ensures continuous power supply and counterbalances the intermittent behavior of renewable sources [5–10].

Baruah et al. [11] investigated the techno-economic feasibility of a hybrid system which consisted of solar photovoltaic, biogas, wind turbine, syngas and hydrokinetic energy. They proposed this system in order to supply the demand of an academic township using HOMER Pro software. They have used Analytical Hierarchy Process in order to optimize the cost of energy generation and area of system. They have also conducted a sensitivity analysis of the system for the changes of different system parameters. Das et al. [12] minimized the net present cost of a hybrid system consisting of PV array, biogas generator, pumped hydro and battery storage using water cycle algorithm and moth-flame algorithms. They have also performed the comparison of statistical characteristics of the net present cost results obtained by water cycle algorithm, moth-flame algorithm and genetic algorithm.

El-houari et al. [13] designed a solar energy, wind energy and biomass-based hybrid system for ten houses located in remote villages in the Moroccan Fez-Meknes region. They found that the proposed hybrid system offered a reduction of 26.48 tons and 28.814 tons of carbon emission in comparison to the utility grid and diesel generator, respectively.

Elkadeem et al. [14] suggested the different combinations of PV array, wind turbines, diesel-based generator and converter for agriculture and irrigation in Dongola, Sudan. They have also performed the sensitivity analysis to evaluate the effect of wind speed, diesel price, interest rate and solar radiation on system economic performance such as net present cost and cost of generation.

Kumar et al. [15] considered three types of battery such as lead acid, nickel iron and lithium ion during the design of hybrid system. They minimized annual cost using Salp Swarm Algorithm and compared the results with other algorithms in obtaining the best optimum solution. Ma et al. [16] proposed the sizing of a hybrid energy system comprised of PV array, wind turbines and battery with consideration of the saturation of renewable sources. They used the saturation factor changing from 0 (only wind system) to 1 (only solar system) in the step of 0.02. Jahangir et al. [17] investigated the economic and environmental assessment of a hybrid system comprised of PV array, wind turbines and biomass generator. They also performed the sensitivity analysis for the changing biomass price, inflation rate and biomass input and evaluated its impacts on the cost of energy and annualized system cost.

Many studies have been performed and investigated the size optimization of the hybrid system. However, many researchers have not considered the battery degradation model during the design of the hybrid system. Additionally, many authors have not accounted for the seasonal changes in the demand and sizes available in market. Renewable fraction and carbon emission in the hybrid system have also not been taken into account by many authors.

The system presented in the work involves a college premises, containing loads of the three hostels and one Sewage Treatment Plant (STP). The design focuses on developing a grid-independent system comprising only SPV and Battery, while diesel generator is working as a backup generator. The maximum number of SPV module is limited depending on the roof area available. The greater size of the SPV system helps to charge the selected size of batteries. The SPV selected through BBO helps to size batteries, which depend on the fulfillment of load primarily during the evening hours. Even then the cumulative usage of the SPV and the batteries are unable to supply the demand due to size constraints of SPV. Hence, sizing of diesel generator is selected to supply the deficit load in case of both load following and cycle charging strategies. In addition, in cycle charging strategy the excess amount of power is fed to charge the batteries. This further reduces the COE of the overall hybrid energy system. The results signify the importance of selecting batteries in place of a diesel generator during the overall operation of the system. This in turn will improve the renewable fraction and reduce the CO_2 emission.

The work also considers number of charging-discharging cycles of the batteries and investment cost, maintenance and operation cost and replacement cost. The parameters that are considered in the work are emission of CO_2, variation of load during summer and winter seasons, Energy index ratio (i.e., a measure that load is fulfilled at all times), renewable fraction, fuel consumption, net present cost.

2. Hybrid System Configuration

In the present paper, three hostels and one sewage treatment plant (STP) of the institute Rajkiya Engineering College Banda of India has been taken as the study area. This area is located at the latitude of 25.29° N and longitude of 80.57° E.

This area receives the yearly average daily solar radiation of 5.262 kWh/m^2. Accordingly, a hybrid system comprised of photovoltaic array and diesel generator has been considered as presented in Figure 1. This system is proposed to supply the electrical energy requirements of three hostels and one STP of the institute. Power output of solar panels is connected to direct current bus, and power production of the diesel generator is linked to alternating current bus. A set of battery bank is also used in the system to store the additional electricity produced from the generating sources and supply the shortage of load demand. A bidirectional converter has also been used in the system in order to convert AC to DC and vice-versa.

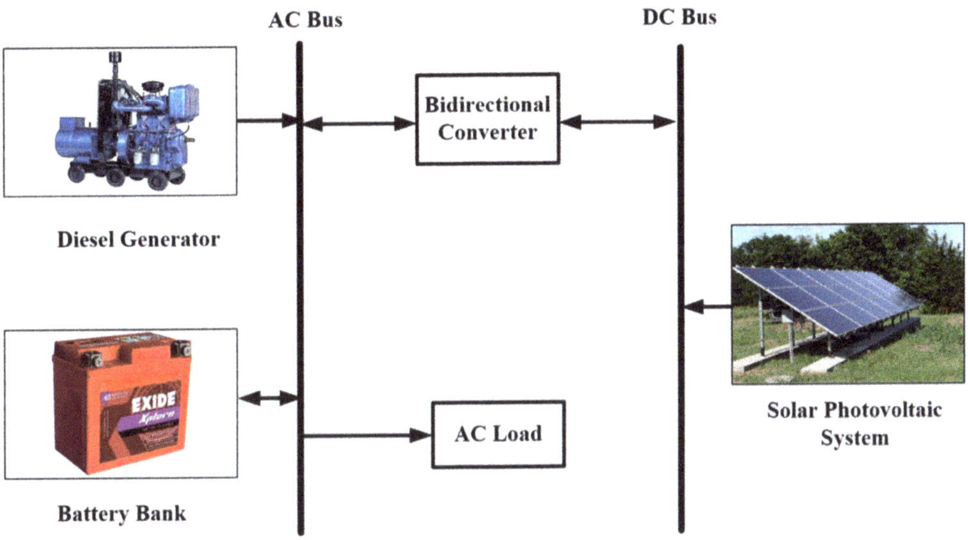

Figure 1. Schematic of Photovoltaic/Diesel Generator/Battery Bank based hybrid system.

A step wise method is required for the performance investigation of the hybrid energy system. It confirms the uninterrupted power supply at the consumer end at minimum system cost. The methodology develops an optimal hybrid system model by addressing all the operational constraints imposed by the user. A step wise description of methodology is summarized as follows:

Step 1: Estimate the electrical energy consumption of each appliance and further, calculate the hourly demand of the selected area.

Step 2: Develop the mathematical model of power output of each generating source and storage system.

Step 3: Choose different configurations of sources.

Step 4: Formulate a framework of objective function and operational constraints of the hybrid system.

Step 5: Take dataset of system components.

Step 6: Perform the simulation of the developed model of the hybrid system for a year.

Step 7: Selection of power dispatch strategy.

Step 8: Show the best optimal configuration of the hybrid system with system sizes and cost parameters.

Step 9: Performance investigation of the best configuration.

A flowchart of methodology is prepared with the steps and depicted in Figure 2.

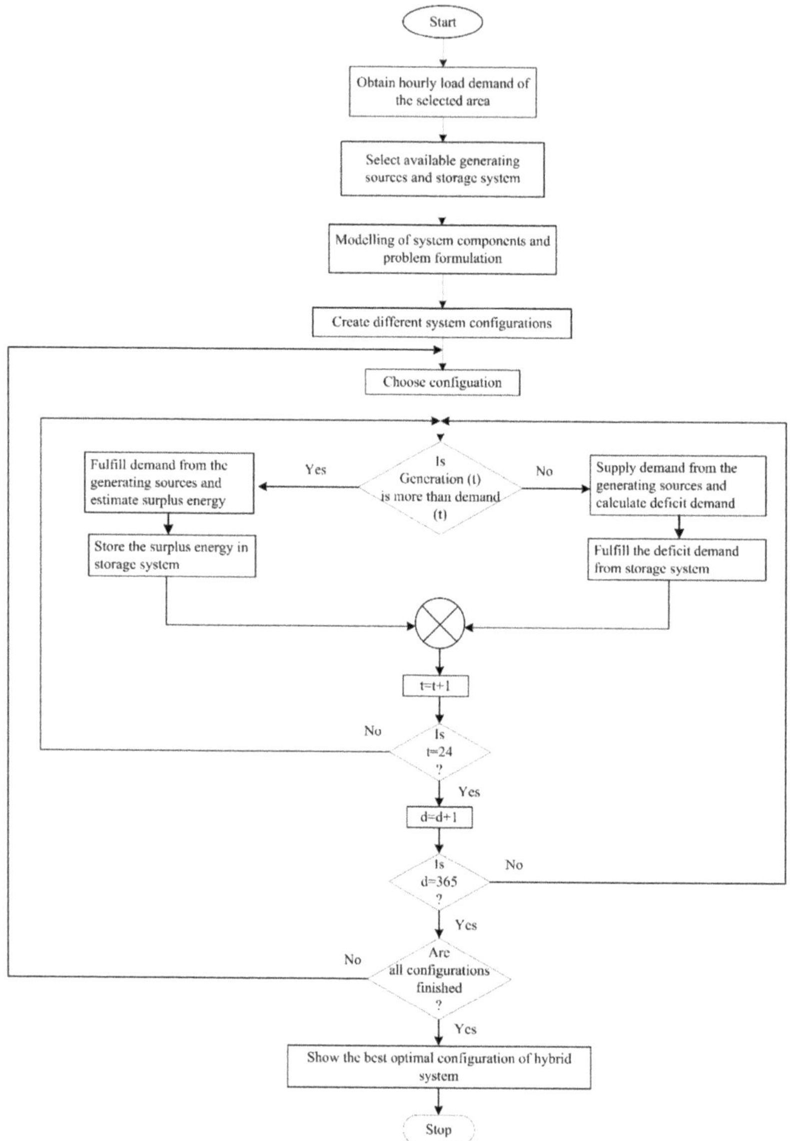

Figure 2. Flowchart of methodology adopted.

3. Mathematical Model

Modelling is an important step before optimization as it gives the static and dynamic characteristics of the component. It relates to the output of the system component in terms of many input variables. The modelling of the hybrid system components is explained as below:

3.1. Model of Solar PV Array

The selected area receives a good amount of daily average solar radiation and therefore PV array has the capability to meet the electricity demand of the area. Power output of a

PV module is the function of open circuit voltage (V_O), short circuit current (I_s) and filling factor (F). It can be modeled as follows [18,19]:

$$P^d_{PV}(t) = V^d_O(t) \times I^d_S(t) \times F(t) \quad (1)$$

Further, the open circuit voltage and short circuit current of PV module depend upon the different module parameters as provided by the manufacturer, and these can be estimated as follows:

$$I^d_S(t) = \left\{I_{S,STC} + C_i\left[T^d_{cell}(t) - 25^0\right]\right\}\frac{\beta^d(t)}{1000}. \quad (2)$$

$$V^d_O(t) = V_{O,STC} - C_v \times T^d_{cell}(t). \quad (3)$$

where $I_{S,STC}$ and $V_{O,STC}$, respectively, represent the short circuit current and open circuit voltage of PV cell at standard test conditions, C_i and C_v, respectively, represent the temperature coefficient of short circuit current and open circuit voltage, β is solar radiation and T_{cell} is PV cell temperature.

PV module cell temperature can be calculated with following equation as follows:

$$T^d_C(t) = T^d_A(t) + \frac{NOCT - 20^0C}{800} \times \beta^d(t). \quad (4)$$

where T_A is ambient temperature and NOCT is nominal operating cell temperature.

3.2. Model of Diesel Generator

Diesel generator is operated to serve the peak demand of the area. Fuel required (Q) in order to operate DG depends upon rated power of DG and can be modeled as follows [20,21]:

$$Q_t(t) = \alpha_{DG}P_{DG}(t) + \beta_{DG}P_{DG,rated}. \quad (5)$$

where P_{DG} is power yield of DG at a particular time, $P_{DG,rated}$ is rated power of DG, β_{DG} (0.08145 l/kWh) and α_{DG} (0.246 l/kWh) are constants of DG.

3.3. Model of Storage System

The storage system is essential for energy balance in the hybrid system. It acts as a tool to mitigate the gap between energy generation and demand. In the present system, battery bank has been considered as a storage system. Battery capacity at a particular instant depends upon the previous capacity and difference between total generation and load demand.

A battery mostly works in two states viz. charging and discharging. In charging state, the battery stores surplus power supplied by sources and the current state of battery $E_B(t)$ can be calculated as follows [22–24]:

$$E_B(t) = E_B(t-1) + E_{CCO}(t) \times \eta_{CHG} \quad (6)$$

In discharging state, demand is more than the total electricity generation and the battery bank storage at hour 't' can be estimated as:

$$E_B(t) = (1-\sigma) \times E_B(t-1) - E_{Required}(t) \quad (7)$$

$$E_{Required}(t) = \frac{E_{NL}(t)}{\eta_{INV} \times \eta_{DCHG}} \quad (8)$$

$$E_{NL}(t) = E_{Demand}(t) - [E_{SPVS}(t) \times \eta_{Inv} + E_{DG}(t)] \quad (9)$$

where $E_{Required}(t)$ is hourly energy required from the battery to meet the load (kWh), $E_{NL}(t)$ is net shortfall load, σ is hourly self discharge rate, E_{CCO} is charge controller output, η_{CHG} and η_{DCHG}, respectively, represent charging efficiency and discharging efficiency of battery.

The n_B^S is the series connected batteries, which depends on the nominal DC bus voltage (V_{BUS}) and individual nominal voltage of battery (V_{nom}). Here the V_{BUS} is considered to be 48 V.

$$n_B^S = \frac{V_{BUS}}{V_{nom}} \tag{10}$$

The battery bank nominal capacity (C_n) is directly proportional to the number of batteries (N_{BAT}) and nominal capacity of each battery (C_B), indirectly related to the number of series connected batteries.

$$C_n = \frac{N_{BAT}}{n_B^S} C_B \tag{11}$$

Number of cycles of failure over minimum average depth of discharge for a period of battery bank is depicted in Figure 3. The battery bank replacement hours (N_{BR}) can be calculated by determining the total number of cycles in which a battery (N_{cycles}) can be operated. Here, De is the yearly average minimum capacity of the battery bank achieved over a day and O_{Batt} is the number of days for which the battery should be operated [25]. It is used to determine the life of the battery. A curve fitting toolbox is used to generate the coefficient values of the equation. Battery degradation model equations are described as:

$$O_{Batt} = a_1 \times (D_e)^{b_1} + c_1 \tag{12}$$

$$\begin{gathered} N_{BR} = n \times 365\text{Ncycles\_DOD\%} \\ a1 = 1.582 \times 105,\ b_2 = -0.9964,\ c_1 = -997.1 \end{gathered} \tag{13}$$

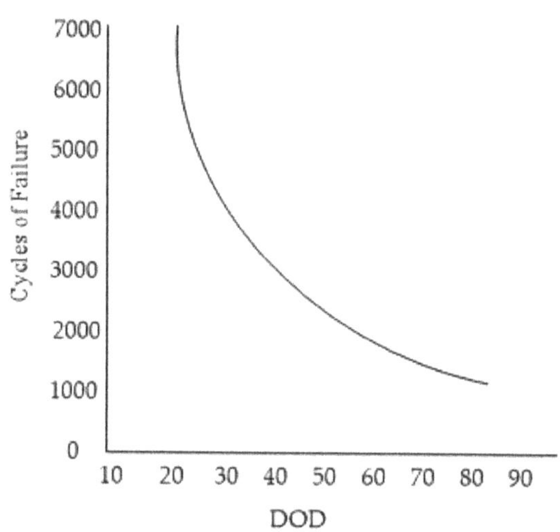

Figure 3. Number of cycles of failure over minimum average depth of discharge for a period of battery bank [25].

3.4. Mathematical Model of Charge Controller

The charge controller makes energy balance among different system components and its model is described as:

$$E_{CCO}(t) = E_{EE}(t) \times \eta_{CC} \tag{14}$$

where E_{CCO} (t) and E_{CCI} (t), respectively, represent the hourly output and hourly input to charge controller (kWh), E_{EE} (t) is amount of excess energy from sources (kWh) after serving the demand and η_{CC} is charge controller efficiency.

4. Problem Formulation

Minimization of system cost of energy is formulated and considered as an objective function for the present paper. Various constraints such as expected energy not supplied, carbon emission, renewable energy fraction and total net present cost have been incorporated during system optimization.

4.1. Objective Function

Cost of generation is the fundamental financial parameter in order to evaluate the techno-economic feasibility of the hybrid system. It can be calculated with annual system cost (ASC) and energy generation (E_{Gen}) as:

$$\text{COE} = \frac{\text{Annual System Cost (ASC)}}{\sum_{t=1}^{8760} E_{Gen}(t)} \tag{15}$$

Annual system cost of hybrid system is a function of net present cost (NPC) and can be estimated as [26]:

$$\text{ASC} = \text{NPC} \times \left[\frac{dr(1+dr)^{\xi}}{(1+dr)^{\xi} - 1}\right] \tag{16}$$

where dr is discount rate and ξ is project lifetime.

4.2. Operational Constraints

System optimization has been investigated under operational economic and reliability constraints which are summarized as follows:

4.2.1. Power Reliability Constraint

At any time, when part of the load demand has not been met from the available generation, energy not supplied is calculated. It is the function of demand not met (NL) and duration of the period (T) as follows [27]:

$$\text{EENS} = \sum_{i=1}^{8760} (\text{NL} \times \text{T}) \tag{17}$$

4.2.2. Economic Parameter Constraint

Constraints of economic parameters have been taken during system analysis. Total net present cost, total recurring cost (TC_{rec}) and non-recurring cost ($TC_{non\text{-}rec}$) of the system have been evaluated. These parameters can be calculated as follows [28]:

$$\text{NPC} = C_{Inv} + TC_{rec} + TC_{non-rec} \tag{18}$$

Total recurring cost of system changes with escalation rate (er) and discount rate. It can be estimated as follows [28]:

$$TC_{rec} = C_{rec}\frac{\left[\frac{1+er}{1+dr}\right]\left\{\left[\frac{1+er}{1+dr}\right]^{\xi} - 1\right\}}{\left[\frac{1+er}{1+dr}\right] - 1} \tag{19}$$

Total non-recurring cost of system can be calculated by using Equation (19) as follows:

$$TC_{non-rec} = \sum_{y=1}^{y=n_{rep}} C_{Inv}\left[\frac{1+er}{1+dr}\right]^{y*n_{frep}} \tag{20}$$

where n_{rep} is total replacement of system component (in Nos.) and n_{frep} is year number of first replacement of system component.

4.2.3. Renewable Energy Fraction

Renewable energy fraction in total system generation ensures the sustainable generation of a hybrid system. It depends on energy generated by diesel generated (E_{DG}) and total energy generation (TE_{Gen}). It can be calculated as:

$$\text{Renewable Energy Fraction}(\%) = \left[1 - \frac{\sum E_{DG}}{\sum TE_{Gen}}\right] \times 100 \tag{21}$$

4.2.4. Total Carbon Emissions

Carbon emission is generated from the use diesel generator in the considered system. Total carbon emission (TE_{Carbon}) can be estimated as:

$$TE_{Carbon} = \sum_{t=1}^{T} \sum_{t=1}^{T} E_{Carbon} \times P_{DG}(t) \tag{22}$$

where E_{Carbon} is carbon emission produced by 1 kWh electricity generation by DG.

5. Energy Management Strategy

5.1. Load Following Strategy

The load following strategy is initiated by determining the demand and solar power generation available in an hour. If the generated solar power in an hour is more than the demand, then the battery state of charge is checked. If the battery bank state of charge is found to be less than the maximum battery state of charge (SOC), the battery bank is charged, otherwise it is fed to the dump loads. In case of demand exceeding the generated solar power, the battery bank is operated until its SOC maximum value is not reached. If the battery bank is unable to fulfill the demand, the diesel generator is operated to supply only the net demand, without charging the battery bank. If any one of these conditions is matched, the iteration as unit time in an hour is updated and the process continuous until the year is complete. Figure 4 shows the methodology of operating load following strategy.

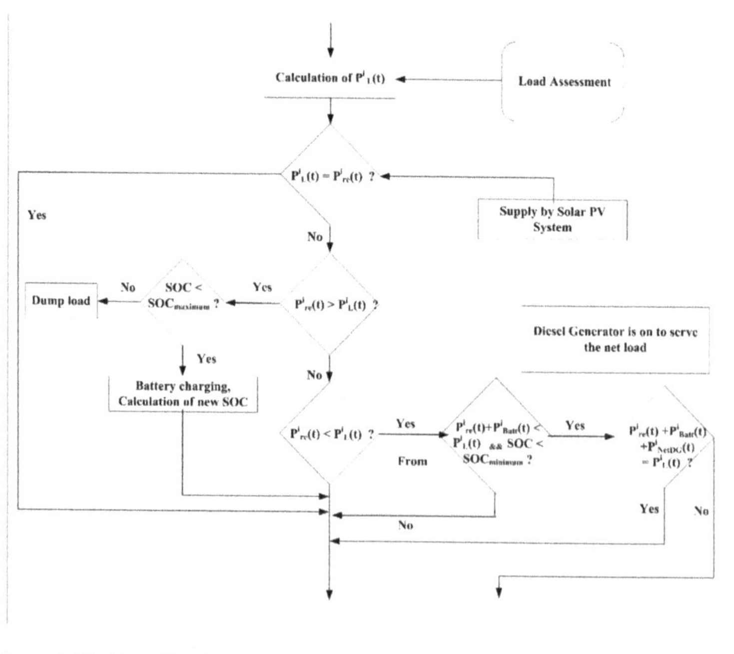

Figure 4. Working of load following strategy.

5.2. Cycle Charging Strategy

In the cycle charging strategy, the hourly load is compared to the available solar power output. If the generated power becomes equal to the demand then the time is updated. If the solar power in an hour is more than the demand then the battery state of charge is checked. If the battery bank state of charge is found to be less than the maximum SOC, the battery bank is charged, otherwise it is fed to the dump loads. In case of demand exceeding the solar power generation, battery bank is operated to fulfill the net demand. If the battery bank is unable to supply the demand, diesel generator is operated. Here the diesel generator is operated to supply the demand as well as charge the battery bank, until the SOC maximum value has not been reached. If any one of these conditions is matched the iteration as unit time in hour is updated and the process continuous until the year is complete. The Figure 5 shows the methodology of operating cycle charging strategy.

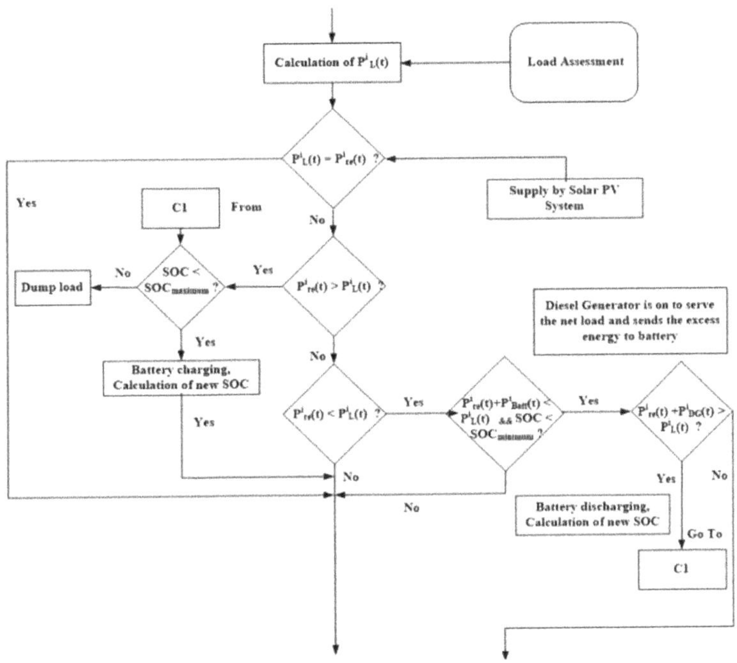

Figure 5. Working of cycle charging strategy.

6. Biogeography-Based Optimization (BBO) Algorithm

The biogeography-based optimization technique has unique features of longevity in the solutions survival nature as solutions in genetic algorithm die after each iteration. It also adds the merit of mutation, which removes the clustering effect seen while implementing BBO [29,30]. Due to these advantages, BBO is selected over GA and PSO for optimizing the cost of energy of the hybrid energy system.

BBO optimization procedure is as follows:

Step 1: Initialize the generation limit, population size and mutation rate.

Step 2: The fitness function values of all the individuals are evaluated.

Step 3: If the termination criterion is not met.

Step 4: Best habitats are saved in the temporary array.

Step 5: HSI (Habitat suitability index) is mapped using the number of ì, € e and S species for each habitat.

Step 6: Probabilistically choose the immigration island.

Step 7: SIV (Suitability index variables) are randomly migrated based on the island selected in Step 6.

Step 8: The population is randomly mutated.

Step 9: The fitness functions of all the individuals are evaluated.

Step 10: The population is sorted and arranged from best to worst.

Step 11: The best values of habitat as stored in temporary array replaces the worst values.

Step 12: Step 3 is again followed for continuing the next iteration.

Step 13: End the algorithm.

A flowchart of process of BBO algorithm is shown in Figure 6.

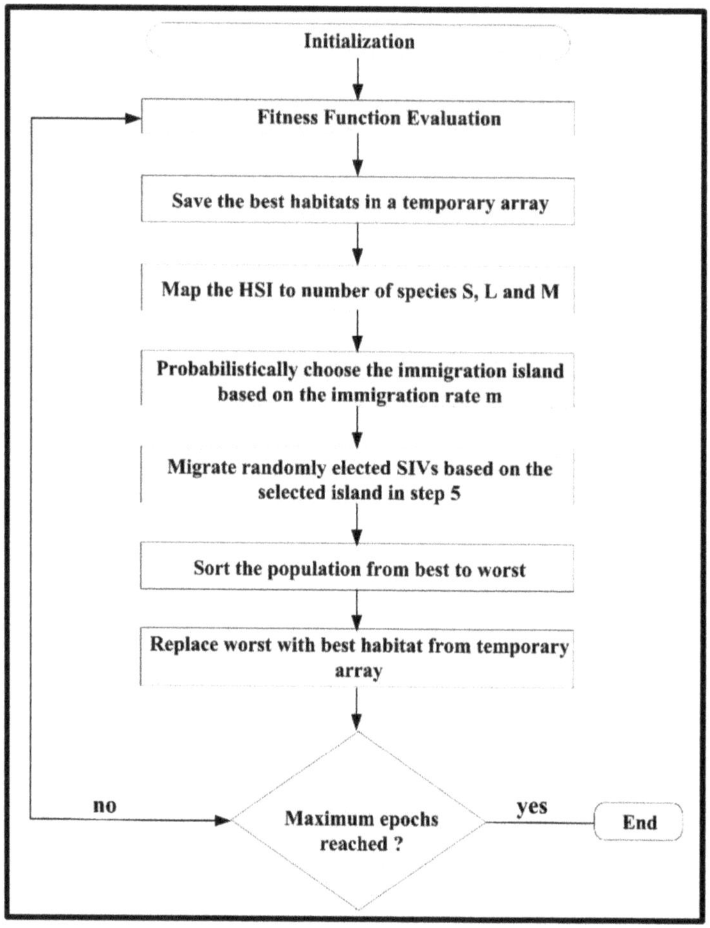

Figure 6. Biogeography based optimization process.

The parameters-chosen BBO technique have mutation probability of 0.4, population size is 50 selected due to better convergence of the objective function, and problem dimension is 3, while the available rooftop area limits the maximum number of solar modules to 500 modules. The variations of the total number of batteries and diesel generator to be selected are limited by 500 and 50, respectively.

7. Results and Discussions

7.1. Input Technical and Economical Dataset

Study area receives the annual average solar radiation of 5.26 kWh/m^2 per day. Peak solar radiation of 7.05 kWh/m^2 is found during the month of May and minimum solar radiation of 4.08 kWh/m^2 has been received during the month of December. A maximum temperature of 45 °C is recorded in the month of May and the lowest temperature of 12 °C is observed during the month of January. Monthly solar radiation and monthly temperature distribution of the selected site are depicted in the Figures 7 and 8, respectively.

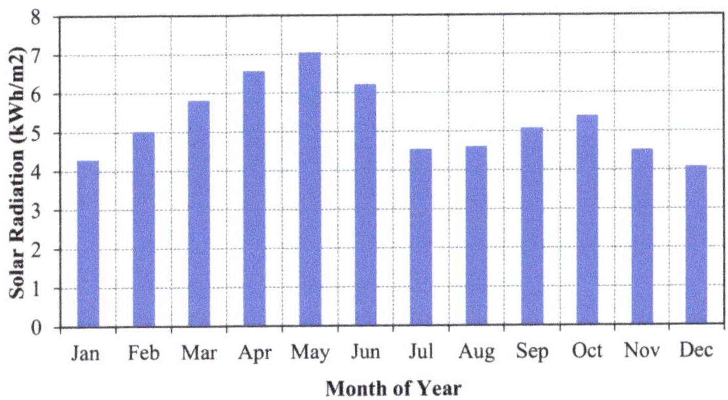

Figure 7. Monthly solar radiation of the selected site.

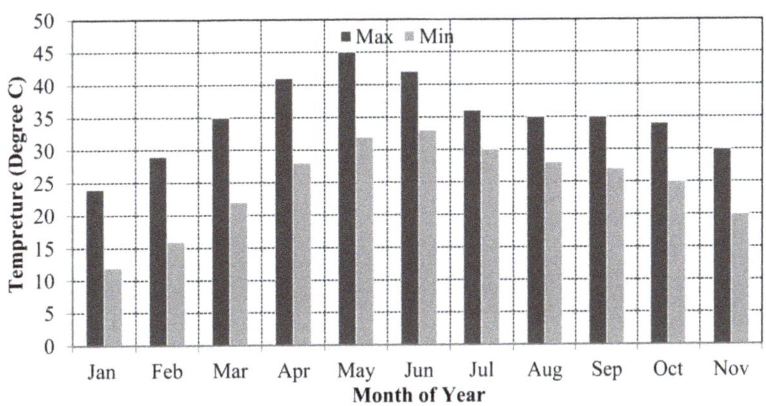

Figure 8. Monthly temperature of the selected site.

Load profile on an hourly basis is shown in Figure 9. Peak electrical load of 89.80 kW and 59.60 kW are estimated during summer season and winter season, respectively. Meanwhile, a minimum load demand of 14.5 kW and 5.43 kW is observed during the summer season and winter season, respectively.

Detailed specifications of system components available in market have been considered in the present paper as given in Table 1. Three types of DG (20 kVA, 30 kVA and 50 kVA) have been taken during the analysis. However, diesel price of $1.05 per Litre is same for all three DG. Two types of PV module (0.375 kW and 0.32 kW) have been used in the study. Capital cost of these modules is $273.79 and $205.34, respectively. Three types of battery sizes, 100 Ah, 150 Ah and 250 Ah have been considered during the performance analysis. Capital cost of these battery types is $96, $130 and $219, respectively. Specifications of charge controller and bidirectional converter are also given in Table 1.

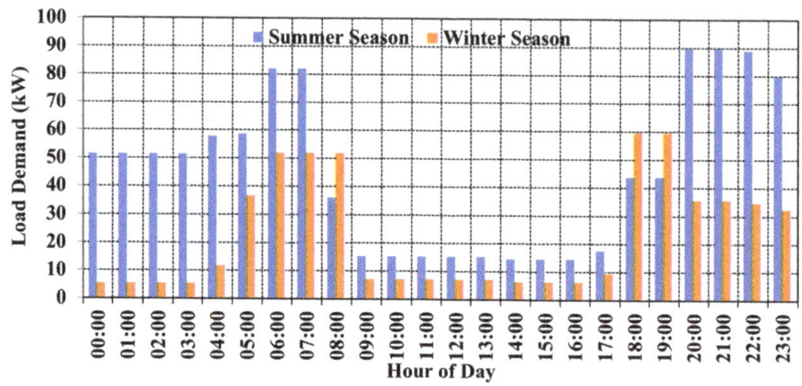

Figure 9. Hourly load demand of the study area.

Table 1. Detail specifications of system components.

(a)	**Diesel Generator**						
Type	**Power Rating DG (kVA)**		**Capital Cost ($)**		**Diesel ($/Ltr)**		
1	20		3450		1.05		
2	30		5133		1.05		
3	50		5818		1.05		
(b)	**Solar Photovoltaic Module**						
Type	**Voc (V)**	**Isc (A)**	**Vmax (V)**	**Imax (A)**	**Pmax (kW)**	**NCOT (°C)**	**Capital Cost ($)**
1	44.5	9.9	39	9.62	0.375	48	273.79
2	45	9.17	36.2	8.84	0.320	48	205.34
(c)	**Battery**						
Type	**Nominal Capacity (Ah)**		**Voltage (V)**		**DOD (%)**		**Capital cost ($)**
1	100		12		80		96
2	150		12		80		130
3	250		12		80		219
(d)	**Charge Controller**						
Type	**n1 (%) PV Battery Charger**		**n2 (%) PV battery Charger**		**Power Rating (kW)**		**Capital Cost ($)**
1	95		70		0.24		68.45
(e)	**Bidirectional Converter**						
Type	**Efficiency of Inv (%)**			**Power rating (kW)**		**Capital cost ($)**	
1	90			5.5		205.34	

7.2. Results and Discussions

The hybrid system model has been developed in MATLAB (R2019b, MathWorks, Natick, MA, USA) covering the specifications of individual components. Further, 18 combinations have been modelled and simulated for a year. The optimum configuration of these combinations is obtained using the BBO algorithm in MATLAB. Further, these combinations have been compared based on cost of energy, carbon emission, fuel consumption, renewable fraction and total operating hours. Descriptions of results are explained as follows.

7.2.1. Cost of Energy

For each combination of battery, solar photovoltaic module and diesel generator, the cost of energy (COE) is evaluated using cycle charging (CCS) and load following strategies (LFS). A comparison between the cost of energy for both the strategies is shown in Figure 10. Cycle charging strategy shows the minimum COE for each combination of components. The 13th combination provides the least COE of 0.225 $/kWh, with COE of 17th combination at 0.229 $/kWh and 7th combination as 0.231 $/kWh.

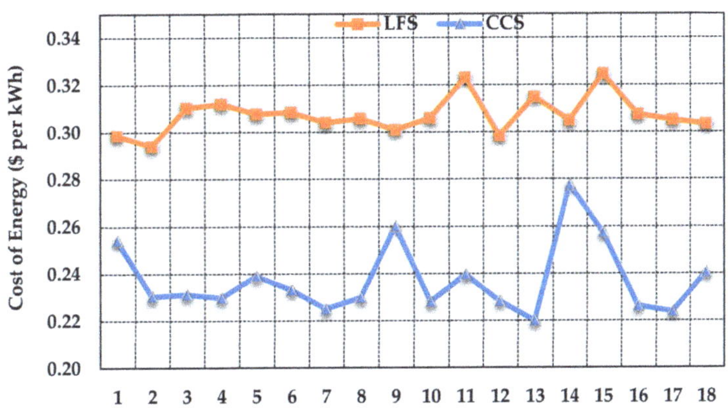

Figure 10. COE for each combination of specifications using BBO with CCS and LFS.

7.2.2. Optimum Size for Each Combination

Each combination has been simulated and optimized using BBO algorithm in MATLAB. Results for each combination are given in Table 2. From the Table 2, it has been found that combination 13 offers minimum cost of energy generation. The net present cost of this combination is calculated as $1,204,972. Combination optimum sizes consist of five numbers of DGs, 519 numbers of PV modules and 307 numbers of batteries.

7.2.3. Carbon Emission

The total CO2 emission is found to be at minimum about 155,184 kg/yr in case of the 13th combination as compared to the 7th and 17th combinations. This is due to the number of operating hours of diesel generator, about 732 for 13th combination as compared to 776 in case of the 7th combination. The carbon emission for each emission has been calculated and given in Table 3.

7.2.4. Renewable Fraction

Renewable fraction is directly dependent on the total output power of diesel generator and the renewable power output. This further depends on the overall fuel consumption by the diesel generator. Hence, higher fuel consumption of DG means minimum renewable fraction, the 13th combination has a maximum renewable fraction of 55% compared to 53% and 52% for the 7th and 17th combinations, respectively. Renewable energy for each combination is given in Table 3.

7.2.5. Total Fuel Consumption

The total fuel consumption of DG is minimal in the case of the 13th combination of 60,418 l, versus 7th and 17th combinations.

Table 2. Optimum size for each combination of hybrid system.

Combination	Device Type			Number of DGs	Number of Batteries	Number of PV Modules	Cost of Energy ($/kWh)	Total Net Present Cost ($)
	DG	SPV	Battery					
1	1	1	1	5	254	466	0.261	1,201,939
2	1	1	2	10	260	490	0.236	1,214,814
3	1	1	3	10	162	489	0.237	1,204,933
4	1	2	1	11	353	495	0.235	1,226,900
5	1	2	2	9	152	524	0.245	1,322,258
6	1	2	3	8	104	495	0.239	1,207,519
7	2	1	1	8	134	493	0.231	1,224,802
8	2	1	2	7	258	488	0.235	1,215,611
9	2	1	3	8	79	493	0.267	1,621,356
10	2	2	1	7	380	495	0.234	1,195,032
11	2	2	2	6	139	495	0.245	1,300,750
12	2	2	3	7	136	483	0.234	1,218,686
13	3	1	1	5	307	519	**0.225**	1,204,972
14	3	1	2	3	65	519	0.284	1,655,641
15	3	1	3	5	78	519	0.264	1,642,181
16	3	2	1	4	386	495	0.232	1,184,115
17	3	2	2	4	261	495	0.229	1,170,417
18	3	2	3	2	104	495	0.246	1,139,395

Note: Bold represents minimum cost of energy generation.

Table 3. Carbon emission, total fuel consumption and renewable fraction for each combination.

Combination	Emissions of CO_2 by DG (kg/yr)	Total Fuel Consumption by DG (Litre)	Renewable Fraction (%)	Energy Index Ratio
1	165,106	64,280	0.38	1
2	154,845	60,285	0.52	1
3	149,587	58,239	0.54	1
4	160,815	62,610	0.51	1
5	195,227	76,007	0.37	1
6	173,399	67,509	0.42	1
7	157,932	61,487	0.53	1
8	154,395	60,110	0.53	1
9	260,302	101,343	0.15	1
10	154,395	60,110	0.52	1
11	195,227	76,007	0.35	1
12	165,258	64,340	0.48	1
13	155,184	60,418	0.55	1
14	270,936	105,483	0.004	1
15	264,576	103,007	0.16	1
16	155,693	60,616	0.52	1
17	155,523	60,550	0.52	1
18	162,562	63,290	0.4	1

7.2.6. Total Operating Hours of DG and Battery Bank

The diesel generator needs to be operated 732 h and battery bank of 4772 h over a year in case of the 13th combination. The generators are operated for the minimum amount of time in this case while using cycle charging dispatch strategy. Operating hours for each combination are given in Table 4. Additionally, it has been found that replacement of battery bank is found to be minimal in case of CCS compared to LFS. The battery bank is to be replaced in every 3 years with CCS while battery banks are replaced after 2 years with LFS.

Table 4. Total operating hours of DG and battery bank.

Combination	Energy Index Ratio	Total Operating Hours (hr/yr)	
		DGs	Batteries
1	1	1947	3680
2	1	913	4683
3	1	882	4714
4	1	862	4734
5	1	1279	4286
6	1	1278	4318
7	1	776	4789
8	1	867	4729
9	1	1279	4286
10	1	867	4729
11	1	1279	4317
12	1	928	4699
13	1	732	4772
14	1	2130	3374
15	1	1248	4256
16	1	918	4678
17	1	917	4679
18	1	1917	3679

7.2.7. Convergence of BBO Algorithm

Convergence plot using BBO algorithm is shown in Figure 11. The results attained using BBO for the 7th, 13th and 17th combinations, by considering COE as the minimization objective function. For the 13th combination the COE is 0.225 $/kWh compared to 0.229 $/kWh and 0.231 for the 17th and the 7th combinations, respectively.

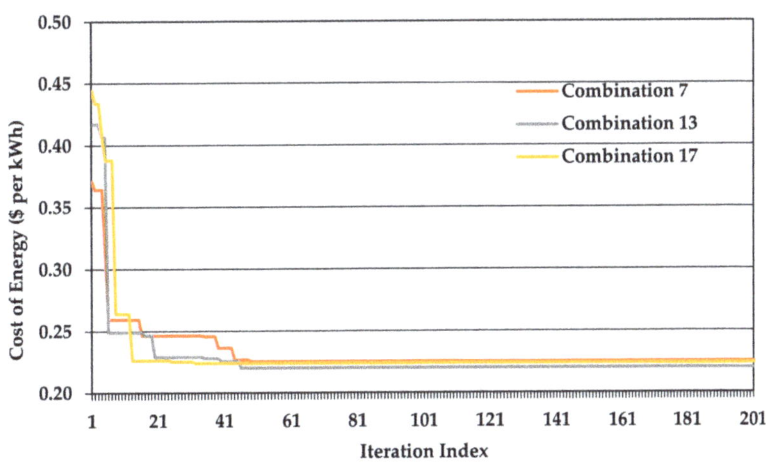

Figure 11. COE convergence plot with BBO for the 7th, 13th and 17th combinations.

8. Conclusions

In the paper, performance investigation of Solar Photovoltaic/DG/Battery-based hybrid system has been performed for the energy access of three hostels and one STP of an institute. In total, 18 combinations of system components sizes available on the market have been considered during the analysis. Cost of energy generation of hybrid system has been optimized using BBO algorithm.

After simulation, it has been found that the cycle charging strategy offers a lower cost compared to the load following strategy. All considered combinations have been compared based on COE, renewable fraction, carbon emission and operating hours of DG's. The optimum configuration offers minimum COE of 0.225 $/kWh, a renewable fraction of 55% and carbon emission of 155,184 kg/yr. Further, government subsidy on the PV module can also further reduce the system cost. In future works, the addition of utility grid and waste utilization, collected from the campus to energy are proposed to make a more sustainable hybrid system.

Author Contributions: Conceptualization, A.C., S.U., M.T.K., S.M.S.H. and T.S.U.; methodology, A.C., S.U., M.T.K., S.M.S.H. and T.S.U.; software, A.C. and S.U.; validation, A.C., S.U., M.T.K. and S.M.S.H.; formal analysis, A.C., S.U. and M.T.K.; investigation, A.C. and S.U.; data curation A.C. and S.U.; writing—original draft preparation, A.C. and S.U.; writing—review and editing, M.T.K., S.M.S.H. and T.S.U.; visualization M.T.K. and S.M.S.H.; funding acquisition, T.S.U. All authors have read and agreed to the published version of the manuscript.

Funding: This research received no external funding.

Institutional Review Board Statement: Not applicable.

Informed Consent Statement: Not applicable.

Data Availability Statement: Not applicable.

Conflicts of Interest: The authors declare no conflict of interest.

References

1. Korkeakoski, M. Towards 100% Renewables by 2030: Transition Alternatives for a Sustainable Electricity Sector in Isla de la Juventud, Cuba. *Energies* **2021**, *14*, 2862. [CrossRef]
2. Balžekienė, A.; Budžytė, A. The Role of Environmental Attitudes in Explaining Public Perceptions of Climate Change and Renewable Energy Technologies in Lithuania. *Sustainability* **2021**, *13*, 4376. [CrossRef]
3. Aleem, S.A.; Hussain, S.M.S.; Ustun, T.S. A Review of Strategies to Increase PV Penetration Level in Smart Grids. *Energies* **2020**, *13*, 636. [CrossRef]
4. Kagimu, V.; Ustun, T.S. Novel business models and policy directions based on SE4ALL global framework for minigrids. In Proceedings of the IEEE International Conference on Emerging Technologies and Innovative Business Practices for the Transformation of Societies (EmergiTech), Mauritius, 1–6 August 2016; pp. 251–256.
5. Sharafi, M.; Elmekkawy, T.Y. Multi-objective optimal design of hybrid energy systems using PSO-simulation based approach. *Renew. Energy* **2014**, *68*, 67–79. [CrossRef]
6. Ustun, T.S.; Hussain, S.M.S.; Kikusato, H. IEC 61850-Based Communication Modeling of EV Charge-Discharge Management for Maximum PV Generation. *IEEE Access* **2019**, *7*, 4219–4231. [CrossRef]
7. Javed, K.; Ashfaq, H.; Singh, R.; Hussain, S.M.S.; Ustun, T.S. Design and Performance Analysis of a Stand-alone PV System with Hybrid Energy Storage for Rural India. *Electronics* **2019**, *8*, 952. [CrossRef]
8. Dey, P.P.; Das, D.; Latif, A.; Hussain, S.; Ustun, T. Active Power Management of Virtual Power Plant under Penetration of Central Receiver Solar Thermal-Wind Using Butterfly Optimization Technique. *Sustainability* **2020**, *12*, 6979. [CrossRef]
9. Latif, A.; Hussain, S.M.S.; Das, D.C.; Ustun, T.S. Double stage controller optimization for load frequency stabilization in hybrid wind-ocean wave energy based maritime microgrid system. *Appl. Energy* **2021**, *282*, 116171. [CrossRef]
10. Ramesh, M.; Saini, R.P. Dispatch strategies based performance analysis of a hybrid renewable energy system for a remote rural area in India. *J. Clean. Prod.* **2020**, *259*, 120697. [CrossRef]
11. Baruah, A.; Basu, M.; Amuley, D. Modeling of an autonomous hybrid renewable energy system for electrification of a township: A case study for Sikkim, India. *Renew. Sustain. Energy Rev.* **2021**, *135*, 110158. [CrossRef]
12. Das, M.; Singh, M.A.K.; Biswas, A. Techno-economic optimization of an off-grid hybrid renewable energy system using metaheuristic optimization approaches—Case of a radio transmitter station in India. *Energy Convers. Manag.* **2019**, *185*, 339–352. [CrossRef]

13. El-houari, H.; Allouhi, A.; Rehman, S.; Buker, M.S.; Kousksou, T.; Jamil, A.; El Amrani, B. Feasibility evaluation of a hybrid renewable power generation system for sustainable electricity supply in a Moroccan remote site. *J. Clean. Prod.* **2020**, *277*, 123534. [CrossRef]
14. Elkadeem, M.R.; Wang, S.; Sharshir, S.W.; Atia, E.G. Feasibility analysis and techno-economic design of grid-isolated hybrid renewable energy system for electrification of agriculture and irrigation area: A case study in Dongola, Sudan. *Energy Convers. Manag.* **2019**, *196*, 1453–1478. [CrossRef]
15. Kumar, P.P.; Saini, R.P. Optimization of an off-grid integrated hybrid renewable energy system with different battery technologies for rural electrification in India. *J. Energy Storage* **2020**, *32*, 101912. [CrossRef]
16. Ma, T.; Javed, M.S. Integrated sizing of hybrid PV-wind-battery system for remote island considering the saturation of each renewable energy resource. *Energy Convers. Manag.* **2019**, *182*, 178–190. [CrossRef]
17. Jahangir, M.H.; Cheraghi, R. Economic and environmental assessment of solar-wind-biomass hybrid renewable energy system supplying rural settlement load. *Sustain. Energy Technol. Assess.* **2020**, *42*, 100895. [CrossRef]
18. Ustun, S.T.; Nakamura, Y.; Hashimoto, J.; Otani, K. Performance analysis of PV panels based on different technologies after two years of outdoor exposure in Fukushima, Japan. *Renew. Energy* **2019**, *136*, 159–178. [CrossRef]
19. Misra, A.; Sharma, M.P. Development of Hybrid Energy System for a Remote Area in Kutch District of Gujarat State, India. *Energy Sources Part A Recovery Util. Environ. Eff.* **2020**. [CrossRef]
20. Ismail, M.S.; Moghavvemi, M.; Mahlia, T.M.I. Techno-economic analysis of an optimized photovoltaic and diesel generator hybrid power system for remote houses in a tropical climate. *Energy Convers. Manag.* **2013**, *69*, 163–173. [CrossRef]
21. Tazvinga, H.; Xia, X.; Zhang, J. Minimum cost solution of photovoltaic–diesel–battery hybrid power systems for remote consumers. *Sol. Energy* **2013**, *96*, 292–299. [CrossRef]
22. Chauhan, A.; Saini, R.P. Size Optimization and Demand Response of A Stand-Alone Integrated. *Renew. Energy System. Energy* **2017**, *124*, 59–73.
23. Fodhil, F.; Hamidat, A.; Nadjemi, O. Potential, optimization and sensitivity analysis of photovoltaic-diesel-battery hybrid energy system for rural electrification in Algeria. *Energy* **2019**, *169*, 613–624. [CrossRef]
24. Ayodele, T.R.; Ogunjuyigbe, A.S.O.; Akpeji, K.O.; Akinola, O.O. Prioritized rule based load management technique for residential building powered by PV/battery system. *Eng. Sci. Technol.* **2017**, *20*, 859–873. [CrossRef]
25. Lopez, R.D.; Agustin, J.L.B. Techno-economic analysis of gridconnected battery storage. *Energy Convers. Manag.* **2015**, *91*, 394–404. [CrossRef]
26. Upadhyay, S.; Sharma, M.P. Selection of a suitable energy management strategy for a hybrid energy system in a remote rural area of India. *Energy* **2016**, *94*, 352–366. [CrossRef]
27. Kanase-Patil, A.B.; Saini, R.P.; Sharma, M.P. Sizing of integrated renewable energy system basedon load profile and reliability index for the state of Uttarakhand in India. *Renew. Energy* **2011**, *36*, 2809–2821. [CrossRef]
28. Kumar, S.; Kaur, T.; Arora, M.K.; Upadhyay, S. Resource estimation and sizing optimization of PV/micro hydro-based hybrid energy system in rural area of Western Himalayan Himachal Pradesh in India. *Energy Sources Part A Recovery Util. Environ. Eff.* **2019**, *41*, 2795–2807. [CrossRef]
29. Simon, D. Biogeography-based optimization. *IEEE Trans. Evol. Comput.* **2008**, *12*, 702–713. [CrossRef]
30. Kumar, R.; Gupta, R.A.; Bansal, A.K. Economic analysis and power management of a stand-alone wind/ photovoltaic hybrid energy system using biogeography based optimization algorithm. *Swarm Evol. Comput.* **2013**, *8*, 33–43. [CrossRef]

MDPI
St. Alban-Anlage 66
4052 Basel
Switzerland
Tel. +41 61 683 77 34
Fax +41 61 302 89 18
www.mdpi.com

Sustainability Editorial Office
E-mail: sustainability@mdpi.com
www.mdpi.com/journal/sustainability

www.ingramcontent.com/pod-product-compliance
Lightning Source LLC
Chambersburg PA
CBHW061819210926
51850CB00006B/751

* 9 7 8 3 0 3 6 5 6 3 7 1 8 *